RECYCLING OF USED OILS

Recycling of Used Oils

Second European Congress on the Recycling of Used Oils
Zweiter Europäischer Gebrauchtöl-Recycling Kongress
Deuxième Congrès Européen sur le Recyclage des Huiles Usagées
Secondo Congresso Europeo sul Riciclaggio degli Oli Usati

Supported by
the Commission of the European Communities
and organized by the
European Commission for Regeneration
of the
European Union of Independent Lubricant Manufacturers

Paris, September 30 - October 2, 1980

D. Reidel Publishing Company
Dordrecht, Holland / Boston, U.S.A. / London, England

Library of Congress Cataloging in Publication Data

Recycling of Used Oils (1980: Paris, France)
  Zweiter Europäischer Gebrauchtöl-Recycling Kongress = Second European Congress on the Recycling of Used Oils = Deuxième Congrès Européen sur le Recyclage des Huiles Usagées = Secondo Congresso Europeo sul Riciclaggio degli Oli Usati.

  Contents in English, French, German, and Italian.
  1. Petroleum waste–Recycling–Congresses.
I. Commission of the European Communities. II. European Commission for Regeneration.
TP687.E95   1980   665.5'389   81-19945
ISBN 90-277-1369-3   AACR2

Publication arranged by
Commission of the European Communities,
Directorate-General Information Market and Innovation, Luxembourg

EUR 7547 DE, EN, FR, IT
Copyright © 1981 ECSC, EEC, EAEC, Brussels and Luxembourg, 1981

LEGAL NOTICE

Neither the Commission of the European Communities nor any person acting on behalf of the Commission is responsible for the use which might be made of the following information.

Published by D. Reidel Publishing Company
P.O. Box 17, 3300 AA Dordrecht, Holland

Sold and distributed in the U.S.A. and Canada
by Kluwer Boston Inc.,
190 Old Derby Street, Hingham, MA 02043, U.S.A.

In all other countries, sold and distributed
by Kluwer Academic Publishers Group,
P.O. Box 322, 3300 AH Dordrecht, Holland

D. Reidel Publishing Company is a member of the Kluwer Group

All Rights Reserved
No part of the material protected by this copyright notice may be reproduced or utilized in any form or by any means, electronic or mechanical, including photocopying, recording or by any informational storage and retrieval system, without written persmission from the copyright owner

Printed in The Netherlands

INHALTSVERZEICHNIS

Begrüssung der Teilnehmer
    H.-W. STUDDERS, Präsident der Europäischen Union der
    unabhängigen Schmierstoffverbände   1

Eröffnung des Kongresses
    M. L. NATALI, Vizepräsident der Kommission der Europäischen
    Gemeinschaften   5

Die Bedeutung der Zweitraffinationsindustrie für die europäische
    Wirtschaft
    P. BRASSART, Präsident der C.E.R.   9

C. A. VANIK, U.S. Representative - House of Congress (D-Ohio)
    presented by B. GONCHAR, Legislative Aide to Congressman
    C. A. VANIK   18

The American Re-Refining Industry
    Yesterday and Today
    J. A. McBAIN, Association of Petroleum Re-refiners,
    Washington, U.S.A.   24

United States Re-Refining - 1980
    K. L. MORRIS, President Association of Petroleum Re-refiners   29

Dr. J. SCHVARTZ
    Rapporteur du Budget de l'Industrie à la Commission de la
    Production et des Echanges de l'Assemblée Nationale, Membre
    du Conseil Supérieur du Pétrole   36

Anwendung der EG-Ratsrichtlinie vom 16. Juni 1975 über die
    Beseitigung von Altölen
    B.W.K. RISCH, Dienststelle für Umwelt- und Verbraucherschutz
    der EG-Kommission   47

Das Gebrauchtöl-Recycling in den verschiedenen Mitgliedsstaaten
    der EG

Gebrauchtöl-Recycling in Belgien
    J. RENGUET, Präsident der Abteilung "Schmierstoffe" in der
    I.H.M.B. Industrie des Huiles Minérales de Belgique, Brüssel   80

Die gesetzlichen Grundlagen sowie die wirtschaftliche Situation
    der Zweitraffinerien in der Bundesrepublik Deutschland
    Dr. H. KOEHN, Vorsitzender der AMMRA, Hamburg   88

Gebrauchtöl-Recycling in Frankreich
    R. KACHLER, Direktor der S.N.F.R.H.G., Paris   99

Zweitraffination von Gebrauchtölen in verschiedenen Mitglied-
    staaten der EG - Grossbritannien
    C. MITCHELL, Chairman der Century Oils Group Ltd.,
    Grossbritannien   111

Recycling in Irland
    F.M.J. DUFFY, Präsident der Association of the Irish
    Independent Lubricant Industry, Dublin   120

Gebrauchtöl-Recycling in Italien
  Dr. R. SCHIEPPATI, Präsident der Gruppo Aziende Indipendenti Lubrificanti (ASCHIMICI), Milano ... 132

Die Preisbildung für Gebrauchtöl in Europa
  P. GRARD, Ehrenpräsident der I.H.M.B. Industrie des Huiles Minérales de Belgique, Brüssel, Vizepräsident der U.E.I.L. ... 143

Das wirtschaftliche Erfordernis einer Produktflexibilität in der Zweitraffinationsindustrie
  R. HAVEMANN, Stellvertretender Vorsitzender der AMMRA, Hamburg ... 150

Die Position der staatlichen Stellen in den USA gegenüber Zweitraffinaten
  M.E. LEPERA, T.C. BOWEN, US Army Mobility Equipment Research & Development Command ... 162
  J.F. COLLINS, US Department of Energy ... 166
  D.A. BECKER, US National Bureau of Standards ... 171

Die Technologie der Zweitraffination
  Dr. C. LAFRENZ, Delegierter der Europäischen Kommission der Zweitraffineure beim technischen Verband der europäischen Schmierstoffindustrie (ATIEL), Bundesrepublik Deutschland ... 175

Qualität von Zweitraffinaten, hergestellt nach modernen Verfahren
  C.J. RATCLIFFE, Operations Director, Anglo-Pennsylvanian Oil Comp., Ltd., Grossbritannien ... 198

Ergebnisse einer vergleichenden Qualitätsuntersuchung von Erst- und Zweitraffinaten
  Dipl.-Ing. E. WEDEPOHL, Institut für Erdölforschung, Hannover
  Prof. Dr.-Ing. W.J. BARTZ, Techn. Akademie Esslingen, Ostfildern (Nellingen)
  Dr.-Ing. K. MÜLLER, Ciba Geigy AG, Basel ... 212

Die Verwendung von Zweitraffinaten nach dem Solventverfahren bei der Formulierung von Motorölen
  Ing. A. MODENESI, Chef des technischen Dienstes der Clipper Oil Italiana, Italien ... 225

Der europäische Schmierstoffmarkt und die Rolle der unabhängigen Firmen
  Dr. M. FUCHS, Geschäftsführender Gesellschafter der Rudolf Fuchs GmbH & Co., Mannheim ... 239

Energieeinsparung durch Wiederaufarbeiten von Gebrauchtöl
  C.J. THOMPSON, D.W. BRINKMAN, US Department of Energy, Bartlesville, Oklahoma, USA ... 253

Emulsionsbehandlung
  J. DUMORTIER, Vizepräsident der S.N.F.R.H.G., Paris ... 267

Umweltbelastung bei Verbrennung von Gebrauchtölen
  W.B. WALKER, Director, Braybrooke Chemical Services, Grossbritannien ... 275

Aspekte des Immissionsschutzes beim Betrieb von Zweitraffinerien
  Dipl.-Ing. G. STEINMETZGER, Haberland Engineering GmbH, Dollbergen ; Arbeiten von F.K. KRENTEL, Leitender Gewerbedirektor, Bezirksregierung Hannover ... 290

Umweltschutz bei der Raffination von Gebrauchtölen.
   Verfahren bei der Wiederverwendung von Säureteeren
     C. VERSINO, Istituto di Chimica-Fisica, Università
     di Torino
     C. MOLINO, O.M.A., Rivalta di Torino   310

Warum Wiederaufarbeitung von Gebrauchtölen ?
   Energetische und ökologische Aspekte des Problems
     C. RICHARD, Compagnie Française de Raffinage, Gonfreville-
     l'Orcher, France   332

Schlusswort
   P. BRASSART, Präsident des Kongresses   345
   F. DELMAS, Secrétaire d'Etat à l'Environnement   348

Teilnehmerliste   351

TABLE OF CONTENTS

Welcoming Address
    H.-W. STUDDERS, President of the European Union of
    Independent Lubricant Manufacturers   1

Opening of the Congress
    M. L. NATALI, Vice President of the Commission of the
    European Communities   5

The Regeneration Industry in the European Economy
    P. BRASSART, President of the E.C.R.   9

C. A. VANIK, U.S. Representative - House of Congress (D-Ohio)
    presented by B. GONCHAR, Legislative Aide to Congressman
    C. A. VANIK   18

The American Re-refining Industry
Yesterday and Today
    J. A. McBAIN, Association of Petroleum Re-refiners,
    Washington, U.S.A.   24

United States Re-refining - 1980
    K. L. MORRIS, President Association of Petroleum Re-
    refiners   29

Dr. J. SCHVARTZ
    Rapporteur du Budget de l'Industrie à la Commission de la
    Production et des Echanges de l'Assemblée Nationale,
    Membre du Conseil Supérieur du Pétrole   36

Application of the Community Directive of 16th June, 1975 on the
Disposal of Used Oils
    B.W.K. RISCH, Environment and Consumer Protection Service
    of the Commission of the European Communities   47

The Recycling of Used Oils in Different Member Countries of the
C.E.C.

The Recycling of Used Oils in Belgium
    J. RENGUET, President of the Lubricants Section of the
    I.H.M.B. (Industrie des Huiles Minérales de Belgique),
    Brussels   80

The Legal Basis and the Economic Situation of Re-refining Plants
in the Federal Republic of Germany
    Dr. H. KOEHN, President of AMMRA, Hamburg   88

The Recycling of Used Oils in France
    R. KACHLER, Director of S.N.F.R.H.G., Paris   99

The Recycling of Used Oils in Different Member Countries of
the C.E.C. - Great Britain
    C. MITCHELL, Chairman of Century Oils Group Ltd., Great
    Britain   111

Recycling in Ireland
    F.M.J. DUFFY, President of the Association of the Irish
    Independent Lubricant Industry, Dublin   120

The Recycling of Used Oils in Italy
    Dr. R. SCHIEPPATI, President of Gruppo Aziende Independenti
    Lubrificanti (ASCHIMICI), Milan   132

The Establishment of Prices for Used Oils in Europe
    P. GRARD, Honorary President of the I.H.M.B. (Industrie
    des Huiles Minérales de Belgique), Brussels, Vice-President
    of E.U.I.L.   143

The Economic Necessity for Qualitative Flexibility of Production
in the Regenerating Industry
    R. HAVEMANN, Vice President of AMMRA, Hamburg   150

Official Attitudes in the United States towards Regenerated Oils
    M.E. LEPERA, T.C. BOWEN, US Army Mobility Equipment Research &
    Development Command   162
    J.F. COLLINS, U.S. Department of Energy   166
    D.A. BECKER, U.S. National Bureau of Standards   171

The Technology of Regeneration
    Dr. C. LAFRENZ, Delegate from the European Commission for
    Regeneration to the Association of European Industrie of
    Lubricants Technicians (ATIEL), Germany   175

The Quality of Products Regenerated in Accordance with Modern
Techniques
    C.J. RATCLIFFE, Operations Director, Anglo-Pennsylvanian Oil
    Comp. Ltd., United Kingdom   198

Results of Comparative Research into the Respective Qualities of
Regenerated Oils and New Oils
    Dipl.-Ing. E. WEDEPOHL, Institut für Erdölforschung, Hannover
    Prof. Dr.-Ing. W.J. BARTZ, Techn. Akademie Esslingen,
    Ostfildern (Nellingen)
    Dr.-Ing. K. MÜLLER, Ciba Geigy AG, Basel   212

Use of Solvent Re-refined Oils in the Formulation of Motor Oils
    Ing. A. MODENESI, Chief of Technical Services of Clipper Oil
    Italiana, Italy   225

The European Lubricant Market - The Role of Independent Companies
    Dr. M. FUCHS, Geschäftsführender Gesellschafter der Rudolf
    Fuchs GmbH & Co., Mannheim   239

Energy Saving by Re-refining Used Oils
    C.J. THOMPSON, D.W. BRINKMAN, US Department of Energy,
    Bartlesville, Oklahoma, USA   253

Treatment of Emulsions
    J. DUMORTIER, Vice President of S.N.F.R.H.G., Paris   267

Pollution of the Environment by the Burning of Waste Oils
    W.B. WALKER, Director, Braybrooke Chemical Services,
    United Kingdom   275

Aspects of the Protection of the Environment in the Operation
of Used Oil Refineries
    Dipl.-Ing. G. STEINMETZGER, Haberland Engineering GmbH,
    Dollbergen ; Research by F.K. KRENTEL, Leitender Gewerbe-
    direktor, Bezirksregierung Hannover   290

Protection of the Environment During the Process of Re-refining Used
Oils.
The Treatment of Acid Sludge
C. VERSINO, Istituto di Chimica-Fisica, Università di
Torino
C. MOLINO, O.M.A., Rivalta di Torino   310

Why Regenerate Used Oils?
Energy and Ecological Aspects of the Problem
C. RICHARD, Compagnie Française du Raffinage, Gonfreville-
L'Orcher, France   332

Conclusions
P. BRASSART, President of Congress   345
F. DELMAS, Secrétaire d'Etat à l'Environnement   348

List of Participants   351

TABLE DES MATIERES

Accueil des Congressistes
   H.-W. STUDDERS, Président de l'Union Européenne des
   Indépendants en Lubrifiants   1

Ouverture du Congrès
   M. L. NATALI, Vice-Président de la Commission des Communautés
   Européennes   5

L'industrie de la régénération dans l'économie européenne
   P. BRASSART, Président de la C.E.R.   9

C. A. VANIK, U.S. Representative - House of Congress (D-Ohio)
   presented by B. GONCHAR, Legislative Aide to Congressman
   C. A. VANIK   18

The American Re-refining Industry
   Yesterday and Today
   J. A. McBAIN, Association of Petroleum Re-refiners,
   Washington, U.S.A.   24

United States Re-refining - 1980
   K. L. MORRIS, President Association of Petroleum Re-refiners   29

Dr. J. SCHVARTZ
   Rapporteur du Budget de l'Industrie à la Commission de la
   Production et des Echanges de l'Assemblée Nationale, membre
   du Conseil Supérieur du Pétrole   36

Application de la directive communautaire du 16 juin 1975 sur
   l'élimination des huiles usagées
   B.W.K. RISCH, Service de l'Environnement et de la Protection
   des Consommateurs de la Commission des Communautés Européennes   47

Le recyclage des huiles usagées dans différents pays membres de la
   C.C.E.

Le recyclage des huiles usagées en Belgique
   J. RENGUET, Président de la section "Lubrifiants" de l'I.H.M.B.
   (Industrie des Huiles Minérales de Belgique), Bruxelles   80

Les bases juridiques et la situation économique des usines de
   régénération en République Fédérale d'Allemagne
   Dr. H. KOEHN, Président de l'AMMRA, Hamburg   88

Le recyclage des huiles usagées en France
   R. KACHLER, Directeur du S.N.F.R.H.G., Paris   99

Le recyclage des huiles usagées dans différents pays membres de
   la C.C.E. - Grande Bretagne
   C. MITCHELL, Chairman of Century Oils Group Ltd., Grande-
   Bretagne   111

Le recyclage en Irlande
   F.M.J. DUFFY, Président de l'Association of the Irish
   Independent Lubricant Industry, Dublin   120

Le recyclage des huiles usagées en Italie
   Dr. R. SCHIEPPATI, Président du Gruppo Aziende Independenti
   Lubrificanti (ASCHIMICI), Milano   132

La formation des prix des huiles usagées en Europe
P. GRARD, Président Honoraire de l'I.H.M.B. (Industrie des Huiles Minérales de Belgique), Bruxelles, Vice-Président de l'U.E.I.L.   143

Nécessité économique d'une flexibilité qualitative de production dans l'industrie de la régénération
R. HAVEMANN, Vice-Président de l'AMMRA, Hamburg   150

Positions d'organismes officiels des Etats-Unis à l'égard des huiles régénérées
M.E. LEPERA, T.C. BOWEN, US Army Mobility Equipment Research & Development Command   162
J.F. COLLINS, U.S. Department of Energy   166
D.A. BECKER, U.S. National Bureau of Standards   171

La technologie de la régénération
Dr. C. LAFRENZ, Délégué de la Commission Européenne de Régénération à l'Association Technique de l'Industrie Européenne des Lubrifiants (ATIEL), Allemagne   175

Qualités des produits de régénération fabriqués selon les techniques modernes
C.J. RATCLIFFE, Operations Director, Anglo-Pennsylvanian Oil Comp. Ltd., Grande-Bretagne   198

Résultats de recherches comparatives portant sur les qualités respectives des huiles régénérées et des huiles neuves
Dipl.-Ing. E. WEDEPOHL, Institut für Erdölforschung, Hannover
Prof. Dr.-Ing. W.J. BARTZ, Techn. Akademie Esslingen, Ostfildern (Nellingen)
Dr.-Ing. K. MÜLLER, Ciba Geigy AG, Basel   212

Utilisation d'huiles régénérées à l'aide de solvants pour la fabrication d'huiles de moteur
Ing. A. MODENESI, Chef des Services Techniques de la Clipper Oil Italiana, Italie   225

Le marché européen des lubrifiants et le rôle des entreprises indépendantes
Dr. M. FUCHS, Geschäftsführender Gesellschafter der Rudolf Fuchs GmbH & Co., Mannheim   239

Les économies d'énergie procurées par la régénération des huiles usagées
C.J. THOMPSON, D.W. BRINKMAN, US Department of Energy, Bartlesville, Oklahoma, USA   253

Traitement des émulsions
J. DUMORTIER, Vice-Président du S.N.F.R.H.G., Paris   267

La pollution de l'environnement par le brûlage des huiles usagées
W.B. WALKER, Director, Braybrooke Chemical Services, Grande-Bretagne   275

Aspects de la protection de l'environnement dans l'exploitation
   d'usines de régénération
   Dipl.-Ing. G. STEINMETZGER, Haberland Engineering GmbH,
   Dollbergen ; Recherches de F.K. KRENTEL, Leitender Gewerbedirektor, Bezirksregierung Hannover   290

La protection de l'environnement dans la régénération des huiles
   usées.
   Procédé de récupération des boues huileuses
   C. VERSINO, Istituto di Chimica-Fisica, Università di
   Torino
   C. MOLINO, O.M.A., Rivalta di Torino   310

Pourquoi régénérer les huiles usagées ?
   Aspects énergétiques et écologiques du problème
   C. RICHARD, Compagnie Française de Raffinage, Gonfreville-
   l'Orcher, France   332

Conclusions
   P. BRASSART, Président du Congrès   345
   F. DELMAS, Secrétaire d'Etat à l'Environnement   348

Liste des participants   351

INDICE

Accoglienza ai congressisti
  H.-W. STUDDERS, Presidente dell'Union Européenne des
  Indépendants en Lubrifiants 1

Apertura del congresso
  M. L. NATALI, Vice Presidente della Commissione delle
  Comunità Europee 5

L'industria della rigenerazione nell'economia europea
  P. BRASSART, Presidente della C.E.R. 9

C. A. VANIK, U.S. Representative - House of Congress (D-Ohio)
  presented by B. GONCHAR, Legislative Aide to Congressman
  C. A. VANIK 18

The American Re-refining Industry
Yesterday and Today
  J. A. McBAIN, Association of Petroleum Re-refiners,
  Washington, U.S.A. 24

United States Re-refining - 1980
  K. L. MORRIS, President Association of Petroleum Re-
  refiners 29

Dr. J. SCHVARTZ
  Rapporteur du Budget de l'Industrie à la Commission de la
  Production et des Echanges de l'Assemblée Nationale,
  Membre du Conseil Supérieur du Pétrole 36

Applicazione della direttiva comunitaria del 16 giugno 1975
sull'eliminazione degli oli usati
  B.W.K. RISCH, Servizio Condizioni Ambientali e Tutela
  dei Consumatori della Commissione delle Comunità Europee 47

Il riciclaggio degli oli usati nei differenti paesi membri della
C.C.E.

Il riciclo degli oli usati in Belgio
  J. RENGUET, Presidente della sezione "Lubrificanti"
  dell'I.H.M.B. (Industrie des Huiles Minérales de Belgique),
  Bruxelles 80

Le basi giuridiche e la situazione economica delle raffinerie
che riciclano gli oli usati nella RF di Germania
  Dr. H. KOEHN, Presidente dell'AMMRA, Amburgo 88

Il riciclo degli oli usati in Francia
  R. KACHLER, Direttore del S.N.F.R.H.G., Parigi 99

Il riciclo degli oli usati nei diversi paesi membri della
C.C.E. - Gran Bretagna
  C. MITCHELL, Chairman of Century Oils Group Ltd., Gran
  Bretagna 111

Riciclo in Irlanda
  F.M.J. DUFFY, Presidente dell'Association of the Irish
  Independent Lubricant Industry, Dublino 120

Il riciclaggio degli olii usati in Italia
  Dr. R. SCHIEPPATI, Presidente del Gruppo Aziende
  Indipendenti Lubrificanti (ASCHIMICI), Milano   132

La formazione dei prezzi degli oli usati in Europa
  P. GRARD, Presidente Onorario dell'I.H.M.B. (Industrie
  des Huiles Minérales de Belgique), Bruxelles, Vice
  Presidente della U.E.I.L.   143

Le esigenze economiche di una flessibilita di produzione nel
  settore della rigenerazione
  R. HAVEMANN, Vice Presidente dell'AMMRA, Amburgo   150

Comportamento degli organismi ufficiali degli Stati Uniti
  nei riguardi degli oli rigenerati
  M.E. LEPERA, T.C. BOWEN, US Army Mobility Equipment Research &
  Development Command   162
  J.F.   COLLINS, U.S. Department of Energy   166
  D.A. BECKER, U.S. National Bureau of Standards   171

La tecnologia della rigenerazione
  Dr. C. LAFRENZ, Delegato della Commissione Europea della
  Rigenerazione presso l'Associazione Tecnica dell'Industria
  Europea dei Lubrificanti (ATIEL), Germania   175

Qualità dei prodotti della rigenerazione fabbricati secondo le
  tecniche moderne
  C.J. RATCLIFFE, Operations Director, Anglo-Pennsylvanian
  Oil Comp. Ltd., Gran Bretagna   198

Risultati delle ricerche comparate sulle qualità rispettivamente
  degli oli rigenerati e degli oli nuovi
  Dipl.-Ing. E. WEDEPOHL, Institut für Erdölforschung,
  Hannover
  Prof. Dr.-Ing. W.J. BARTZ, Techn. Akademie Esslingen,
  Ostfildern (Nellingen)
  Dr.-Ing. K. MÜLLER, Ciba Geigy AG, Basel   212

Impiego di oli riraffinati con solvente nella formulazione di
  oli motore
  Ing. A. MODENESI, Responsabile dei Servizio Tecnici della
  Clipper Oil Italiana, Italia   225

Il mercato europeo dei lubrificanti e il ruolo delle imprese
  indipendenti
  Dr. M. FUCHS, Geschäftsführender Gesellschafter der Rudolf
  Fuchs GmbH & Co., Mannheim   239

Le economie di energia ottenute attraverso la rigenerazione
  degli oli usati
  C.J. THOMPSON, D.W. BRINKMAN, US Department of Energy,
  Bartlesville, Oklahoma, USA   253

Trattamento delle emulsioni
  J. DUMORTIER, Vice Presidente del S.N.F.R.H.G., Parigi   267

L'inquinamento dell'ambiente attraverso la combustione degli
  oli usati
  W.B. WALKER, Director, Braybrooke Chemical Services,
  Gran Bretagna   275

Aspetti della protezione dell'ambiente nell'esercizio degli
   stabilimenti di rigenerazione
   Dipl.-Ing. G. STEINMETZGER, Haberland Engineering GmbH,
   Dollbergen ; Ricerche di F.K. KRENTEL, Leitender Gewerbedirektor, Bezirksregierung Hannover   290

La protezione dell'ambiente nella riraffinazione degli oli·
   usati.
   Processo di ricupero delle melme oleose
   C. VERSINO, Istituto di Chimica-Fisica, Università di
   Torino
   C. MOLINO, O.M.A., Rivalta di Torino   310

Perchè rigenerare gli oli usati?
   Aspetti energetici ed ecologici del problema
   C. RICHARD, Compagnie Française du Raffinage,
   Gonfreville-l'Orcher, France   332

Conclusioni
   P. BRASSART, Presidente del Congresso   345
   F. DELMAS, Secrétaire d'Etat à l'Environnement   348

Lista dei partecipanti   351

BEGRÜSSUNG DER TEILNEHMER
WELCOMING ADDRESS
ACCUEIL DES CONGRESSISTES
ACCOGLIENZA AI CONGRESSISTI

H.-W. STUDDERS
Präsident der Europäischen Union der unabhängigen Schmierstoffverbände
President of the European Union of Independent Lubricant Manufacturers
Président de l'Union Européenne des Indépendants en Lubrifiants
Presidente dell'Union Européenne des Indépendants en Lubrifiants

Sehr geehrter Herr Präsident Natali, sehr geehrter Herr Präsident Brassart, verehrte Gäste, meine sehr geehrten Damen und Herren, liebe Kolleginnen und Kollegen,

ich habe die Ehre, Sie in meiner Eigenschaft als derzeitiger Präsident der Europäischen Union der unabhängigen Schmierölverbände zum zweiten Europäischen Gebrauchtöl-Recycling-Kongreß sehr herzlich begrüßen zu dürfen.

Diese Ehre wird mir zuteil, weil der Veranstalter des Kongresses, - die Europäische Kommission der Zweitraffinieure -, ein wichtiger Bestandteil unserer Europäischen Schmierölunion ist.

Die Themen des Programms, die sich mit wirtschaftspolitischen und technologischen Fragen befassen, spiegeln die große Bedeutung wieder, die die Aufbereitung von Gebrauchtölen für die gesamte westliche Welt hat. Der zweite Kongreß - das Timing könnte gar nicht besser sein, um die Bedeutung der Zweitraffination in das rechte Licht zu rücken - fällt in eine Zeit, in der sich die westlichen Industrienationen täglich von neuem ihrer Abhängigkeit von dem Rohstoff Öl bewußt werden müssen. Die rasanten Preiserhöhungen der vergangenen Monate für das Öl haben empfindlich in die volkswirtschaftlichen Vorgänge in den einzelnen Ländern eingegriffen. Die Regierungen haben erkennen müssen, wie sehr diese preislichen Bewegungen ihre wirtschaftspolitischen Konzeptionen treffen können. Es liegt daher auf der Hand, daß jede Möglichkeit genutzt werden muß, um den wertvollen Rohstoff Öl so nutzbringend, wie überhaupt nur denkbar, einzusetzen. Hierzu trägt die Regeneration von Gebrauchtölen in erheblichem Umfang bei. Die aufgezeigte Entwicklung hat die Bedeutung der Zweitraffination in beträchtlichem Umfang wachsen lassen. Der Zweitraffination dürfte daher künftig in erhöhtem Maße eine aufgeschlossene Betrachtungsweise sicher sein. Im Verlaufe der Tagung werden Ihnen die Vorträge hierzu sicherlich viele Gedanken, Anregungen und Hinweise vermitteln. Ich möchte den kommenden Referaten nicht vorgreifen. Deshalb beschränke ich mich darauf, Ihnen zur Einstimmung auf die Tagung zur Bedeutung der Zweitraffinate zu sagen, daß ihr Anteil am gesamten Aufkommen von Schmierölen und Schmierfetten in der Bundesrepublik etwa zwischen 12 und 13 % beträgt. In eine absolute Zahl übersetzt sind dies pro Jahr 230.000 Tonnen, - eine durchaus beachtliche Menge !

Die Regeneration des Gebrauchtöls hat aber nicht nur eine zunehmende wirtschaftliche Bedeutung, die sich zwangsläufig aus der aufgezeigten Entwicklung ergeben hat. In den Jahren des Überflusses schenkte man diesem Punkt nur eine sehr geringe Beachtung. Zu dieser Zeit stand die Zweitverarbeitung unter dem Gesichtspunkt eines wachsenden Umweltschutzbewußtseins. Der Umweltschutzgedanke hat in der Bundesrepublik Deutschland schon vor vielen Jahren dazu geführt, die Aufarbeitung von Altölen dadurch nachhaltig zu fördern, daß dem Verbraucher eine sogenannte Altölabgabe auferlegt wurde. Wenn auch hierdurch zunehmend größere Mengen der Wiederverwertung zugeführt werden konnten, so ist doch die Dunkelziffer der Mengen, die in gedankenloser Weise der Wiederaufarbeitung entzogen werden, meines Erachtens immer noch zu hoch. Man wird sich Gedanken darüber machen müssen, was man tun kann, um diesem Zustand besser begegnen zu können, ohne daß dies mit dem Aufbau eines Verwaltungsapparates verbunden wird, was nach leidvollen Erfahrungen auf anderen Gebieten die Gefahr beinhaltet, daß entweder die Wirtschaft mit Kosten belastet wird oder die Kosten für einen solchen Verwaltungsapparat höher sind als die für den eigentlichen Zweck der Erfassung von Gebrauchtöl zur Verfügung stehenden Mittel.

Was für die Bundesrepublik gilt - Umweltschutz und Verringerung der Rohstoffabhängigkeit - gilt nicht nur für die Mitgliedstaaten der Europäischen Gemeinschaft, sondern für alle anderen Länder in gleicher Weise. Deshalb ist es gut und richtig, daß die betroffene Wirtschaftsgruppe im Rahmen eines derartigen internationalen Kongresses eine Selbstdarstellung und Standortbestimmung vornimmt – eine Standortbestimmung sowohl in wirtschaftlicher als auch in technischer Hinsicht.

In diesem Sinne haben wir es besonders begrüßt, daß es gelungen ist, diesen Kongreß zustandezubringen. Mein Dank gilt all denen, die dazu beigetragen haben, daß wir uns heute hier versammeln konnten. Ich hoffe, daß die Erwartungen, die die Veranstalter mit dem Kongreß verbinden, in Erfüllung gehen mögen. In diesem Sinne wünsche ich der Veranstaltung einen erfolgreichen Verlauf.

ERÖFFNUNG DES KONGRESSES
OPENING OF THE CONGRESS
OUVERTURE DU CONGRES
APERTURA DEL CONGRESSO

L. NATALI
Vizepräsident der Kommission der Europäischen Gemeinschaften
Vice President of the Commission of the European Communities
Vice-Président de la Commission des Communautés Européennes
Vice Presidente della Commissione delle Comunità Europee

Monsieur le Président,
Mesdames,
Messieurs,

Je regrette de ne pas être en mesure, en raison d'obligations importantes et urgents, d'ouvrir personnellement ce Congrès et d'y participer.

La Commission a accepté le patronage de ce congrès et lui a accordé son soutien matériel, convaincue qu'elle est, de longue date déjà, de l'intérêt et de l'importance de la récupération des huiles usagées tant du point de vue de la protection de l'environnement que de celui des économies d'énergie.

Le Premier Congrès Européen sur les Huiles usagées, tenu en 1976 à Bruxelles, patronné également par la Commission, a contribué à sensibiliser les responsables politiques et économiques sur l'importance de ce problème.

Je rappelerai également que ce sujet a fait l'objet de la première directive adoptée par le Conseil des Ministres des C.E., le 15 juillet 1975, en application du programme d'action en matière d'environnement.

Cette directive réglemente l'élimination et la régénération des huiles usagées.

Chaque année, en effet - comme vous le savez - environ 2.200.000 tonnes d'huiles usagées sont générées dans la Communauté pour une consommation totale de lubrifiants de 4.300.000 tonnes environ. Un peu plus de 50 % de ces huiles usagées sont ramassées, les autres sont encore souvent rejetées sans contrôle, ce qui ne va pas sans conséquences dommageables pour l'environnement et plus particulièrement pour les eaux.

Quantitativement, il est vrai, les huiles usagées ne représentent qu'une faible partie du flux total des déchets dans la Communauté, mais leur importance est considérable en raison de : la gravité de la pollution qu'elles causent, de leur grande valeur économique, de la nécessité d'importer les matières premières indispensables à la fabrication d'huiles neuves et des économies de devises qui résultent de la récupération des huiles usagées.

Aussi la directive a-t-elle institué un système de contrôles rigoureux portant sur la collecte, le dépôt et l'élimination de ces huiles, tout en soulignant que "dans la mesure du possible, l'élimination des huiles usagées est effectuée par réutilisation (régénération et/ou combustion à des fins autres que la destruction)".

Les chiffres cités ci-dessus montrent qu'un important effort reste à accomplir dans les différents Etats membres pour que davantage d'huiles usagées soient récupérées dans des conditions satisfaisantes pour l'environnement. La crise pétrolière que nous connaissons depuis un certain nombre d'années et qui a trouvé une illustration dramatique dans les événements actuels au Moyen Orient, ainsi que la dépendance de la Communauté pour son approvisionnement en énergie et en matières premières font que l'intérêt porté à la récupération et à la réutilisation des huiles usagées, aussi bien dans les Etats membres que dans les pays tiers, reste entière. Votre Union aura comme par le passé un rôle particulièrement important à jouer à cet égard.

Quant à la Commission, il lui incombe notamment de veiller à ce que les directives communautaires soient transposées et appliquées rapidement et complètement dans tous les Etats membres. La mise en oeuvre de la directive sur les huiles usagées a connu quelque retard dans certains Etats membres. Même si la situation n'est pas encore tout-à-fait satisfaisante, on peut dire que pour l'essentiel le travail d'adaptation nécessaire est à présent accompli.

La Commission a soumis par ailleurs une communication au Conseil des ministres de l'Environnement du 9 avril 1979 dans laquelle elle traitait, dans le cadre de la politique de gestion des déchets, du problème des huiles usagées. Ce document a permis aux ministres de réfléchir aux résultats accomplis et aux moyens mis en oeuvre dans les différents Etats membres dans ce domaine.

Il me paraît, cependant, que deux problèmes doivent encore faire l'objet d'un examen attentif de la part de la Commission.

Il serait utile d'une part, de mieux connaître les avantages comparatifs de la régénération et de la combustion avec récupération de chaleur, compte tenu du bilan "ressources" et du bilan "pollution". Votre Union nous a déjà fourni des éléments intéressants à cet égard. Une étude est en cours pour approfondir cette question.

Il s'agit, d'autre part, du problème de la combustion des huiles usagées sans pré-traitement. En effet, cette utilisation des huiles usagées est en augmentation en raison même de l'accroissement du prix du pétrole depuis 1973. Or, la quantité de métaux lourds contenus dans ces huiles se traduit - comme vous le savez - par une aggravation de la pollution atmosphérique.

*
*   *

Il me reste, Monsieur le Président, Mesdames, Messieurs, à souhaiter à ce Congrès un plein succès.

La Commission attend de ce Congrès une information complète et actualisée sur la législation et l'état de la récupération des huiles usagées dans les Etats membres de la Communauté, un vaste échange d'expériences sur les aspects techniques et économiques de cette récupération ainsi que surtout des impulsions pour l'application accélérée des mesures prévues dans la directive communautaire relative aux huiles usagées, conduisant notamment à l'augmentation des quantités d'huiles usagées ramassées et réutilisées ainsi qu'à la solution d'un certain nombre de problèmes technico-économiques et écologiques dans le domaine du traitement et de la réutilisation des résidus de la régénération d'huiles usagées.

Je vous souhaite d'avoir des discussions fructueuses, constructives et une entière réussite.

DIE BEDEUTUNG DER ZWEITRAFFINATIONSINDUSTRIE FÜR DIE EUROPÄISCHE
WIRTSCHAFT
THE REGENERATION INDUSTRY IN THE EUROPEAN ECONOMY
L'INDUSTRIE DE LA REGENERATION DANS L'ECONOMIE EUROPEENNE
L'INDUSTRIA DELLA RIGENERAZIONE NELL'ECONOMIA EUROPEA

P. BRASSART
Präsident der C.E.R.
President of the E.C.R.
Président de la C.E.R.
Presidente della C.E.R.

L'ensemble des neuf pays de la Communauté Européenne consomme actuellement plus de 4.000.000 tonnes/an d'huiles de graissage de toutes catégories, ce qui se traduit par la nécessité, après utilisation, d'éliminer chaque année 800.000 tonnes d'huiles noires usagées.

A ces 800.000 tonnes, il convient évidemment d'ajouter :
. les huiles dites "claires" (Turbines, Hydrauliques, Transfo) qui sont susceptibles d'être réutilisées, soit en l'état, soit à la suite d'un traitement très simple, qui n'est pas un traitement de régénération proprement dit,
. ainsi que les tonnages très importants d'huiles solubles. Ces dernières, qui peuvent être éventuellement régénérées après cassage des émulsions sont évidemment justiciables au préalable de traitements qui, eux aussi, n'ont rien à voir avec le traitement de régénération tel qu'on doit l'entendre.

Quels que soient les schémas retenus pour organiser une collecte exhaustive de ces huiles usagées et si l'on veut que la collecte soit la plus exhaustive possible - pour des raisons évidentes d'environnement - il est souhaitable de réunir un certain nombre de conditions, que nous allons rapidement passer en revue.

REGENERATION DES PRODUITS

Il est apparu que les huiles noires - qu'elles soient moteurs ou d'origine industrielle - sont le plus souvent parfaitement régénérables (ainsi qu'un certain nombre de communications le mettront en évidence), à la condition qu'un minimum de ségrégations soit assuré à la collecte.

Le mélange d'huiles industrielles et d'huiles moteurs peut, en effet, entraîner entre les différents additifs des réactions chimiques parasites en cours de traitement qui, non seulement compliquent ce dernier, mais encore nuisent gravement à la qualité et donc à la valorisation optima des produits obtenus.

C'est pourquoi il est essentiel que les Administrations chargées éventuellement d'organiser la collecte ne contrarient pas, a priori, cet effort de ségrégation, mais au contraire l'encouragent.

COLLECTE DES PRODUITS

En ce qui concerne l'exhaustivité de la collecte, l'économie bien comprise de cette dernière suppose une organisation rationnelle des tournées.

De même, la recherche des responsabilités en cas de collecte insuffisante ou de défaut de collecte entraînant une pollution, suppose également de la part de la réglementation des dispositions telles que ces impératifs ne se voient pas contrariés.

Bien entendu, ceci ne signifie nullement que la concurrence et le respect des lois du marché ne soient pas respectés, mais il y a plusieurs moyens de faire jouer la concurrence à la collecte, qu'il s'agisse de l'adjudication périodique des zones (comme la législation française le prévoit) ou qu'il s'agisse éventuellement d'autres dispositions.

Ce qui est important, en effet, c'est que l'intérêt des collecteurs ne se trouve point, par le biais de la réglementation, en opposition avec l'intérêt de la collectivité.

Il appartient donc aux Administrations de faire en sorte que leurs prescriptions facilitent l'obtention du but poursuivi.

C'est pour avoir méconnu éventuellement ces nécessités que l'on a pu constater dans l'un des pays membres de notre Fédération la mise en difficulté de l'Industrie de la Régénération, non pas faute de tonnage mais, paradoxalement pourrait-on-dire, par excès de tonnage, les produits qui lui parviennent étant particulièrement difficiles à traiter (faute de ségrégation) et d'un coût anormalement élevé, entraînant corrélativement la nécessité d'une aide publique substantielle, ce levier formidable de progrès économique qu'est la concurrence étant mal utilisé.

Si nous insistons autant sur ces aspects c'est parce que l'Industrie de la Régénération ne pourra apporter à l'ensemble de l'économie des pays de la Communauté tout ce que ces pays sont en droit d'en attendre que si les réglementations tiennent compte de ces différents impératifs.

Au surplus, il ne faut pas perdre de vue le fait que l'élimination proprement dite des huiles usagées peut recevoir deux autres solutions, en dehors de la

régénération, qui sont :
- la destruction par le brûlage.
- ou, pire encore, le rejet pur et simple de ces produits dans le milieu naturel.

I - REJET DES HUILES USAGEES DANS LE MILIEU NATUREL.

Le souci d'assurer la protection de l'environnement a déjà conduit tous les pays de la Communauté à interdire le rejet des huiles usagées dans les sols, les canalisations, les cours d'eau et les eaux côtières.

A cet égard, on peut donc considérer que les Etats Membres ont d'ores et déjà pris les mesures demandées par la Directive édictée le 16 juin 1975 (art. 4) par le Conseil des Communautés Européennes au sujet de l'élimination des huiles usagées.

Il est dès lors inutile d'insister davantage sur un mode d'élimination unanimement écarté.

II - BRULAGE DES HUILES USAGEES.

Deux types de brûlage peuvent être envisagés :

- sans récupération de calories
- avec récupération de calories

A partir du moment où il est possible de récupérer de l'énergie en brûlant des huiles usagées, ne pas le faire serait un non-sens économique et c'est pourquoi nous n'examinerons ensemble que le seul brûlage de ces huiles avec récupération de calories.

Sous certaines réserves développées plus loin, l'huile usagée peut constituer un bon combustible puisque son pouvoir calorifique est assez voisin de celui du fuel lourd. Il serait dès lors possible de l'utiliser pour remplacer un fuel de cette catégorie, soit telle quelle, soit en mélange avec du fuel lourd.

A) Brûlage des Huiles Usagées en l'état.

Le brûlage des huiles usagées dans l'état où elles se présentent au moment de leur récupération présente deux inconvénients majeurs : il est à la fois dangereux et nocif.

1. <u>Dangereux</u> : L'huile usagée est un produit hétérogène qui contient très souvent - toujours lorsqu'il s'agit d'huile usagée moteur - des parties à très bas point d'inflammabilité (essence, gazole, etc...). Les risques d'explosion qui en découlent ont été malheureusement confirmés par un certain nombre d'accidents dont quelques uns très graves.

2. <u>Nocif</u> : Le brûlage des huiles usagées, s'il permet d'assurer la protection des sols et des eaux, ne fait en réalité que rejeter la pollution dans l'atmosphère.

   Les huiles usagées moteur - dont les quantités représentent approximativement 65 % des quantités totales d'huiles usagées à éliminer - contiennent des métaux qui, après brûlage, se retrouvent en totalité dans les fumées.

   On notera par exemple que l'huile usagée moteur a, le plus souvent, une teneur en plomb comprise entre 1.200 et 12.000 ppm.

   Cela veut dire que le brûlage dans l'ensemble des pays de la Communauté, de 5 à 600.000 tonnes d'huiles usagées moteur, reviendrait à rejeter chaque année dans l'atmosphère entre 2.000 et 3.000 tonnes de plomb.

   Cette seule considération devrait suffire pour que soit interdite une telle pratique et, ceci, sans préjudice du fait que cette dernière se traduirait en outre par la combustion de phosphore, de baryum et de zinc, par centaines de tonnes.

B) <u>Brûlage des Huiles Usagées en mélange avec du Fuel lourd</u>.

Parfaitement conscients du caractère dangereux et surtout nocif du brûlage pur et simple des huiles usagées, d'aucuns ont pensé que leur intégration à très faible dose dans le fuel lourd avant brûlage permettrait de résoudre le problème.

Il est vrai qu'une telle incorporation, effectuée selon un dosage "homéopathique", permet au mélange fuel + huile usagée de présenter des caractéristiques très proches des spécifications admises pour le fuel lourd, mais il s'agit là d'un simple camouflage : la dilution ne change en rien l'importance des rejets de métaux et les tonnages indiqués à la fin du paragraphe précédent se retrouvent de toutes façons et dans leur intégralité dans l'atmosphère.

C) Brûlage des Huiles Usagées sans pollution de l'air.

La seule solution permettant de brûler une huile usagée sans polluer l'atmosphère consiste à la prétraiter.

Une simple décantation accompagnée de déshydratation et de filtration ne suffit pas : pour retirer les métaux contenus dans l'huile, il faut en effet :
- ou la traiter également à l'acide sulfurique,
- ou la traiter avec un solvant approprié,
- ou la centrifuger à chaud après déshydratation et distillation,
- ou avoir recours à des techniques actuellement en cours d'étude, plus compliquées encore.

Un tel prétraitement aussi poussé de l'huile usagée ayant sa réutilisation comme combustible revient finalement à un véritable traitement de régénération, ce qui aboutit à transformer l'huile usagée en huile de base pour, en définitive, brûler cette dernière comme un fuel.

En d'autres termes, le fait de procéder à ce prétraitement reviendrait à engager des dépenses du même ordre de grandeur que celles de la régénération proprement dite, sans pour autant communiquer au produit la même valeur ajoutée et sans (à beaucoup près) permettre la même économie d'énergie.

L'intérêt du brûlage des huiles prétraitées s'avère donc finalement comme étant une opération économiquement stérile.

III - REGENERATION DES HUILES USAGEES.

- Intérêt économique de la Régénération.

    L'intérêt économique de la Régénération, tel qu'il apparaît à la suite d'une comparaison, à la fois technique et comptable entre les huiles neuves et les huiles reraffinées, sera mise en évidence au cours de ce congrès par une voix plus autorisée que la mienne.

    Je n'insisterai donc pas sur cet aspect, mais par contre je tiens à attirer l'attention sur toute une série de points que j'estime au moins aussi importants.

    Au cours des trente ou quarante dernières années, l'Industrie Européenne de la Régénération - née de la guerre 1939/45 - n'a cessé de poursuivre ses recherches, de développer ses techniques et d'améliorer ses produits.

En conséquence, l'huile de base produite par l'Industrie de la Régénération est actuellement qualitativement tout à fait comparable à l'huile de base obtenue en Raffinerie.

De cet état de fait, il en résulte les considérations suivantes :

1°) L'industrie de la Régénération est susceptible de fournir environ 15 % des besoins du marché en huiles de base.

Cette proportion très importante permet de comprendre que dans tous les pays où existe une Industrie de la Régénération vigoureuse, la concurrence que celle-ci exerce sur le marché des lubrifiants contribue à contenir les prix, ce qui est tout à fait conforme à l'intérêt de ses clients et donc, finalement, à l'intérêt du consommateur.

2°) Les huiles régénérées étant le plus souvent - pour ne pas dire toujours- consommées dans le pays d'origine, et à tout le moins dans le Territoire de la Communauté, chaque tonne d'huile régénérée produite à partir de cette matière première européenne qu'est l'huile usagée permet d'exporter à l'extérieur de la Communauté une tonne d'huile neuve produite par l'Industrie pétrolière à partir du brut importé.

Cette tonne d'huile neuve produite par l'Industrie pétrolière permet naturellement à cette dernière de valoriser en devises, non seulement la valeur en brut de ce produit, mais la valeur ajoutée par le travail européen et la rémunération considérable des investissements consentis par l'Industrie pétrolière européenne pour la production d'huiles de base lubrifiantes.

3°) Le fait pour la production communautaire indépendante de pouvoir, au départ de ces huiles usagées, couvrir 15 à 20 % des besoins en lubrifiants de la C.E.E. en cas de crise mondiale pétrolière est un point que les événements de Suez en 1956 et les événements actuels depuis la guerre du Kippour ont remarquablement mis en valeur.

On notera à ce propos que, dans la mesure où certains pays membres ont, en matière de stocks obligatoires de réserve, une réglementation qui s'étend également aux lubrifiants, on voit se conforter les objectifs de leur politique, puisque la disponibilité d'une ressource d'huile complémentaire d'origine communautaire ne peut que renforcer la portée d'une telle réglementation, en prolongeant l'autonomie fournie par les stocks.

4°) En cas de conflit armé, les grandes Raffineries pétrolières constituent malheureusement, traditionnellement, des cibles de choix, alors qu'au contraire les usines de régénération (relativement petites, dispersées et ne produisant pas de carburant) ont beaucoup plus de chance d'échapper à une destruction totale et, donc, de pouvoir continuer à fournir les produits nécessaires à la lubrification des matériels militaires ou industriels.

5°) L'avance technique considérable acquise par l'Industrie européenne de la Régénération depuis la dernière guerre et qui lui est reconnue dans le monde entier, lui a déjà permis et lui permettra encore plus dans l'avenir d'exporter son know how et ses techniques, jusques et y compris vers les pays les plus développés d'autres continents.

Il y a là, dans ce domaine, une source très appréciable d'entrées de devises et une augmentation non négligeable de la capacité des pays industrialisés à résister à une diminution de leurs ressources en lubrifiants.

CONCLUSIONS.

Entre le brûlage et la régénération qui sont les deux seuls modes d'élimination restant envisageables pour les huiles usagées à partir du moment où leur rejet dans la nature n'est plus concevable, il apparaît bien que la REGENERATION est le meilleur du point de vue de l'intérêt général des pays de la Communauté Européenne puisque, tout en assurant la protection de l'environnement, ce mode d'élimination :
- joue un rôle non négligeable dans la régulation des prix et de la lutte contre l'inflation dans le domaine des lubrifiants,

- procure un accroissement sensible de la ressource européenne en matière de lubrifiants en assurant une autonomie partielle non négligeable dans ce domaine,

- contribue très efficacement à la réalisation de la plus grande économie de devises.

Dans sa directive du 16 juin 1975, le Conseil des Communautés Européennes prescrivait une réglementation de la collecte et des dépôts d'huiles usagées mais laissait aux Gouvernements Nationaux le soin de choisir entre le brûlage et la régénération pour leur élimination.

Ce qui précède montre tout d'abord que le seul choix possible est la Régénération puisque même le brûlage nécessite une "régénération" préalable.

L'alternative devient dès lors la suivante : régénération vers le combustible ou régénération vers l'huile ?

La réponse est maintenant évidente :

La seule solution est la
REGENERATION VERS L'HUILE

Cette conclusion en entraine une autre, apparemment accessoire, mais en fait primordiale : la régénération vers l'huile implique que toutes les huiles régénérables soient effectivement régénérées. Or, seuls les professionnels de la régénération peuvent déterminer quelles sont ces huiles. Les ramasseurs n'ont, ni les connaissances techniques, ni les moyens technologiques de distinguer les huiles régénérables de celles qui ne le sont pas.

Il faut donc que la collecte des huiles usagées soit organisée sur les indications, voire sous le contrôle technique des régénérateurs.

D'autre part, certaines techniques de destruction des huiles solubles entraînent un gaspillage d'énergie - qu'il s'agisse d'incinération ou de traitement par évaporation - et ceci sans préjudice de la perte des huiles pour l'économie générale.

Or, ainsi qu'on le verra au cours de nos débats, il existe actuellement une technique de destruction de ces solutions qui, non seulement entraîne une dépense d'énergie minimum dans le cadre de l'opération de l'élimination, mais encore permet le recyclage de la partie huile obtenue vers la régénération.

Là encore, bien évidemment, les réglementations intervenues ou à intervenir se doivent d'exalter les techniques et les options économiques les plus favorables à la Communauté Européenne.

Charles A. VANIK
U.S. Representative - House of Congress (D-Ohio)

presented by Brian GONCHAR
Legislative Aide to Congressman Charles A. VANIK

It is an honor to present this paper to a delegation of world leaders in the field of rerefining used lubricating oil. Unfortunately, because of an unusually heavy congressional schedule, I am unable to present this paper myself. I have asked Jim McBain of the Association of Petroleum Rerefiners to make my regrets and offer this in my place.

I am not a technical expert in the field of petroleum refining. But I am completing twenty six years as a representative in the U.S. Congress, a period which enables me to speak on the development of a national used oil recycling policy. Events which have led to the momentous occasion of both houses of Congress close to approving recycled oil legislation have been torturously slow. The Federal Government seems to take two steps backward for each step forward. I believe by examining some of the actions of the past we-and you-can learn the difficulties of forging a political consensus on this issue, and use this information to make up for lost time so that the rerefining industry will be vigorous and healthy in the 1980's.

An Evolving National Policy

The legislation which has been adopted by the Senate and the House Interstate and Foreign Commerce Committee (S. 2412) has been a long time in the making. The players and backdrops for this legislative drama have changed with time, but the goals have not. Many of your strongest supporters have left the halls of Congress long ago. On the other hand, some of the world's greatest foot-draggers have also left.

In 1967 former Senator Boggs held a hearing before a subcommittee of the Senate Public Works Committee to investigate the waste oil problem. The need for a comprehensive solution to the problem was abundantly clear then. Some of the problems evident to us 12 years ago have been rectified, others have grown worse. As an outgrowth of these 1967 hearings, Senators Magnuson, Muskie and Boggs wrote to the Department of Commerce asking the National Bureau of Standards to perform quality tests on rerefined oil. The Commerce Department requested $ 300.000 for the task, but the request was denied by the Bureau of the Budget.

In 1970, the Council on Environmental Quality began a study of waste oil disposal, but the CEQ dropped its study largely because its staff could not collect adequate information on the topic. Before the CEQ forgot its investigation, however, it did ask the Defense Department to look into the possibility of procuring recycled oil.

Against this backdrop I introduced the first National Oil Recycling Act in Congress on December 2, 1971. After some revision, the legislation was reintroduced on June 14, 1972 ; and it is this bill, H.R. 15502, which has become the model for succeeding legislation. Let me dissect this bill for a moment.

The magnitude of the problem was as recognizable then as it is now. Approximately 2.5 billion gallons of lubricating oils were sold in the U.S. in 1970. The Association projects demand for lubricating oils to be 2.8 billion gallons in 1980. Of the billion plus waste oil generated by these sales more than 75 percent was disposed of in an environmentally harmful way. Today more than 95 percent is being dumped, burned, or disposed of in some other way that harms the environment or endangers our health.

The original legislation, rather than proposing new Federal laws making the dumping and incineration of used oils illegal, would have created positive incentives which would have led to the recycling of a much greater proportion of the used oils being wasted. While I have championed subsequent laws which tightened controls over water and air pollution and solid waste disposal, my bill sought to minimize the need for an administrative police force. First, it would have corrected the defect caused by the Excise Tax Reduction Act of 1965 and Revenue Ruling 68-108. By resubjecting new lubricating oil to the 6¢/gallon excise tax under section 4091, and by removing the tax rebate currently given off-highway users of lubricating oil under section 6424, the bill would have raised $ 75 million in additional tax revenue, more than enough to pay for the other provisions of the bill. The Energy Tax Act of 1978 corrected part of this deficiency by removing the excise tax on rerefined lubricating oil.

The other major Federal defect, the FTC labeling requirement was to be removed. The word "recycled" was to be placed prominently on the face of all recycled oil containers rather than the words "previously used" as is presently required. I optimistically stated then that, "... once the public is educated on the relative merits of recycled oil, and methods for quickly and economically comparing it with new oil are developed, the use of recycled oil should increase significantly."

The bill would have also changed the Federal procurement practice of the time to not allow the use of recycled products. When such oil was available at competitive prices, the bill required that it be purchased and used by all agencies of the Federal government, including the military and private contractors.

Another provision to encourage public acceptance of recycled oil was to require convenient, leakproof containers to return used oil to the marketers. These returnable containers would have carried a 10¢ deposit. Not a high price for the enormous benefits to be gained automobile owners recycling their oil. All marketers would be required to provide ample disposal facilities and would be encouraged to deliver or sell this collected used oil to you, the oil recyclers.

The costs of not adopting this policy in 1972 are difficult to assess, but I am convinced that if a comprehensive oil recycling policy had been adopted in 1972, we would have saved approximately 122 million barrels of oil. This oil has cost America more than 1.7 billion dollars. Other costs to the environment and health are much more difficult to estimate, but have been tremendous. We could have made great progress in reducing our oil import bill, which will approach the $ 90 billion mark this year. This energy savings takes into account both the energy used in refining and the energy lost by not burning that portion of used lubricating oil that is presently being used for that purpose.

It was late in 1972 when the Defense Supply Agency issued the first comprehensive government report on waste oil. The DSA study made recommendations to encourage recycling of waste oil and the purchase of rerefined lube oils by the Defense Department. Although the proposal went to the Department of the Army in early 1973, it was not until recently that the Defense Department changed its procurement policy.

In 1972 Congress passed the Water Pollution Control Act amendments which mandated EPA to conduct a two year waste oil study. After two years of study and despite specific policy recommendations developed by the EPA researchers, the agency issued a call for further research and study. At that time I restated my position that, "It is clear that if the FEDERAL government is ever going to confront the waste oil problem, Congress is going to have to legislate a solution. All that is required is the will to recognize the problem, confront it, and overcome the bureaucratic stagnation that has thus far plagued efforts at improving waste oil recovery practices."

Since the disappointing EPA Study, several advances have been made. First, the Federal Energy Administration embarked on a waste oil program which has led to a greater understanding of the problem and increased local participation in solving the problem. Early in 1975, the FEA published a Waste Oil Fact Sheet which clearly stated the problems of the rerefining industry and the government obstacles which must be overcome. In an effort to improve access to used oil for rerefiners through better collection at the local level, the FEA drafted a Model Used Oil Recovery Act for use by state and local governments. The model law was introduced about one year after the FEA began its waste oil program. The FEA then began, together with the EPA, a national oil recycling program targeted to the individual who changes his own oil and dumps the drainings. The Oil Recycling Kit offered in the fall of 1976 is one tangible result of this education program.

Your Association can make a real difference at the grass roots level. Many states have already undertaken ambitious recycling programs. In California more than 36 million gallons of automotive and industrial lubricating oil were recycled last year. To encourage used oil recycling, the state has established a program entitled SOAR (Save Oil America, Recycle). Under the program, more that 2,400 gasoline stations, stores and recycling centers have been formed into a statewide network of collection points.

Illinois is another state injecting new life into old oil. The Institute of Natural Resources, a state agency, is now running a statewide campaign operating in over 300 communities to monitor the 1,500 collection stations. To educate oil wasters, the INR has publized its program in local newspaper articles, over radio and television, and on bumper stickers and posters. San Diego, California and West Virginia use rerefined oil in their official vehicles. Taxi fleets and many bus, truck and railroad lines have operated on used oil for decades.

The Association of Petroleum Rerefiners has made great progress in the adoption of a national oil recycling policy ; you must now concentrate your energy and resources on getting your message to America. Results of the National Bureau of Standards equivalency test for virgin and rerefined oil are several years off. However, this last vestige of discrimination by the Federal Government will soon disappear. Donald A. Becker, Chief of the Bureau of Standard's Recycled Oil program, recently stated, "An adequately rerefined oil can be as good as a high quality virgin lubricating oil ... there is no question that oil recycling is here to stay." The Associations must now expand their program of political education, which has been successful, to include public education.

Reasons for Rerefining Industry Decline

Of course, the U.S. Government is not entirely to blame for the decline of the rerefining industry. An uncertain supply of used oil and a shrinking profit margin as refiners compete with heating oil dealers for a limited supply have forced many rerefining companies out of business. These problems have been exacerbated by the OPEC crisis. A lack of innovation and a falling behind of the changes in the lubricating oil business as more and more additives are invented have also contributed to the decline.

The premium being paid by fuel oil dealers, more expensive collection because the largest single source of used oil, automobiles, is more dispersed as people change their own oil and the disposal of the acid sludge by-product have all drive your costs skyhigh. These added costs have forced old and inefficient plants, many operating at less than capacity design, out of business. The needed economic and technical changes can only come about through the combined investment of government and business in new technologies and processes. In response to the mandate contained in the Energy Policy and Conservation Act, the Department of Energy is engaged in on-going research at their facilities in Bartlesville, Oklahoma. I will continue to support a strong Federal participation in this project. I understand that technologies exist which produce a rerefined oil of high quality, but do not produce hazardous waste. One such state of the art technology is the Phillips Re-refined Oil Process or PROP. They have begun construction of a demonstration facility in a North Carolina community.

I hope this example of a close-working relationship between business and government can be repeated throughout the country.

The dramatic almost daily changes in the energy picture require a constant and close vigilance by your industry of economic, political and technical changes. More than most industries in America, you are compelled to monitor these changes to survive and prosper. The good you do deserves support from all sectors of our society and I wish you a successful and prosperous future.

THE AMERICAN RE-REFINING INDUSTRY
YESTERDAY AND TODAY

James A. McBAIN
Association of Petroleum Re-refiners, Washington, U.S.A.

Thank you. It is indeed a pleasure for the Association of Petroleum Re-refiners and for me to be here representing the re-refining industry in the United States. I would like to express my personal gratitude and the gratitude of the members of the APR to Patrick Brassart for his kind invitation and attention and help in making our visit to your country such a pleasure.

The role of this panel will be to inform you on the important elements in the development of the U.S. re-refining industry, the industry's status and it's outlook for future growth. My role will be to bring you up to just a little bit before yesterday. Kimball Morris, President of the Association of Petroleum Re-refiners, will discuss today's technology and it's outlook for tomorrow. We hope the information we provide will be useful to all of us in determining areas of common experience and problems. Then, perhaps together we can develop common solutions.

HISTORY

The conservation and recovery of petroleum products is not a new phenomenon. In the United States, the process dates back to 1915. At that time, used oils went through a process of heating, settling, and centrifuging, and then on to renew their lubricating functions. The Armed Forces used this renewed oil all during World War I and after. Commericial airlines began using renewed oil in 1932, and during the Second World War, virtually all of the used oils in the U.S. were re-refined. By 1950 almost one quarter of all Air Force engine lubricants were re-refined. However, with the advent of the jet age, more completely formulated lubricants were needed, and the use of re-refined oils declined.

But other markets were opening. America was fast becoming a mobile society, and our love affair with the automobile stimulated automotive lubricant markets to grow rapidly. The re-refining industry was able to compete favorably with the producers of new oil. By 1960 the industry claimed well over 150 companies producing over 300 million gallons of re-refined lubricants - almost 18% of the nation's total lubricant consumption. The stage was set for an industry with dramatic growth.

Today our industry is composed of less than 10 companies producing less than 100 million gallons of re-refined lubricating oils. There are many reasons that the industry in the United States did not grow as expected. We can point to a few.

FIRST, the use of used oil as a road dust suppressant and as a fuel increased dramatically. The increased competition for used oil reduced the flow of re-refiner's feedstock, and led to drastically higher feedstock prices for the feedstock they did get.

SECOND, improvement of automotive technology led to the development of higher performance engines. Lube oil formulation became more complex in order to meet the specifications for these newer and faster engines. The more sophisticated additives were much more difficult to remove in the re-refining process and higher levels of re-refining technology were necessary. The industry began the process of technological change, but product quality did not grow as fast as the growth in automotive and lubricant technology. This cast a stigma of product inferiority, which still plagues the industry today.

THIRD, overcapacity in the fresh oil market brought lube oil prices down sharply. The lower market prices prohibited re-refiners from passing increased cost of changing technology on to the consumer. This cost/price squeeze continued to tighten creating financial losses for many re-refiners, and a shrinking of the industry.

All of this is not unique. Every industry, vibrant and active today has gone through its own process of technological change. Some companies lost their places of prominence, only to be replaced by newer companies with new ideas and advancing technology. It is the process of industrial growth. But for the United States re-refining industry, this process of growth was stunted by several important government actions.

The Department of Defense (DOD) eliminated re-refined oils from its procurement list. DOD specifications were used as guidelines by most federal, state, and local governments, and therefore the entire governmental market was closed to the re-refining industry.

The Federal excise tax of 6¢ a gallon was imposed unfairly on re-refined lubricants. This had to be absorbed by the re-refiners.

One more decimating blow to the re-refining oil market was the Federal Trade Commission's labeling rule imposed in 1964. In essence, the rule required that re-refiners indicate prominently on the label, "Made From Previously Used Oil". This, connotated an inferior product.

These government actions increased the cost/price squeeze further, eliminating more and more re-refiners from the industry, and thus almost bringing to a complete halt, the process of technological change that is just now beginning to occur today.

The advent of the oil embargo and the continued increase in energy and lubricant prices led to a national demand for conservation of all of our energy resources. Re-refining of used oil can save the U.S. up to 70 thousand barrels of oil a day, and for 365 days a year that adds up to a very significant number.

Congress has already recognized the importance of re-refining to the country's energy conservation efforts. Congress has exempted re-refined oil from the 6¢ a gallon excise tax. The DOD has changed its specifications to allow re-refiners to qualify their oils for government purchase. Just last month the Federal Trade Commission announced that the words "recycled petroleum product" fulfills the requirements of its labeling rule. At this writing Congress is considering legislation that would completely eliminate the need for such labeling and establish a grant program to expend 10 to 25 million dollars in promoting the recovery of used oil.

The Environmental Protection Agency is expected to promulgate its used oil regulations under the Resource Conservation and Recovery Act (RCRA) in the latter part of this year. While we can not be completely certain as to what these regulations will do, we do expect that they will restrict the use of used oil as a road dust suppressant and limit the use of untreated used oil as a fuel. This will mean more feedstocks for the re-refiner.

The DOD and FTC actions already are beginning to expand the markets, and financial analysts forecast a favorable return on capital investment. Already we have seen significant investment in new technology.

In 1978 at our Third International Conference on Waste Oil Recovery and Reuse, entitled "The Road to Recovery", we said that the industry would soon be taking steps forward on it's own road to recovery. In the past two years we have seen new growth. Four new installations utilizing new technology are in operation. We expect the next 2-3 years to be more dramatic. Government barriers have for the most part been eliminated and we are working on new technology. There are new vistas for re-refining and reclaiming oil throughout all of the industry. I think you would be interested to see a view of the future from one of our new re-refiners. Let me present Kimball Morris, President of the Association of Petroleum and also President of Cam-Or, Inc., operating Westville Oil Division in Indiana.

UNITED STATES RE-REFINING - 1980

KIMBALL L. MORRIS
President
ASSOCIATION of PETROLEUM RE-REFINERS

Thank you Jim. First of all I, too, would like to thank you for having the opportunity to speak with you today. Before I get into my talk with you this morning, I would like to tell you a little about CAM-OR, INC. CAM-OR, INC., is a small publically held company in the United States which own a re-refinery in Westville, Indiana. This re-refinery until recently was an acid-clay re-refinery In 1978 we switched to a proprietary high vacuum distillation process. Our capacity is approximately 30,000 gallons of material a day. The majority of our finished product is marketed on a custom recycling basis where a customer furnishes oil to us and we return the material back to him, blended to his specifications and needs. In addition, we are also in the waste oil pickup business with operations both out of our Westville Refinery and Indianapolis, Indiana.

Now that you know something about the company I represent, I would like to examine what has happened in the US re-refining industry in the last 12 months and to give you my feelings on the direction and future growth prospects for re-refining in the United States.

The re-refining industry in our country is entering a period of rapid technological change. Our three largest re-refineries - Motor Oils Refining, CAM-OR, INC., and Ekotech have each made the change away from the acid-clay process to some form of distillation. There is very little in the printed literature concerning the processes that each of these companies are using, but, I do believe that each company feels it's process is somewhat propriatary. The product coming out of each of these refineries is a high quality rerefined base stock which is suitable for compounding or blending into a variety of both motor oil and industrial products. In addition to the above three companies, Booth Oil of Buffalo New York, recently announced it had secured financing to build a plant in Buffalo, New York area. The Booth family has not announced the process they will be employing in the new facility. I believe it is not one of those processes that are commercially available from Phillips or Resource Technology. I do know the plant will employ some form of vacuum distillation.

If a new process is to succeed in the United States, it must produce little or no waste by-products. The passage of the Resource Reclamation & Recovery Act of 1976 has resulted in the Environmental Protection Agency in our country enacting thousands of pages of rules and regulations in regard to the disposal of waste by-products. In order for waste material to be disposed of in a landfill in the United States, the material must undergo a series of difficult and complex leeching tests. Waste products passing those tests can go into any normal sanitary landfill. However, if a waste product does not pass those tests, it is branded as hazardous, in most cases, and must go to a special landfill with substantially higher disposal costs resulting. The principals of each of the three companies that have switched away from acid-clay to new processes indicate they have little or no by-product waste disposal problems.

The recent development of new processes that are available for sale or license both in the United States and Europe has stimulated renewed interest in rerefining in the United States. There are two process technologies currently available from United States companies, for either direct sales or license. One is PROP - the Phillips Petroleum Re-refining Process and the other is a process developed by Resource Technology of Kansas City, Missouri. The first commercial PROP plant is in its final shakedown stages in North Carolina. This plant is owned by the State of North Carolina and will recycle oil from state vehicles and also act as a research project on rerefined oil for the State University. The PROP plant, installed by Mohawk Lubricants in Canada, has also begun commercial production. I understand that there are people from Mohawk attending this conference and I am sure they will be glad to answer any questions about PROP with you individually. In addition, a company called Clayton Chemical was recently announced as the purchaser of a PROP facility in St. Louis, Missouri. This is to be a 5 million gallon PROP plant according to the press release on the facility and should be built within the next 14-15 months. There are also a lot of rumors pending on other PROP installations in the United States including the possibility of facilities by major oil companies such as Gulf and Texaco

A research company in Kansas City, called Resource Technology has also announced a commercially available process which has the unique advantage of being able to be retrofitted into existing rerefining plants. This process is presently being used commercially in an old Kansas City Re-refinery with a reported high degree of success. Our CAM-OR labs have reviewed the oil coming from this process and I thought you might be interested to hear the base specifications of the sample that we received from Mr. Sparks of the Resource Technology organization. The sample had a Viscosity of 200 Sec at 100°F, 46.6 Sec at 210°F for a Viscosity Index of 97.42. The API Gravity was 26.7, Flash was 420°F, the Fire Point was 465°F and the total acid number was .420. The process also produces a #6 like residue which our labs calculated at an API Gravity of 10.6, Viscosity at 210° of 644 and a Flash Point of 580. Resource Technology indicates this material may be commercially desirable by certain types of asphalt manufacturing facilities.

To sum up, there is a great deal of activity going on in the United States in terms of developing new technologies and new processes for re-refining oil. I am sure that we will continue to see more and more companies studying the re-refining industry in the United States as a potential market and looking at the variety of processes and perhaps coming up with some new ones, on their own, that might be used to successfully enter the market place.

In the next few minutes I would like to talk about the market factors currently affecting re-refined oil in the United States. Jim McBain has already indicated to you there have been many Government imposed restrictions on the use of rerefined oil in this country. These restrictions artificially prohibited the growth of a viable re-refining business in the United States and those of us who remained in the business remained in it as speciality companies servicing a particular market or markets. As you have heard, one of the largest United States Markets, the automotive market, was literally shut out to the re-refiner by taxation, Government policy and labeling restrictions.

The Association of Petroleum Re-refiners has taken an aggressive stance in the last 15 months to try to do something about some of the market restrictions that have affected our business. In April of this year, we petitioned the Federal Trade Commission to either remove the labeling restriction altogether or to change the label requirements from "made from previously used oil" to "this is a recycled product". The Association felt that the phrase "made from previously used oil" had strong negative connotations in American society. The word USED in America generally means defective or not as good as new. The first hope was to remove the labeling entirely, but, we did recognize that Government regulations, specifically the Magnuson-Moss Act, required the Federal Trade Commission to go through a long and cumbersome hearing process in order to eliminate a trade regulation. Consequently, we suggested the alternative "recycled" oil as a more desirable label than the current label in use.

On August 18th, the Federal Trade Commission issued a ruling modifying their long standing labeling regulations eliminating the requirements of the large display letters on a can of rerefined motor oil indicating the oil was "made from previously used oil". In it's place the FTC did accept the use of the word "recycled" oil product. While the labeling requirement was not completely removed from motor oil products, I believe the change APR helped effect with the FTC will allow more rerefined oil to go towards the motor oil market in the future.

At the same time we were petitioning the FTC for a rule change, Congress was looking for ways for the United States to decrease it's dependence on foreign oil. Among one of these solutions was the encouraging development of a viable rerefining industry in the United States. The Senate, under the leadership of Senator Domenici from ARizona, passed Senate Bill 2412 in July of this year. This legislation addressed many of the problems that the rerefining industry in the United States had faced in previous years. It specifically delt with the labeling problem that our Association and industry had been fighting for so many years. Basically, this legislation eliminated all restrictions on rerefined oil in the motor oil market but did require the

Environmental Protection Agency to study whether purchasers of the re-refined oil were informed of the quality of the product and it's fitness for intended use. The API/SAE labeling codes are in widespread use and we at APR feel these codes would certainly provide sufficient consumer information under the terms of this Bill.

As it takes both the House and Senate of the Congress of the United States to make a Bill law, a similar Bill was introduced in the House of Representatives by Representative Florio of New Jersey, and Representative Dingell of Michigan. We believe that this Bill will entirely eliminate the need for labeling of rerefined motor oil in the United States, provided that the API/SAE automotive quality codes are used by the rerefiner as a statement of the product's fitness for use. I hope at the time this speech is given, we will have had some news on whether that Bill has passed the House. Perhaps, more importantly, on what form the Bill will take.

Assuming the House Bill passes, a Conference Committee between House and Senate Representatives will meet to compromise on the differences between the two pieces of Legislation and both Acts will become Law when the President of the United States signs them.

While growth of rerefined motor oil in the automotive market has been limited by Government restrictions, acceptance of rerefined oil by industries in the United States is increasing. The Resource Recovery Conservation Act (RECRA) has caused industry in the United States to take a hard look at what it previously considered to be it's waste products. I believe we will see more industries trying to utilize their used oil by using in-plant reclaiming techniques and by rerefining. Since the publication of RECRA regulations, my own company has had inquiries from such widely diverse industries as the steel mills on one hand to companies who brew America's beer on the other hand. The last example is a very interesting one. A high percentage of beer in America is sold in cans and now brewers have installed lines to extrude aluminum cans next to, or even in their breweries. These lines use a high cost extrusion oil, which while difficult to re-refine, can offer a new market to America's rerefiners. To sum up, I believe the industrial

market for rerefined oils in this country should grow rapidly in the
next few years because of industry's need to conserve and turn what
industry previously viewed as waste or by-products into assets. The
Association of Petroleum Re-refiners will be sponsoring an International
WasteOil Conference in Las Vegas, Nevada on September 28 to October 1,
1981, with one of the primary objectives of looking at the way industry
can turn waste oil into valuable and recoverable resource. We invite
all of you to attend. (By the way, both Jim McBain and I have
reservation forms.) We, at APR in the United States, have recognized
a need for our Association to broaden it's scope and to act as a focal
point for the recovery and continued reuse of all waste oil products in
our country. With that in mind, APR recently enlarged its membership
to include those companies that reclaim oil by mechanical means for
further use as a lubricant. We have realized that the interest of the
reclaiming industry in the United States coincides closely with that of
the rerefiner. The enlargement of our Association should allow us to
have a greater voice in convincing industry and government that used
oil should be reused for lubricant purposes.

Yet, in spite of all the things that APR and the rerefining industry
in the United States has accomplished in the last year, the amount of
oil that is rerefined in the United States still is less than 5% of
total lubricant consumption. A lot needs to be done to increase that
percentage to a meaningful level of our nation's lubricant's needs.
I believe the technologies that are on the market place and the
Governmental developments of 1979 and 1980 should provide our industry
with a base to grow that percentage to a more respectable level in the
coming years. I certainly hope that the next time the President of APR
has a chance to address you, he will report that 1979 and 1980 years
were the turning point for rerefiners in the United States.

I would like to thank you today for giving me the opportunity to
talk to you and will be happy to answer any questions you might have.

Dr. J. SCHVARTZ
Rapporteur du Budget de l'Industrie à la Commission de la Production et des Echanges de l'Assemblée Nationale, Membre du Conseil Supérieur du Pétrole

Mesdames, Messieurs,

C'est la deuxième fois que je participe à un congrès international consacré à l'industrie de la régénération des huiles usées. En 1978, lorsque j'ai assisté brièvement à une partie du congrès de Houston, je dois dire que l'avenir de l'industrie française de la régénération me paraissait bien menacé. Aujourd'hui, deux ans après, il me semble que cette industrie peut regarder l'avenir avec un certain optimisme et aborder la prochaine décennie avec confiance.

Compte tenu du programme général de ce congrès et compte tenu des circonstances au cours desquelles je suis amené à prendre la parole, je pense que chacun attend de moi que je sois bref et que par ailleurs il me revient de sortir quelque peu des problèmes techniques pour lesquels de surcroît je suis tout à fait incompétent. Je me contenterai donc très succinctement de rappeler en quelques mots les conditions générales dans lesquelles l'industrie de la régénération a eu à vivre depuis une dizaine d'années et à rappeler également quels ont été les obstacles à son essor.

Lorsque, en 1974, j'étais amené pour la première fois à m'intéresser aux problèmes de la régénération des huiles usées, j'ai été à la fois saisi par la simplicité des données de base au regard de l'intérêt général et par l'extraordinaire complication de la démarche de l'Etat en la matière.

En effet, à la suite des désordres monétaires et de leurs conséquences immédiates, à savoir l'augmentation brutale des prix du pétrole, les pouvoirs publics français avaient défini les quelques grandes lignes déterminant le cadre des décisions économiques qu'il était nécessaire de prendre. Il s'agissait d'assurer à la France une balance des paiements en équilibre, de développer sur notre sol les activités à forte valeur ajoutée, d'économiser l'énergie et les matières premières et de redéfinir une croissance jusqu'alors trop peu soucieuse de l'environnement.

Il était clair que le recyclage des huiles usées était un bon exemple, certes mineur mais cependant éclairant, des dossiers qu'il convenait de pousser : la régénération des huiles permet d'économiser l'énergie et les matières premières ; c'est une activité qui localise sur notre sol la valeur ajoutée ; elle permet de moindres importations et elle peut jouer un rôle décisif dans la diminution de la pollution puisque le rejet des huiles usées dans la nature a représenté à une certaine époque entre 25 et 35% de la pollution des nappes phréatiques dans certaines régions de notre territoire. C'est pourquoi j'ai été surpris - et le mot est faible - de l'extraordinaire timidité de l'administration française dans ce domaine. Pendant très longtemps je n'ai rencontré dans les milieux dont dépendait l'instauration en France d'une réglementation nécessaire au développement de l'industrie de régénération qu'incohérence, confusion, hypocrisie, laxisme, hésitation, arrière-pensées, j'arrête là cette énumération qui n'est pas cependant limitative.

Si l'on cherche à comprendre pourquoi l'industrie de la régénération française a été pendant si longtemps bridée dans son développement, je crois que l'explication provient du fait que le dossier se situe à un carrefour. D'une part, il y avait la politique pétrolière française qui est une affaire d'Etat et d'autre part il y avait les tares de notre, ou plutôt, si l'on veut être plus exact, de nos administrations dont le comportement paraît parfois à l'observateur comme particulièrement opaque dans ce sens que l'on peut difficilement discerner son mobile et ses buts.

En quelques mots je pense qu'il faut resituer l'affaire de l'industrie de la régénération dans le contexte de la longue et difficile histoire pétrolière française. On sait en effet que la France dispose de ressources d'hydrocarbures peu importantes sur son sol et que les différents gouvernements ont tenté de mettre en place, d'une part un système réglementaire permettant d'obtenir une sécurité d'approvisionnement optima et, d'autre part, de permettre le développement des firmes pétrolières contrôlées par des capitaux français. Cette politique s'est longtemps symbolisée par la fameuse loi de 1928 et par la constitution des deux compagnies pétrolières françaises, à savoir la C.F.P. et Elf-Aquitaine. Pendant très longtemps également les pouvoirs publics ont désiré que ces deux compagnies puissent contrôler à l'extérieur la production d'hydrocarbures équivalant à la consommation annuelle de la France et puissent contrôler au moins la moitié du marché français. Il faut bien voir que ces objectifs ont nécessité d'établir une sorte de contrôle des prix sur le marché intérieur, de telle sorte que le jeu naturel du marché n'aboutisse pas à contrarier l'objectif que j'ai signalé tout à l'heure, à savoir, réserver la moitié du marché national aux compagnies françaises. Ceci a signifié que les prix français des produits pétroliers, premièrement ont moins varié que les prix dans les autres Nations, que deuxièmement vraisemblablement les filiales des majors ont gagné plus d'argent en France que le marché ne l'aurait permis et que, troisièmement, compte tenu du fait que les gouvernements successifs ne pouvaient pas faire payer aux industriels français le pétrole plus cher que leurs concurrents des autres pays, il était nécessaire de subventionner les compagnies nationales dans leurs efforts de recherche tandis que, dans le même temps, la France mettait peu à peu en place un dispositif fiscal qui, à l'image du dispositif fiscal américain, dégrevait en fait d'impôts les compagnies pétrolières françaises.

C'est donc un cadre dirigiste qui a prévalu mais, à l'intérieur de cette construction compliquée, on a ménagé quelques "fenêtres" que l'on peut d'ailleurs appeler des "fenêtres de profit". L'une de ces fenêtres de profit était constituée par le marché des lubrifiants. Sur ce marché, les prix ont été très largement fixés par les compagnies pétrolières elles-mêmes qui, pour employer une expression un peu triviale mais imagée, faisaient leur "gras" sur la vente des huiles moteur notamment. Il est clair que cette fenêtre de profit ne pouvait rester largement ouverte que si aucune autre huile ne venait concurrencer celles mises en vente par les compagnies pétrolières dominant, à l'époque, le marché des lubrifiants. C'est pourquoi celles-ci se sont arrangées pour freiner au maximum le développement de la régénération des huiles usées.

Or, dans le même temps où le gouvernement acceptait cette fenêtre de profit à l'intérieur d'un cadre très largement contrôlé, les pouvoirs publics, conscients de l'intérêt de la régénération des huiles usées, interdisaient aux détenteurs de ces huiles de les brûler. Il s'agissait, en interdisant le brûlage des huiles, d'orienter celles-ci vers l'industrie de la régénération. Cependant, cette réglementation contrariait la volonté des compagnies pétrolières de vendre leurs lubrifiants au prix maximum, et certaines compagnies pétrolières ont pensé à se doter d'un moyen éventuel de contrôle du ramassage des huiles usées.

Je me dois de vous signaler, pour compléter ma description de cette grande spécialité française qu'est la contradiction dans la démarche des pouvoirs publics que, alors que les administrations étaient tout à fait au courant des pratiques anti-concurrentielles qui avaient été mises en place à l'instigation de certaines compagnies pétrolières, ces pouvoirs publics continuaient, dans le même temps, à accorder une aide à l'industrie de la régénération sous la forme d'une détaxe fiscale : les huiles neuves payaient en effet une taxe intérieure alors que les huiles régénérées ne la payaient pas.

Telle était la situation en 1974, date à laquelle les retombées de la crise monétaire, de la guerre du Kippour et de la hausse du prix du pétrole ont nécessité un redéploiement de notre industrie mais également une redéfinition de toute une série de comportements et de réglementations, parmi lesquels ceux et celles qui nous intéressent. Je souhaite aujourd'hui passer sur les différentes péripéties, pourtant nombreuses et à bien des égards passionnantes pour le sociologue politique, qui ont marqué les six dernières années. 1980 en effet a vu la mise en place d'une nouvelle réglementation qui fait coexister d'une part un système de ramassage des huiles usées qui, espérons-le, permettra une collecte aussi exhaustive que possible de la matière première et d'autre part un système d'aide financière à l'industrie de la régénération devant favoriser des investissements de modernisation permettant le traitement d'un volume sans cesse croissant d'huiles usées. Je crois qu'il est inutile d'entrer dans le détail de ces systèmes qui vous a été ou vous sera exposé par ailleurs. Je crois également inutile d'essayer aujourd'hui de prévoir ce que donnera finalement cette réglementation assez complexe car le recul nous manque encore. Je voudrais simplement essayer de résumer en quelques mots ce qui me donne quelque espoir de voir ce dossier enfin sorti des ornières dans lesquelles il s'était embourbé.

Tout d'abord, les augmentations générales et continuelles de prix qu'a connues le marché des produits pétroliers depuis 1974 ont dû à mon sens donner mauvaise conscience aux pouvoirs publics en matière d'économie d'énergie.

Deuxièmement, j'ai noté au cours des dernières années une évolution favorable de l'attitude des compagnies pétrolières françaises qu'il s'agisse de la S.N.E.A. ou de la C.F.P. qui ont adopté vis-à-vis de l'industrie de la régénération une attitude positive contrastant heureusement avec certains agissements anciens. Ce changement de politique a certainement des causes complexes qu'il est très difficile pour un observateur extérieur de démêler, mais je pense qu'à des considérations d'ordre commercial dont le poids est certainement important, la persévérante action du Syndicat français des régénérateurs a permis de faire triompher au sein des états majors des grands groupes pétroliers français une appréciation objective et réfléchie des problèmes posés.

La troisième raison du déblocage final de ce dossier tient à la fermeté dont a fait preuve, dans les difficiles négociations interministérielles, le Ministre de l'Industrie, M. André Giraud. Je profite de l'occasion qui m'est donnée pour rendre hommage publiquement à son fair play et à son sens de la parole donnée.

Certes, des difficultés surgiront encore, notamment en raison d'un dialogue qui sera toujours difficile entre les régénérateurs et les administrations. Non seulement parce que le dialogue avec l'administration française est consubstantiellement difficile, mais parce que quatre ministères sont compétents - j'emploie ce mot avec des guillemets - et que dans ces conditions il est habituel que plutôt que d'essayer de construire, chacun essaie d'imposer son point de vue pour des raisons de défense d'un pré-carré administratif ou d'une souveraineté sur tel ou tel aspect de la réglementation. C'est pourquoi il me semble que l'industrie de la régénération doit rester vigilante et surtout unie tout en maintenant la ligne raisonnable qu'a su définir son Syndicat.

J'ai finalement peut-être été un peu trop long et je vous prie de m'en excuser.

Ladies and Gentlemen,

This is the second time that I have taken part in an international congress of the used-oil regeneration industry. In 1978, when I was briefly present during a part of the Houston congress, I must admit that I felt the French regeneration industry was in a sorry state. Now, two years later, it looks as if this industry can face the future with some optimism and start the new decade with confidence.

In view of the general programme for this conference and the circumstances that have led me to be speaking to you today, I am sure you are all hoping that I shall be brief. I also think it appropriate that I should stand aside a little from technical problems, in which I am quite incompetent into the bargain. I shall thus restrict myself to a short summary of the general conditions under which the regeneration industry has had to operate for the last ten years, and a reminder of the obstacles it has had to face.

When I first became interested in the problems of used-oil regeneration, in 1974, I was struck both by the simplicity of the basic data in terms of the national interest, and also by the extraordinary complication of government attitudes to them.

Faced with the situation of monetary disorder and its immediate consequences, in other words the sharp increase in oil prices, the French authorities laid down basic guidelines for the economic decisions that had to be taken. The object was to keep the French foreign payments account in balance, to develop activities providing high added value within the country, to save energy and raw materials and to redefine the policy for growth, which hitherto had taken too little account of the environment.

It was obvious that recycling used oils was an excellent example, of minor importance no doubt but nevertheless pointing in the right direction, of the kind of action that should be encouraged. For oil regeneration saves energy and raw materials ; it is an activity that keeps added value at home, it leads to reduced imports and it can play a decisive role in pollution control since at one point in time between 25 and 35% of the pollution of the water table in certain regions of France was attributable to used oil being poured out into the ground. This is why I was so surprised - and I am underestimating what I felt - at the extraordinary timidity of the French authorities in this field. For a very long period I found, in the circles responsible for drawing up the regulations that were necessary if the regeneration industry was to develop, nothing but incoherence, confusion, hypocrisy, laxity, hesitation, and reservations. I could say more but I will stop there.

If we try to understand why the French regeneration industry was shackled for so long, I think we shall find the explanation in the fact that policy on it was at a cross-roads. On the one hand, oil decisions in France come under the heading of affairs of state, and on the other there were the shortcomings of our civil service, or to be more accurate services, whose behaviour sometimes appears to the observer as particularly obscure, in the sense that it is hard to discern either its motives or its goals.

In a few words, I think that we have to resituate the problem of the regeneration industry in the context of the long and difficult history of the French petroleum industry. It is well known that France has very limited hydrocarbon resources on its national territory and that the various governments have tried to set up, firstly, a system of regulations that would provide maximum security of supply and, secondly, a situation that would permit the development of oil companies controlled by French capital. This policy was for long symbolized in our celebrated law of 1928 and in the creation of two French oil companies, the C.F.P. and Elf-Aquitaine. For long also, the authorities have desired that these two companies should control hydrocarbon production outside the country equivalent to French annual consumption, and should also control at least one half of the French market. It must be realized that these objectives have made it necessary to establish some kind of control over prices on the domestic market, with the object of ensuring that the natural play of market forces should not prevent achievement of the objective I have just mentioned - that half of the national market should be reserved to the French companies. This has meant that French prices for oil products have varied less than those in other countries, that the French subsidiaries of the majors have made more money in France than they would have done on a free market, and that finally, since the successive governments could not expect French industry to pay more for its oil than its competitors in other countries, it was necessary to subsidize the exploration efforts of the national companies while at the same time France was progressively setting up fiscal arrangements on the lines of those in America which led in practice to a reduction in the tax burden for the French oil companies.

It is thus a controlled situation that has prevailed ; but within this complicated construction a number of "windows" have been cut out, which may in fact be described as "profit windows". One of these profit windows was the lubricant market. On this market, prices have been very largely fixed by the oil companies themselves which, to use a somewhat trivial but eloquent expression, have made their "pickings" on the sale of motor oils in particular. It is clear that this profit window could only remain wide open if no other oil could come and compete with those sold by the oil companies that dominated the lubricant market at the time. This is why these companies took steps to discourage as much as possible the development of used-oil regeneration.

And yet, concurrently with government approval of this profit window in a very largely controlled situation, the authorities, because they were aware of the advantage offered by oil regeneration, forbade those in possession of used oils to burn them. The aim in prohibiting the burning of such oils was to steer them towards the regeneration industry. However, this regulation went against the desire of the oil companies to sell their lubricants at the maximum price, and some companies considered setting up means by which they could control the collection of used oils.

I must now tell you, to complete my description of this great French speciality of contradiction in official policies, that while the government departments were perfectly aware of the practices in restraint of competition that had been established on the instigation of certain oil companies, these same departments continued to grant aid at the same time to the regeneration industry in the form of tax relief : new oils paid a tax on the domestic market, while regenerated oils did not.

This was the situation in 1974, a year when the aftermath of the monetary crisis, the Yom Kippur war and the rise in the price of oil made it necessary to redeploy our industry and also to redefine a whole series of behaviours and regulations, among which those that concern us. I will not go into all the different vicissitudes, numerous and of passionate interest to the political sociologist as they were, that have marked

the last six years. We have now in 1980 new regulations that permit
the coexistence of a system for collecting used oils which, it is to
be hoped, will allow as much as possible of the raw material to be
recovered, and also one of financial aid to the regeneration industry
which should encourage investments for modernization that will make it
possible to process a growing volume of used oils.

I do not think it necessary to go into these systems in detail ; other
people have described or will describe them to you. I also think that
no purpose will be served by trying today to forecast the final effect
of this rather complex pattern of regulation ; it is still too close to
us for reasoned judgement.

I would just like to summarize in a few words why I have some hope that
this problem may at last be coming out of the morass in which it has
been bogged down.

Firstly, the continual and general increases in the prices of oil
products since 1974 have, I believe, given the authorities a bad
conscience in respect of energy saving.

Secondly, I have in recent years noted a favourable change in the
attitudes of the French oil companies, both S.N.E.A. and C.F.P., which
have adopted a positive attitude to the regeneration industry that provides a happy contrast to certain of their earlier actions. This
policy change certainly has complex causes which for an outside observer
it is extremely difficult to analyse ; however I think that apart from
commercial considerations, whose influence is clearly considerable, we
must thank for it the perseverance of the French Association of Oil
Regenerators, which has succeeded in making an objective and reasoned
appreciation of the problems prevail in the counsels of the leading
French oil groups.

The third reason why the situation has got out of its rut is the firm-
ness shown by the Minister for Industry, Mr. André Giraud, in the difficult interministerial negotiations. May I take this opportunity to
pay public homage to his sense of fair play and of keeping his word.

There will no doubt be further difficulties, especially as the dialogue
between the regeneration industry and government departments is unlikely
to be easy. This is not only because dialogue with the French civil

service is difficult per se, but also because four different ministries are "competent" in this matter - and you will note my use of inverted commas ; in such conditions the normal practice is that each one should, rather than try to be constructive, seek to impose its own point of view for reasons of defending its administrative stronghold or sovereignty over some aspect of the regulations.

This is why I think the regeneration industry will have to remain vigilant and above all united, while keeping to the reasonable policy line which the Association has laid down.

After all, I have perhaps taken a little too much time, and I hope you will forgive me.

ANWENDUNG DER EG-RATSRICHTLINIE VOM 16. JUNI 1975 ÜBER DIE BESEITIGUNG
VON ALTÖLEN
APPLICATION OF THE COMMUNITY DIRECTIVE OF 16TH JUNE, 1975 ON THE
DISPOSAL OF USED OILS
APPLICATION DE LA DIRECTIVE COMMUNAUTAIRE DU 16 JUIN 1975 SUR
L'ELIMINATION DES HUILES USAGEES
APPLICAZIONE DELLA DIRETTIVA COMUNITARIA DEL 16 GIUGNO 1975
SULL'ELIMINAZIONE DEGLI OLI USATI

B.W.K. RISCH
Dienststelle für Umwelt- und Verbraucherschutz der EG-Kommission
Environment and Consumer Protection Service
of the Commission of the European Communities
Service de l'Environnement et de la Protection des Consommateurs
de la Commission des Communautés Européennes
Servizio Condizioni Ambientali e Tutela dei Consumatori
della Commissione delle Comunità Europee

Die Anwendung der EG-Ratsrichtlinie vom 16. Juni 1975 über die Beseitigung von Altölen

KURZFASSUNG

Die Erkenntnis von der wirtschaftlichen Bedeutung des Altöls ist mit den Energiepreisen und der Versorgungsabhängigkeit in den letzten Jahren stark gestiegen. Angesichts der sich abzeichnenden weltweiten Verknappung der Verfügbarkeit von Rohöl wird das Altöl somit eine volkswirtschaftlich unentbehrliche zweite Versorgungsquelle für die europäische Wirtschaft. Die Rückführung des gesamten Schmierstoffverbrauchs in den Mineralölkreislauf und dies mit der Möglichkeit der Wiederholung und auch der Mengensteigerung stellt eine ausserordentlich wichtige Entlastung des Rohölbedarfs dar.

Aus 1000 Tonnen Altöl können über 500 Tonnen Schmieröl gewonnen werden. Der Rohölbedarf für die gleiche Menge Schmieröl beträgt ca. 5.000 Tonnen. In der Europäischen Gemeinschaft fallen jährlich 2 - 2,5 Millionen Tonnen Altöl an. Werden diese Altöle regeneriert, dann kann die Rohöleinfuhrbilanz der Europäischen Gemeinschaft um rund 20 Mio. Tonnen Rohöl entlastet werden.

Die vollständige Sammlung und Nutzung des Altöls wird somit zu einem volkswirtschaftlichen Gebot der Stunde und der Zukunft. Keine Tonne Altöl darf mehr verschwendet werden! Es liegt im Interesse der europäischen Wirtschaft, dass Altöl in umfassender Weise zur Sicherung der Gemeinschaft mit der Versorgung von Schmierstoffen in allen Situationen herangezogen wird.

Das Nordseeöl aus den Gemeinschaftsländern hat keine Schmierstofffraktion. In der Rohstoffversorgung zur Herstellung von Schmierstoffen ist die Gemeinschaft also völlig abhängig. Die Altöle stellen danach die einzige sichere Versorgungsquelle für die Schmierstoffherstellung in der Gemeinschaft dar.

Die Altölwirtschaft in der Gemeinschaft, die die Sammlung und Verwertung von Altöl besorgt, hat also eine wichtige volkswirtschaftliche Funktion zu erfüllen.

Die vom Ministerrat am 16. Juni 1975 verabschiedete Altölrichtlinie enthält ein Gebot zur Sammlung aller Altöle und zu ihrer wirtschaftlichen Verwertung.

Aber die Sammlung und Verwertung der in der Ge einschaft anfallenden Altöle ist heute - 5 Jahre nach Verabschiedung der Altölrichtlinie durch den Rat - in keiner Weise befriedigend.

Noch immer werden heute nicht mehr als rund 50 % der anfallenden Altöle gesammelt, verwertet und umweltunschädlich beseitigt. Die Situation hat sich in den letzten 5 Jahren kaum weiter entwickelt.

Dies rührt entscheidend daher, dass die Richtlinie, die seit dem
16. Juni 1977 voll in Kraft ist, noch immer nicht von allen Miegliedstaaten der Gemeinschaft vollständig in nationales Recht umgesetzt worden ist. In einer Reihe von Mitgliedstaaten fehlen heute noch immer die notwendigen rechtlichen und wirtschaftlichen Instrumente, um den Bestimmungen der Richtlinie und den wirtschaftlichen Erfordernissen zu entsprechen.

Seit letztem Jahr ist die Entwicklung aber wieder etwas in Bewegung geraten. Mehrere Mitgliedstaaten haben ihre Gesetzgebung über die Sammlung, Verwertung und umweltschädliche Beseitigung, sowie über die Kontrolle der Altöle vervollständigt.

Die Kommission appelliert an alle Mitgliedstaaten, die verschiedenen Bestimmungen der Richtlinie vom 16. Juni 1975 rasch und vollständig anzuwenden. Dabei soll insbesondere auch darauf gesehen werden, dass bei der Altölverwertung keine zusätzlichen Umweltbelastungen entstehen. Nach den der Kommission aufgrund eigener Untersuchungen und von Daten aus mehreren Mitgliedstaaten vorliegenden Informationen führt eine direkte Verbrennung ohne Vorbehandlung und Rauchabgaswäsche zu einer zusätzlichen Luftverschmutzung. Die Kommission beabsichtigt einen Richtlinienentwurf über die Verbrennung von Altöl ohne Vorbehandlung vorzulegen, um die Ratsrichtlinie vom 16. Juni 1975 entsprechend zu ergänzen.

Wenn alle wiederaufbereitungsfähigen Altöle zur Herstellung von Zweitraffinaten eingesetzt werden, könnte der Anteil der Zweitraffinate am Schmierstoffmarkt, der heute in der Gemeinschaft bei etwa 15 % liegt, 25 - 30 % des Marktes mittelfristig erreichen. Die Altöle stellen damit eine wichtige zweite Versorgungsquelle für die Herstellung von Schmierstoffen dar. Wenn dagegen alle anfallenden Altöle der Verbrennung mit Energieverwendung zugeführt würden, so könnten damit nur etwa 0,2 % des Energieverbrauchs in der Gemeinschaft gedeckt werden. Im Sinne einer rationellen Abfallwirtschaft sollte daher im Rahmen einer Altölwirtschaftspolitik aus volkswirtschaftlichen Gründen der Rohstoffversorgung - unter der Voraussetzung freier Wirtschaftsbedingungen - grundsätzlich der Vorrang vor der Energieverwertung eingeräumt werden. Schliesslich sollte die öffentliche Hand als Nachfrager in stärkerem Masse Zweitraffinate für ihren Bedarf einsetzen.
Der Verpflichtung zur vollständigen Altölsammlung und zur Altölverwertung müsste die Selbstverpflichtung der öffentlichen Hand zur Verwendung von Recyclingprodukten, hier von Zweitraffinaten, entsprechen.

Application of the Community Directive of 16 June 1975 on the disposal of used oils

SUMMARY

Awareness of the economic importance of used oils has increased appreciably in recent years following the increase in energy prices and the painful experience of dependence on third parties for oil supplies. The obvious scarcity of crude oil makes used oils an indispensable second source of supply for the European economy.
Incorporation of the entire lubricant consumption into the mineral oils cycle, on a repetitive basis and even with an increase in quantity, would represent a considerable reduction in the Community's **crude oil needs.**
One thousand tonnes of used oil can be converted into 500 tonnes of lubricant. The crude oil demand for the same quantity of lubricant amounts to around 5 000 tonnes. Each year between 2 million and 2 500 000 tonnes of used oil are generated in the Community. If these used oils were regenerated, the Community's balance of **imports in crude oil could be reduced** by about 20 million tonnes.
Full-scale collection and recovery of used oils is thus essential both now and in the future. Not a single tonne of used oil should be wasted. It is in the interests of the Community's economy for full use to be made of used oils so as to guarantee the Community's supply of lubricants under all circumstances. Without lubricants the economy cannot run smoothly.
North Sea oil has no lubricant fraction, and therefore the Community is completely dependent on third parties for raw material supplies for the manufacture of lubricants.
Used oils are thus the only certain source of supply the Community has for the manufacture of lubricants.
The used oil sector in the Community, which handles the collection and recovery of used oils, thus has a very important economic role to play.
The directive of 16 June 1975 stipulates that all used oils generated must be collected.
However, the situation regarding the recovery of used oils in the Community - five years after the adoption of the directive on used oils by the Council - is anything but satisfactory.
Little more than 50 % of used oils are collected, recovered and eliminated under satisfactory environmental conditions. Thus, the situation has practically remained at a standstill over the past five years.
This is without doubt due to the fact that the directive of 16 June 1975, which has been in force since 1977, has still not been completely incorporated into the laws of all the Member States. In general, the incorporation of directive provisions into national laws is too slow.
Over the past 12 months the Commission has finally noticed a certain positive development, several Member States having further developed or established regulations relating to the collection, recovery, inoffensive disposal and control of used oils.
The Commission insists that Member States apply the various provisions laid down in the directive of 16 June 1975 as quickly and as comprehensively as possible and that they make an active contribution to the collection and recovery of used oils. The Commission is particularly anxious that all additional pollution stemming from the recovery of used oils should be avoided.
The information available to the Commission indicates that the combustion of used oils without pretreatment aggravates air pollution.
Several Member States have already taken measures designed to cope with this situation. The Commission intends to submit a draft directive to the

Council in the near future concerning the burning of used oils without pre-treatment with a view to completing the directive of 16 June 1975.

If all used oils were used to manufacture regenerated lubricants the volume of these products, which is currently around 15 % in the Community, could be increased to 25 and 30% in the medium term. If, on the other hand, all the used oils generated were burnt with energy recovery it would amount to no more than +/- 0.2 % of the Community's energy consumption. To make for rational management of waste, priority must be given to the supply of raw materials rather than to the recovery of energy.

Finally, public authorities should use far more recycled products and thus provide a good example as consumers.

The commitment to collect and recover should be completed by a commitment on the part of public authorities to give priority to recycled and regenerated products.

L'application de la Directive Communautaire du 16 juin 1975 sur l'élimination des huiles usagées

RESUME

La prise de conscience de l'importance économique des huiles usagées s'est sensiblement développée depuis quelques années suite à l'augmentation des prix énergétiques et à l'expérience douloureuse de la dépendance d'approvisionnement en pétrole. La rareté de la disponibilité de pétrole brut qui se dessine concrètement fait des huiles usagées une deuxième source d'approvisionnement indispensable pour l'économie européenne. La réconduction de toute la consommation de lubrifiants dans le cycle des huiles minérales et ceci d'une façon répétitive et même avec une augmentation quantitative représente donc une diminution considérable et importante des besoins communautaires en pétroles bruts.

Mille tonnes d'huiles usagées peuvent être transformées en 500 tonnes de lubrifiants. Le besoin en pétrole brut pour la même quantité de lubrificants se chiffre à environ 5.000 tonnes. Chaque année entre 2.000.000 et 2.500.000 tonnes d'huiles usagées sont générées dans la Communauté. Si toutes ces huiles usagées étaient régénérées, le bilan d'importation en pétrole brut de la Communauté pourrait être diminué d'environ 20 millions de tonnes.

La collection exhaustive, et la récupération des huiles usagées devient donc une exigence impérative pour aujourd'hui et demain. Plus une seule tonne d'huile usagée ne doit plus être gaspillée. Il est de l'intérêt de l'économie communautaire que les huiles usagées soient employées d'une manière intégrale pour la sécurité d'approvisionnement de la Communauté en lubrifiants en toutes circonstances. Sans lubrifiants l'économie ne peut pas tourner !

Le pétrole de la mer du Nord n'a pas de fraction lubrifiante. La Communauté est donc entièrement dépendante de son approvisionnement de matières premières pour la fabrication de lubrifiants.

Les huiles usagées représentent par conséquent l'unique source propre et sûre d'approvisionnement de la Communauté pour la fabrication de lubrifiants.

Le secteur des huiles usagées dans la Communauté, qui s'occupe de la collecte et de la récupération des huiles usagées, a donc un rôle économiquement très important à jouer.

La directive du 16 juin 1975 fixe une obligation de collecte de toutes les huiles usagées générées.

Mais la situation de la récupération des huiles usagées dans la Communauté – cinq ans après l'adoption de la directive sur les huiles usagées par le Conseil – n'est pas tout à fait satisfaisante.

Il n'y a toujours pas beaucoup plus que 50 % d'huiles qui soient collectées, récupérées et éliminées dans des conditions satisfaisantes pour l'environnement. La situation n'a donc pas beaucoup évolué depuis les cinq dernières années.

Ceci est certainement la conséquence du fait que la directive du 16 juin 1975, qui est d'application depuis 1977, n'est toujours pas transposée d'une manière complète dans tous les Etats membres. La transposition des dispositions de la directive en lois nationales se fait généralement trop lentement.

Depuis 12 mois, la Commission a pu constater enfin une certaine évolution. Plusieurs Etats membres ont approfondi ou mis sur pied une réglementation relative à la collecte, la récupération, l'élimination inoffensive et le contrôle des huiles usagées.

La Commission insiste auprès des Etats membres d'appliquer des différentes dispositions de la directive du 16 juin 1975 aussi rapidement et complètement que possible et de promouvoir activement la collecte et la récupération des huiles usagées. Elle est d'avis qu'il faut notamment éviter toute pollution supplémentaire provenant de la récupération des huiles usagées.

Selon les informations dont dispose la Commission, la combustion des huiles usagées sans pré-traitement conduit à une aggravation de la pollution atmosphérique.

Plusieurs Etats membres ont déjà pris des mesures pour faire face à cette situation. La Commission a l'intention de soumettre prochainement au Conseil une proposition de directive concernant le brûlage des huiles usagées sans prétraitement afin de compléter la directive du 16 juin 1975.

Si toutes les huiles usagées étaient utilisées pour la fabrication de lubrifiants régénérés, la part de ces produits, qui est actuellement de l'ordre de 15 % dans la Communauté pourrait être portée à moyen terme à 25 et 30 %. Si au contraire toutes les huiles usagées générées étaient brûlées avec récupération de l'énergie ceci ne contribuerait qu'à $\pm$ 0,2 % à la consommation de l'énergie de la Communauté. Au sens d'une gestion rationnelle des dechets une priorité devrait être accordée en principe à l'approvisionnement en matières premières par rapport à la récupération d'énergie.

Finalement, les autorités publiques devraient employer beaucoup plus de produits recyclés et donner ainsi un bon exemple en tant que consommateur.

L'obligation de collecte et de récupération devrait être complétée par une obligation pour les autorités publiques d'utiliser par priorité des produits recyclés et régénérés.

L'applicazione della direttiva comunitaria del 16 giugno 1975 concernente
l'eliminazione degli oli usati

RIASSUNTO

Da qualche anno ci si è resi conto dell'importanza economica degli oli usati, soprattutto in seguito all'aumento dei prezzi dei prodotti energetici ed alle grovi difficoltà derivanti dalla dipendenza per quanto riguarda gli approvvigionamenti petroliferi. La rarefazione della disponibilità di greggio che si profila chiaramente, fa degli oli usati una seconda fonte, indispensabile per l'economia europea. La riconduzione di tutto il consumo di lubrificanti nel ciclo deli oli minerali in modo ripetitivo e persino con un aumento quantitativo costituisce pertanto una riduzione considerevole e importante del fabbisogno comunitario di petrolio greggio.

Mille tonnellate di oli usati possono essere trasformati in 500 t di lubrificante. Il fabbisogno di petrolio greggio per la stessa quantità di lubrificanti è di circa 5.000 t. La Comunità produce annualmente dai 2.000.000 ai 2.500.000 t di oli usati che se venissero rigenerati totalmente potrebbero far scendere le importazioni di petrolio greggio di circa 20 milioni di t.

Una raccolta generalizzata e il recupero degli oli usati è e sarà pertanto una esigenza assoluta. Nemmeno una tonnellata di questi oli dovrà andare perduta. Per l'economia comunitaria è importante che gli oli usati vengano utilizzati integralmente per la sicurezza dell'approvvigionamento di lubrificanti in ogni circostanza. Senza lubrificanti l'economia si ferma.

Dal petrolio del Mare del Nord non si possono trarre lubrificanti e quindi la Comunità dipende interamente per la fabbricazione di tali prodotti dal suo approvvigionamento di materie prime.

Gli oli usati costituiscono quindi l'unica fonte pulita e sicura di approvvigionamento.

Nella Comunità, il settore degli oli usati che si occupa della raccolta e del recupero di tali oli, ha quindi una funzione economicamente molto importante.

La direttiva del 16 giugno 1975 fissa l'obbligo di raccogliere tutti gli oli usati disponibili.

La situazione in questo settore rimane tuttavia, cinque anni dopo l'adozione della direttiva da parte del Consiglio, del tutto insoddisfacente.

Soltanto il 50% circa degli oli usati viene raccolto, recuperato ed eliminato in condizioni soddisfacenti per l'ambiente. La situazione non ha pertanto subito evoluzioni tangibili negli ultimi cinque anni.

Ciò dipende certamente dal fatto che la direttiva del 16 giugno 1975, in applicazione dal 1977, non viene sempre applicata in modo completo in tutti gli Stati membri. La trasformazione delle norme della direttiva in leggi nazionali viene fatta generalmente con troppa lentezza.

Da 12 mesi la Commissione ha potuto però constatare una certa evoluzione. Numerosi Stati membri hanno approfondito o attuato una regolamentazione relativa alla raccolta, al recupero e all'eliminazione non inquinante e controllata degli oli usati.

La Commissione insiste presso gli Stati membri affinché vengano applicate le varie disposizioni della direttiva del 16 giugno 1975 il più rapidamente e completamente possibile e siano attivamente incentivati la raccolta e il recupero degli oli usati. Essa ritiene che sia necessario peraltro evitare qualsiasi forma di ulteriore inquinamento proveniente dal recupero di tali oli. Secondo le informazioni di cui dispone la Commissione, la combustione degli oli usati non preventivamente trattati porta ad un aggravamento d'inquinamento atmosferico.

Vari Stati membri hanno già preso delle misure per far fronte a tale situazione. La Commissione si propone di trasmettere in un prossimo futuro al Consiglio una proposta di direttiva concernente la combustione degli oli usati senza trattamento preventivo, per completare la direttiva del 16 giugno 1975.

Se tutti gli oli usati venissero utilizzati per la fabbricazione di lubrificanti rigenerati, la quota di tali prodotti, che è attualmente dell'ordine del 15% nella Comunità, potrebbe essere portata a medio termine al 25-30%. Se invece tutti gli oli usati generati venissero bruciati con ricupero dell'energia, il contributo al consumo energetico della Comunità sarebbe soltanto dello 0,2%. Per una razionale gestione dei residui dovrebbe essere data la precedenza all'approvvigionamento delle materie prime rispetto al ricupero energetico.

Infine, le autorità pubbliche dovrebbero impiegare maggiori quantità di prodotti riciclati e dare quindi il buon esempio in qualità di consumatori.

L'obbligo di raccolgiere e di recuperare gli oli residui dovrebbe essere completato da un obbligo per le autorità pubbliche di utilizzare in primo luogo i prodotti riciclati e rigenerati.

Herr Vorsitzender,
meine sehr verehrten Damen und Herren !

Lassen Sie mich zu Beginn meiner Ausführungen meiner Freude
darüber Ausdruck geben, daß vier Jahre nach dem ersten so er-
folgreichen Europäischen Altölkongreß 1976 in Brüssel nun im
Jahre 1980 in Paris ein 2. Europäischer Gebrauchtöl- und
Recycling-Kongreß stattfindet, und daß damit nunmehr eine
Tradition begründet ist.

Ihr Verband kann auf dieses Ergebnis mit Recht stolz sein.
Und die Kommission ist befriedigt, daß sie hierzu in enger
Zusammenarbeit mit Ihnen beigetragen hat.

Vizepräsident Natali hat in seiner Botschaft an den Kongreß
ja bereits darauf hingewiesen, welche fachliche und politische
Bedeutung die Kommission auch Ihrer diesmaligen Veranstaltung
beimißt, da die heute im Vergleich zu 1976 noch kritischere
energiepolitische und rohstoffpolitische Situation das Altöl
zu einem wichtigen Wirtschaftsgut hat werden lassen, das für
die Wirtschaft der Gemeinschaft einen hohen Stellenwert hat.

In der Tat findet der 2. Europäische Altölkongreß wiederum zu
einem Zeitpunkt statt, der dieser Veranstaltung ein zusätzli-
ches Gewicht verleiht.

Der erste Europäische Altölkongreß im März 1976 fand ein Drei-
vierteljahr nach der Verabschiedung der Altölrichtlinie durch
den Ministerrat statt.

Sie haben dem 2. Europäischen Altölkongreß eine besondere Ausrichtung gegeben, indem Sie die Themen und Diskussionen auf das Recycling konzentrieren. Dies ist aber nicht nur von grundsätzlicher Bedeutung, sondern darüber hinaus auch wegen besonderer Umstände von einer außerordentlichen aktuellen Brisanz. Der Krieg zwischen dem Irak und dem Iran, zwei der wichtigsten Erdölerzeugerländer der Welt, macht blitzartig die kritische Versorgungslage der europäischen Länder mit Energie und Rohstoff deutlich. Die rohstoff- und energiepolitische Abhängigkeit Europas ist ein permanentes Datum. Auf das Ende des Tunnels kann nicht gewartet werden. Wir müssen uns etwas einfallen lassen.

Hier gewinnt die Abfallwirtschaftspolitik als übergeordneter Rahmen der altölwirtschaftlichen Tätigkeiten ihren besonderen volkswirtschaftlichen Stellenwert. Das rohstoffarme Europa braucht die Abfälle als wichtige Zweitrohstoffe. Der eigentliche Kampf gegen die Verschwendung muß erst noch beginnen. Abfälle und Rückstände müssen als eine der wichtigsten und sichersten Rohstoffquellen Europas betrachtet werden. Wir können es uns in den achtziger und in den neunziger Jahren nicht leisten, noch weniger als heute, auch nur eine Tonne Abfall ungenutzt zu verschwenden und die in ihm steckenden Wertstoffe ungenutzt zu lassen.

Es müssen Anstrengungen technischer, rechtlicher und wirtschaftlicher Art gemacht werden, damit die Zweitrohstoffe und die Recyclingprodukte den natürlichen Wettbewerb mit Primärrohstoffen und den aus ihnen hergestellten Produkten aufnehmen können.

Die öffentliche Hand hat hierbei als Nachfrager, als Marktmacht, eine besondere Aufgabe zu erfüllen. Sie kann nicht Recycling von der Wirtschaft fordern - und dies vielleicht noch zu unwirtschaftlichen Bedingungen und gar unter Verlusten - wenn sie selbst z.B. Verschwendung betreibt, indem sie z.B. für sogenannte niedrige Verwendungszwecke Hochqualitätsprodukte gebraucht wie z.B. Hochglanzpapier zum Fotokopieren, als Notizpapier, für Papierhandtücher, für Toilettenpapier usw., Erstraffinate anstelle von Zweitraffinaten als Schmierstoffe, Neureifen anstelle von runderneuerten Reifen für die öffentlichen Fuhrzeugparks usw.

Wir stehen hier einer Größenordnung und einem komplexen Problem gegenüber, das in den nächsten Jahren verlangt, daß über die beschränkte umweltbezogene Betrachtungsweise der Abfallwirtschaft hinausgegangen wird, und daß eine globale Abfallwirtschaftspolitik entwickelt und durchgeführt wird. Die Abfallwirtschaft muß zu einem integrierten Bestandteil einer europäischen Ressourcenpolitik ausgebaut werden. Schwerpunkte müssen dabei die Verringerung und Verwertung der Abfälle sein bei gleichzeitiger Sicherstellung einer umweltgerechten Beseitigung der nicht vermeidbaren und nicht verwertbaren Abfälle. Abfallwirtschaftspolitik ist und darf kein Anhängsel der Umweltpolitik sein, sondern ist integraler Bestandteil der Wirtschafts- und Gesellschaftspolitik. Sie hat eine wichtige Funktion zu erfüllen als Klammer zwischen der Umweltpolitik, der Rohstoffpolitik, der Energiepolitik, der Industriepolitik, der Handelspolitik usw.

In der Europäischen Gemeinschaft fallen jährlich bereits zwei Milliarden Tonnen Abfälle aller Art an. Das jährliche Wachstum beträgt zwischen 2 und 5 Prozent.

Man geht heute davon aus, daß zwischen 70 und 90 Prozent aller anfallenden Abfälle und Rückstände in der einen oder anderen Form wieder verwertet werden können. Dennoch werden bisher immer noch 80 bis 90 Prozent der Abfälle vernichtet oder durch Ablagerung lediglich beseitigt. Das ist eine enorme Verschwendung potentieller Wertstoffe.

Wenn sie als sekundäre Rohstoffe oder energetisch genutzt werden, so hat dies große wirtschaftliche und politische Vorteile für die Europäische Gemeinschaft und ihre Mitgliedstaaten, insbesondere angesichts der zunehmenden Rohstoffknappheit und der erhöhten Energie- und Rohstoffpreise und zwar :

- erhebliche Deviseneinsparungen bei Importen in der Größenordnung mehrerer hundert Millionen Dollar;

- Verringerung der Abhängigkeit der Gemeinschaft von Rohstoff- und Energieeinfuhren und gleichzeitige Verbesserung der Rohstoff- und Energieversorgung, unter anderem bei einer Reihe strategischer Rohstoffe;

- erhebliche Energieeinsparungen (allein der Einsatz von Zweitrohstoffen anstelle von Primärrohstoffen kann bis zu 5 Prozent Energieeinsparung des Gesamtenergieverbrauchs der EG bringen, in der Reifenproduktion 20 bis 40 Prozent und bei der Aluminiumherstellung fast 90 Prozent);

- langfristig rationellere Nutzung der nicht oder nur nach einer gewissen Zeit erneuerbaren natürlichen Ressourcen;

- Verringerung der Umweltbeeinträchtigungen;

- Verringerung der Kosten für Beseitigung und Ablagerung von Abfällen;

- Schonung der Landschaft und des verfügbaren knappen Bodens durch geringere Anforderungen nach Deponieflächen.

Abfallwirtschaft ist also ein sehr wichtiger und zentraler Bestandteil der Wirtschafts-, Rohstoff- und Energiepolitik. Zweitrohstoffe und Zweitenergien stellen eine der wichtigsten und vor allem sichersten eigenen Versorgungsquellen der Europäischen Gemeinschaft dar.

Die Abfallwirtschaftspolitik ist ein noch sehr junger Politikbereich : als nationale, als europäische und internationale Politik gibt es sie erst seit etwa sechs bis zehn Jahren. Sie hat sich auf allen Ebenen in den letzten Jahren zunächst darauf konzentriert, Rahmenbedingungen für eine geordnete Abfallwirtschaft zu schaffen, die insbesondere als Teil der Umweltpolitik eine geordnete Beseitigung ohne Umweltschäden gewährleisten muß und jetzt auch kann.

Nun stehen wir vor einer zweiten Phase der Abfallwirtschaftspolitik, deren Aufgabe darin bestehen wird, die Abfallwirtschaft zu entwickeln und zu einem zentralen Bestandteil der Wirtschaftspolitik zu machen. Beziehungen müssen hergestellt werden zur Energiepolitik, zur Rohstoffpolitik, zur Industriepolitik, zur Handelspolitik usw. In den achtziger Jahren wird nach unserer Auffassung in größerem Stil der Schritt von der Abfallbeseitigung und Abfallentsorgung zur umfassenden Abfallverwertung und Abfallwirtschaft getan werden.

Das bedeutet aber auch, daß die künftige Abfallwirtschaftspolitik sich von der reinen umweltpolitischen Betrachtungsweise entfernen muß. Sie muß, um ihre volle Rolle spielen zu können, aus dem Ghetto der reinen Umweltpolitik herausgeführt werden. Die Abfallwirtschaftspolitik als Ressourcenpolitik muß langfristig ausgerichtet sein. Sie darf von unterschiedlichen und wechselnden konjunkturellen Situationen nicht wieder kurzfristig in Frage gestellt werden. Sie muß in erster Linie nicht von wechselnden Preisen sondern von dem Datum der immer größer werdenden Versorgungsabhängigkeit der Gemeinschaft bestimmt werden. Die Schonung der Ressourcen, der Aufbau und die Erhaltung leistungsfähiger Recyclingindustrien, die Sammlung, Behandlung, Umwandlung und Wiederverwertung der Abfälle als Zweitrohstoffe und Recyclingprodukte bleibt eine langfristige permanente Aufgabe.

Kernsatz der Abfallwirtschaftspolitik muß es sein, sich vor Augen zu halten, daß für sehr viele Grund- und Wertstoffe die in den Abfällen und Rückständen steckenden Stoffe oft die einzige und die sicherste Versorgungsquelle darstellen.

Mittelpunkt der Abfallwirtschaftspolitik der nächsten zehn bis zwanzig Jahre müssen dabei im Wesentlichen folgende Aufgaben sein :

1. Die Verringerung des Abfallanfalls und die Entwicklung sauberer und recyclingfreundlicher Abfälle, die ohne große Schwierigkeiten in den Wirtschaftskreislauf zurückgeführt werden können. Es handelt sich hier um eine mittel- und langfristige Aufgabe, die insbesondere Anstrengungen auf den Gebieten der Forschung und Entwicklung erfordert, aber auch in den Bereichen der Produktentwicklung, der Verfahrenstechnologie, der Betriebsorganisation usw.

2. Die maximale Verwertung der in den Abfällen steckenden Wertstoffe und Potentiale.

Die Abfallbeseitigungsgesetzgebung, die in den letzten fünf bis zehn Jahren entstanden ist, muß zu einer Abfallwirtschaftsgesetzgebung weiterentwickelt werden. Dazu gibt es bereits Ansätze und Tendenzen in den Mitgliedsländern, aber auch bereits in den Gemeinschaftsrichtlinien, so insbesondere in der Altölrichtlinie und in der Rahmenrichtlinie für Abfälle vom 15. Juli 1975.

In der Bundesrepublik soll jetzt neben die Verpflichtung zur Abfallsammlung und Abfallbeseitigung auch die Pflicht zur Abfallverwertung gestellt werden, nämlich als ein Gebot, das besagt, daß bei der Abfallbeseitigung zunächst vorrangig geprüft werden muß, ob der Abfall wirtschaftlich sinnvoll verwertet werden kann, bevor er eventuell auf eine Deponie verbracht wird.

Die Kommission der Europäischen Gemeinschaft prüft zur Zeit sehr sorgfältig, ob auch auf Gemeinschaftsebene dieser Schritt getan werden soll. Wahrscheinlich muß er schon aus Gründen einer einheitlichen wirtschaftlichen Entwicklung und der Wettbewerbsneutralität in dem einheitlichen Wirtschaftsgebiet des Gemeinsamen Marktes getan werden.

x

x   x

Lassen Sie mich im zweiten Teil meiner Ausführungen nun zu den speziellen Problemen der Altölbeseitigung und Altölwirtschaft kommen, und insbesondere auch zum Stand der Anwendung der Altölrichtlinie vom 16. Juni 1975.

Auf die besondere rechtliche und faktische Situation in den einzelnen Mitgliedstaaten der Gemeinschaft gehe ich hier nicht ein. Darüber wird in Kurzreferaten im Anschluß an diesen Vortrag berichtet werden. Ich werde mich hier darauf beschränken, die Gesamtlage zu diesem Zeitpunkt aus der Sicht der EG-Kommission darzustellen.

Wie es Herr Vizepräsident Natali in seiner Botschaft an den
Kongreß unterstrichen hat, und wie auch ich es in meiner Einleitung bereits betont habe, so sieht die Kommission in den
Altölen einen wichtigen und wertvollen Rohstoff, der umfassend zur Rohstoff- und Energieversorgung der Gemeinschaft
herangezogen werden muß. Eine der wesentlichen Bestimmungen
der Altölrichtlinie, auf die sich auch die künftige Politik
der Kommission in diesem Bereich stützen wird, sieht vor,
daß die Beseitigung von Altölen durch Wiederverwendung in
Form der Aufbereitung zu Zweitraffinaten und durch Verbrennung mit Energienutzung erfolgen soll. Eine bloße Vernichtung des Altöls, wie dies früher geschehen ist, ist nach
Artikel 3 der Altölrichtlinie nicht mehr statthaft.

Die Altölrichtlinie hat nach Auffassung der Kommission daher eine Bedeutung, die über die Umweltpolitik und die Altölbeseitigung weit hinausgeht. Sie wird von uns verstanden
als ein Instrument einer Altölwirtschaftspolitik im Rahmen
einer gemeinschaftlichen Abfallwirtschafts- und Ressourcenpolitik. Dies war bereits unsere Auffassung im Jahre 1973,
als wir diese Altölrichtlinie konzipiert haben. Heute aber
hat angesichts der Energiepreise und der bitter empfundenen
Versorgungsabhängigkeit vom Rohöl sowie der sich abzeichnenden weltweiten Verknappung der Verfügbarkeit von Rohöl die
wirtschaftliche Bedeutung des Altöls noch einen weit höheren
Stellenwert erhalten. Die vollständige Sammlung und Nutzung
der Altöle in der Gemeinschaft wird somit zu einem volkswirtschaftlichen Gebot der Stunde und der Zukunft. Keine
Tonne Altöl darf mehr verschwendet werden! Altöl muß in
umfassender Weise zur Sicherung der Gemeinschaft mit der Versorgung von Schmierstoffen in allen Situationen herangezogen werden. Die Rückführung des gesamten Schmierstoffverbrauchs in den Mineralölkreislauf und dies mit der Möglichkeit der Wiederholung und auch der Mengensteigerung stellt
eine außerordentlich wichtige Entlastung der Import- und der
Devisenbilanz dar. Da das Nordsee-Öl praktisch zur Zeit
keine Schmierstoff-Fraktion enthält, ist das in der Gemein-

schaft anfallende Altöl die einzige und sicherste eigene Versorgungsquelle der Gemeinschaft für die Schmierstoffherstellung. Der Aspekt der Rohstoffversorgung hat dabei für uns eindeutig Vorrang vor der Energieverwertung der Altöle.

Die Zweitraffinate haben bereits heute, wo nur knapp etwas mehr als die Hälfte des anfallenden Altöls gesammelt und verwertet wird, einen Marktanteil von rund 15 % am Schmierstoffmarkt. Wenn die Gemeinschaftsrichtlinie voll greift und die Altöle zu nahezu 100 % wiederverwertet werden, wobei die wiederaufbereitungsfähigen Altöle der Schmierstoff-Herstellung in der Zweitraffination zufließen sollten, so könnten die Zweitraffinate durchaus mittelfristig 25 bis 30 % und mehr des Schmierstoffmarktes erreichen. Die Altöle stellen damit eine wichtige zweite Versorgungsquelle für die Herstellung von Schmierstoffen in der Gemeinschaft dar. Ohne Schmieröl läuft nichts!

Wenn dagegen alle anfallenden Altöle der Verbrennung mit Energieverwertung zugeführt würden, so könnte damit noch nicht einmal 0,5 % des Energieverbrauchs in der Gemeinschaft gedeckt werden. Das ist eine relativ zu vernachlässigende Größe!

Die Kommission ist davon überzeugt, daß die Altölwirtschaft in der Gemeinschaft, die die Sammlung und Verwertung von Altöl besorgt, eine sehr wichtige volkswirtschaftliche Funktion wahrnimmt. Daher hat sie die Durchführung dieses Zweiten Europäischen Altölkongresses auch sehr begrüßt und insbesondere seine thematische Ausrichtung auf die verschiedenen Aspekte des Altölrecyclings. Wir sind über die Dynamik und die gute Organisationsstruktur Ihres Wirtschaftszweiges außerordentlich befriedigt, weil sie eine der wesentlichen Voraussetzungen für die Erfüllung dieser wichtigen volkswirtschaftlichen Aufgabe darstellt. Außerdem ist Ihre Branche für uns eine Art Modell, das beispielhaft für Tätigkeiten sein kann, die wir für erforderlich halten und fördern wollen zur Nutzbarmachung von Abfallstoffen auch in anderen Bereichen.

Aber die Sammlung und Verwertung der in der Gemeinschaft anfallenden Altöle ist heute, im Jahre 1980, fünf Jahre nach Verabschiedung der Altölrichtlinie durch den Rat, dennoch in keiner Weise befriedigend. In der Gemeinschaft fallen jährlich zwischen 2 und 2,5 Millionen Tonnen Altöle an. Aber noch immer werden heute nicht viel mehr als 50 % der anfallenden Altöle gesammelt, verwertet und umweltunschädlich beseitigt, sowie dies bereits vor fünf oder sieben Jahren der Fall war. Die Situation auf dem Altölmarkt hat sich in den letzten fünf Jahren quantitativ kaum weiterentwickelt, obwohl die Technik Fortschritte gemacht hat.

Das rührt nach Auffassung der Kommission entscheidend daher, daß die Altölrichtlinie, die seit dem 16. Juni 1977 voll in Kraft ist und damit einklagbares Recht darstellt, noch immer nicht von allen Mitgliedstaaten der Gemeinschaft vollständig in nationales Recht umgesetzt worden ist. In einer Reihe von Mitgliedstaaten fehlen noch immer die notwendigen rechtlichen und wirtschaftlichen Instrumente, um den Bestimmungen der Richtlinie und den volkswirtschaftlichen Notwendigkeiten voll zu entsprechen.

Es wäre unehrlich, hier Dinge beschönigen zu wollen, die noch nicht völlig in Ordnung sind. Auf jeden Fall ist die Altölrichtlinie in der Kommission noch nicht in der Ablage als abgehakt gelandet.

Es zeigt sich hier, daß sich in der praktischen Umsetzung der Gemeinschaftsrichtlinien, und dies ist nicht typisch für den Bereich der Abfallwirtschaft, oft viel längere Fristen ergeben als dies in den Bestimmungen der Richtlinien vorgesehen ist. Das kann vielfache Gründe haben : rechtliche, technische, wirtschaftliche und politische. Häufiger Regierungswechsel verzögert allemal und oft den Gang der Rechtsprechung.

Aber unabhängig von der Umsetzung in den Mitgliedstaaten in nationales Recht ist die Richtlinie vom 16. Juni 1975 seit Juni 1977 voll in Kraft. Sie stellt daher ein unmittelbar einklagbares Recht dar, sowohl für die Gemeinschaftsinstitutionen als aber auch für jeden einzelnen Wirtschaftsbürger in der Gemeinschaft.

Am Tage des Inkrafttretens der Richtlinie, am 16. Juni 1977, war kein einziges Mitgliedsland den Bestimmungen der Richtlinie vollkommen nachgekommen. Nur einige Mitgliedstaaten hatten eine wenigstens faktisch befriedigende Situation aufgrund früheren nationaler Regelungen, die wir zum Teil in die Gemeinschaftsrichtlinie integriert hatten.

Die zuständigen Stellen der Kommission haben in unzähligen Gesprächen in den letzten drei Jahren und in Stellungnahmen auf den bisherigen Vollzug der Richtlinie gedrängt. Der Umweltministerrat hat sich am 9. April 1979 mit diesem Fragenkomplex befaßt. Danach sind dann die Dinge innerhalb der letzten zwölf Monate etwas in Fluß gekommen. Mit Befriedigung haben wir insbesondere festgestellt, daß Frankreich eine umfassende neue Altölgesetzgebung verabschiedet hat, deren Maßnahmen am 23. November - also innerhalb von 2 Monaten - voll in Kraft treten müssen.

In Frankreich beruht die neue Gesetzgebung auf den drei grundliegenden Zielen, die die Gemeinschaftsrichtlinie definiert hat, nämlich :

- Verbot der unkontrollierten Altölbeseitigung;
- umfassende und vollständige Sammlung aller anfallenden Altöle;
- Verwertung der Altöle als Rohstoff, der nicht mehr verschwendet werden darf.

Die Niederlande haben ebenfalls Fortschritte bei der Vertiefung der Gesetzgebung über die Sammlung, Verwertung, Kontrolle und schadlose Beseitigung der Altöle gemacht.

Außerdem hat Luxemburg ein entsprechendes Gesetz am 26. Juni 1980 in Kraft gesetzt.

Dennoch kann man sagen, daß nur fünf der neun Mitgliedsländer der Gemeinschaft die Bestimmungen der Richtlinie vom 16. Juni 1975 rechtlich weitgehend umgesetzt haben. Vorbehaltlich einer letzten formalen Prüfung der uns erst vor kurzer Zeit zugegangenen neuen nationalen Rechts- und Verwaltungsvorschriften, wird die Akte dieser Länder bei den zuständigen Dienststellen der Kommission hinsichtlich der rechtlichen Aspekte wahrscheinlich bald geschlossen werden können.

Die tatsächliche Lage der Sammlung, Verwertung und umweltunschädlichen Beseitigung der Altöle ist allerdings ein ganz anderes Kapitel. In Frankreich und in Luxemburg ist das Gesetz gerade so eben erst in Kraft getreten bzw. tritt in einigen Wochen in Kraft. In diesen Ländern ist eine Bewertung der tatsächlichen Situation erst in ein bis zwei Jahren möglich. Hinsichtlich der anderen Länder ist eine vergleichende Beurteilung außerordentlich schwierig, da die der Kommission zur Verfügung stehenden Daten pro Mitgliedsland außerordentlich unterschiedlich sind sowohl in ihrer zeitlichen Relevanz als in ihrer Zusammensetzung und Vollständigkeit. Aus diesem Grunde war es der Kommission auch nicht möglich, neuere Daten als die aus dem Jahre 1977 für alle Mitgliedsländer zur Verfügung zu stellen. Diese sind als Anlage zu dem Vortragstext beigefügt. Und auch hier handelt es sich nicht um harte Daten, sondern nur um Schätzwerte. Für den so wichtigen Altölbereich, aber auch für die anderen Bereiche der Abfallwirtschaft, wird dringend eine funktionsfähige Abfallwirtschaftsstatistik innerhalb der EG gebraucht.

Einige der Mitgliedstaaten, die die Richtlinie vom 16. Juni 1975 noch nicht vollständig in nationales Recht umgesetzt haben, haben dennoch gewisse Schritte unternommen, so zum Beispiel in Form eines Verbots der unkontrollierten Ableitung der Altöle in den Boden und in die Gewässer. Dies ist ein sehr wichtiger Aspekt, dennoch wird dadurch nicht die gesamte Richtlinie abgedeckt. Wir sind daher der Auffassung, daß die Richtlinie ohne eine Gesetzgebung über die systematische und umfassende Sammlung und Verwertung der Altöle nicht befriedigend erfüllt werden kann.

Die praktische Erfahrung hat uns in unserer Auffassung bestätigt, daß eine umfassende und strenge gesetzliche Regelung zur Lösung des Altölproblems erforderlich ist.

In den zwei Mitgliedstaaten, die bereits seit längerer Zeit Altölsondergesetze haben, nämlich die Bundesrepublik Deutschland und Dänemark, haben unseres Erachtens diese Gesetze ihre Bewährung voll bestanden. Die gesammelten und wiederverwendeten Altölmengen sind infolge der einschlägigen gesetzlichen Regelungen ständig gestiegen. Auf der anderen Seite ist die unkontrollierte und verschmutzende Beseitigung von Altölen in solch einer Weise zurückgegangen, daß man sagen kann, daß das Umweltproblem durch die unkontrollierte Beseitigung von Altölen in diesen beiden Ländern, wenn auch nicht völlig gelöst, so doch als im Griff befindlich bezeichnet werden kann.

Ein ähnliches Ergebnis streben wir mit der vollständigen Anwendung der Altölrichtlinie vom 16. Juni 1975 auch in den anderen Gemeinschaftsländern an.

Leider ist es nicht möglich, heute zu sagen, was ich in meinem Vortrag während des Ersten Altölkongresses im März 1976 in Brüssel als Hoffnung formuliert habe, daß es nämlich möglich sein möge, anläßlich eines Zweiten Europäischen

Altölkongresses zu sagen, daß das Problem der Umweltverschmutzung durch die unkontrollierte Beseitigung von Altölen für die gesamte Gemeinschaft gelöst worden ist.

Die Kommission bleibt aber optimistisch und hofft, daß eine solche positive Feststellung schließlich spätestens während des Dritten Europäischen Altölkongresses möglich sein wird. Wir hoffen, daß wir dann aber auch gleichzeitig vermelden können, daß nahezu alle wiedervertungsfähigen Altöle gesammelt und einer optimalen volkswirtschaftlichen Nutzung zugeführt worden sind.

Inzwischen sind aber darüber hinaus neue Probleme aufgetaucht, die eine Behandlung und Initiativen durch die Gemeinschaft erforderlich machen.

Gleichzeitig mit der Entwicklung der Heizölpreise konnte nämlich in fast allen Mitgliedstaaten der Gemeinschaft beobachtet werden, daß mehr und mehr Altöl direkt zu Heizzwecken verbrannt wird. Das führt nach uns aus verschiedenen Mitgliedsländern vorliegenden Informationen nicht nur dazu, daß die Luftverschmutzung stieg, sondern daß den Aufarbeitungsunternehmen der Rohstoff entzogen wurde. Dadurch gerieten insbesondere Unternehmen in England, Frankreich und in Italien in echte Schwierigkeiten.

Das Altöl, so wie es anfällt, stellt mit seinen Zusatzstoffen ohne Frage ein gefährliches, schädliches und hinsichtlich seiner Wirkungen auch toxisches Produkt dar. Die Risiken für die Wasserverschmutzung und die Verschmutzung des Bodens sind frühzeitig erkannt worden und standen Pate bei den verschiedenen nationalen und gemeinschaftlichen Versuchen, zu einer Regelung in einem Sondergesetz über Altöle zu gelangen. Fünf Liter Motorenöl verschmutzen eine Wasserfläche von 5.000 qm. Sachverständige haben seinerzeit vor einer entsprechenden gesetzlichen Regelung festgestellt, daß die

unkontrollierte Ableitung von Altöl in die Gewässer und in
den Boden nahezu 20 Prozent der gesamten industriellen Umweltverschmutzung darstellt. Die Altölrichtlinie vom 16.
Juni 1975 war daher die erste Umweltschutzrichtlinie der
Europäischen Gemeinschaft.

Diese Form der unkontrollierten Ableitung in die Gewässer
und in den Boden hat man heute in nahezu allen Ländern der
EG weitgehend im Griff.

Ein neues Problem ist jetzt allerdings die Luftverschmutzung
infolge einer nicht durchgeführten Vorbehandlung. Untersuchungen und Messungen, die in verschiedenen Ländern durchgeführt worden sind, haben ergeben, daß beim Verbrennen von
Altöl erhebliche Emissionen an gasförmigen und staubförmigen
Substanzen auftreten. Unter "Altöl" versteht man dabei landläufig ein recht heterogenes Gemisch von teilweise brennbaren flüssigen Stoffen, welches zum Beispiel Gebrauchtöle aus
Motoren und Getrieben, Wärmeträger- und Transformatorenöle,
ölhaltige Emulsionen und Schlämme und nicht selten auch organische Lösungsmittel enthält. Aufgrund der heterogenen Zusammensetzung liegt der festgestellte durchschnittliche
Schwefelgehalt von Altölen zwischen o,8 und 2 %, wobei bestimmte Spezialöle (zum Beispiel einige Metallbearbeitungsöle) einen Schwefelgehalt von über 10 % erreichen können,
ihr durchschnittlicher Aschegehalt zwischen 1 und 2 % (also
100 bis 200 -mal höher als bei Heizöl EL) und auch weit darüber hinaus. Zusätzlich ins Gewicht fallen die relativ
hohen Anteile an Schwermetallen wie Blei und Zink, aber auch
Kadmium und Barium, welche den Neuölen meist als Additive
zugegeben wurden und bei der Verbrennung als staubförmige
Stoffe emitiert werden. Im Weiteren können bedeutende Chlor-
und Chlorwasserstoffmengen ausgestoßen werden, da Altöle
Chlorverbindungen, in Einzelfällen (zum Beispiel Schneidöle,
Metallbearbeitungsöle) über 5 % Chloranteile enthalten.

Dies gilt bis zu einem gewissen Grad auch für andere Halogene, insbesondere Fluor und Brom.

In 5 Liter Motoröl sind etwa 20 Gramm Blei enthalten. Es ist ausgerechnet worden, daß wenn rund 900.000 Tonnen Altöl direkt ohne Vorbehandlung in der Gemeinschaft verbrannt werden, dann 2.600 Tonnen Blei in die Luft abgeleitet werden. Außer Blei würden zusätzlich noch 700 Tonnen Phosphor, 900 Tonnen Barium und 7.000 Tonnen Zink in die Luft abgeleitet. Der durchschnittliche Bleigehalt von Motorenöl ist mit 2.900 ppm festgestellt worden.

Altöle sind daher eindeutig keine normalen oder genormten Brennstoffe wie leichtes und schweres Heizöl, sondern sie sind problematische Brennstoffe, die deshalb bei ihrer Verbrennung erhebliche zusätzliche Umweltschutzmaßnahmen erfordern. Vor allem eine Verbrennung in normalen kleinen Ölfeuerungsanlagen einschließlich der kleinen Altölöfen mit Verstäubungs- oder Verdampfungsbrennern kann nach allgemeiner Auffassung von Sachverständigen verschiedener Länder nicht mehr länger akzeptiert werden.

Eine Auswertung zahlreicher Untersuchungen und Forschungsarbeiten zeigt, daß die Verbrennung von Altöl in kleinen Ölfeuerungsöfen zu weitaus höheren festen und gasförmigen Schadstoffemissionen führt als die Verbrennung von Öl extraleicht.

Einen besonders hohen Anteil weisen dabei einmal Ruß und andere staubförmige Schadstoffe auf mit einem hohen Prozentsatz an toxischen Bestandteilen wie Zink- und Bleiverbindungen sowie gasförmige Schadstoffe wie Schwefeldioxyd, Chlor und Chlorwasserstoff (Salzsäure), aber auch Verbrennungsrückstände von PCB und PCT.

Die mit der direkten ohne Vorbehandlung erfolgte Verbrennung von Altölen anstelle von Heizöl verbundenen Gefahren der Explosion, der Luft- und Umweltverschmutzung haben dazu geführt, daß mehrere Mitgliedstaaten diese Art der Altölverwertung verboten haben.

Das gilt insbesondere für die Bundesrepublik Deutschland, die am 22. September 1978 eine Änderung der ersten Durchführungsverordnung zum Bundesemissionschutzgesetz verabschiedet hat, durch die ab 1. Oktober 1981 nach einer angemessenen Umstellungs- und Anpassungszeit die Altölverbrennung in kleinen Ölfeuerungsanlagen verboten wird, indem von diesem Zeitpunkt an die Anlagen nur noch mit Heizöl extra leicht nach DIN 51603 zu betreiben sind, das vorher zu keinem anderen Verwendungszweck eingesetzt sein durfte.

Ein ähnliches Verbot ist von Frankreich mit Gesetz vom 21. Mai 1980 erlassen worden.

Auch in Italien besteht grundsätzlich ein Verbot der Altölverbrennung. Dieses schon viele Jahre zurückliegende Verbot - es stammt aus dem Jahre 1940 - ist allerdings in der Praxis der letzten Jahre kaum noch beachtet worden. Italien hat allerdings die Absicht, im Rahmen der Umsetzung der Altölrichtlinie vom 16. Juni 1975 das Verbot der Verbrennung von Altöl zu verschärfen.

Auch die Schweiz hat als ersten Schritt in dieser Richtung in diesem Jahr eine Empfehlung über das Verbot der direkten Altölverbrennung verabschiedet.

Schließlich gibt es auch in den USA eine deutliche Tendenz sowohl bei den zuständigen Behörden wie in der Industrie zugunsten der Altölregenerierung anstelle der Altölverbrennung und zwar im Wesentlichen aus zwei Gründen :

- zu hohe Luft- und Umweltverschmutzung durch
  Altölverbrennung

- bessere und weniger umweltbelastende Verwertung
  von Altöl durch Zweitraffination.

Erste von der Kommission in Auftrag gegebene Untersuchungen haben die nationalen Untersuchungsergebnisse weitgehend bestätigt.

Daher ist die Kommission der Auffassung, daß die Verbrennung von Altöl in kleinen Betriebseinheiten aufgegeben werden sollte. Diese Verwendung von Altöl hat sich insbesondere nach der Energiekrise von 1973 sehr stark verbreitet. Solche Kleinbetriebe wie Garagen, Autoreparaturwerkstätten, Treibhäuser usw. verfügen nicht über die geeigneten Anlagen, die für die Behandlung der giftigen Bestandteile im Altöl erforderlich sind, um die Luft- und Umweltverschmutzung und eine umweltschädliche Verwertung zu vermeiden.

Das gilt insbesondere für die Verbrennung in normalen kleinen Ölfeuerungsanlagen einschließlich der kleinen Altölöfen. Insbesondere Altölöfen mit Zerstäubungsbrennern geben umfangreiche staub- und gasförmige Schadstoffemissionen mit einem hohen Prozentsatz an toxischen Bestandteilen wie Schwermetallverbindungen an die Atmosphäre ab. Diese Verbrennungsöfen können nur Partikel, die größer als 20 Micron sind, beseitigen, während schon Partikel von 0 bis 5 Micron Umweltprobleme aufwerfen. Bei Altölöfen mit Verdampfungsbrennern sind zwar die Schwermetallemissionen relativ geringer, weil sie in der Schlacke gebunden werden, diese müssen allerdings als toxische Abfälle auf Sondermülldeponien verbracht werden, was bei der Vielzahl solcher kleinen Anlagen kaum kontrolliert werden kann. Außerdem haben

Messungen von Ölfeuerungsanlagen mit Verdampfungsbrennern ergeben, daß relativ hohe Rußzahlen erreicht werden, die ebenfalls nicht zulässig sind.

Die Kommission ist daher zur Zeit dabei, aus den in den Mitgliedstaaten gemachten Erfahrungen, eigenen Untersuchungsergebnissen und aufgrund der unterschiedlichen Rechtslage in den Mitgliedstaaten im Hinblick auf die Einschränkung der Altölverbrennung Konsequenzen zu ziehen und eine Richtlinie über die Verbrennung von Altöl ohne Vorbehandlung auszuarbeiten, um auf diese Weise zu einer einheitlichen harmonischen Gesetzgebung in der Gemeinschaft in Ergänzung der Richtlinie vom 16. Juni 1975 zu gelangen.

Das bedeutet nun nicht, daß die Verbrennung von Altöl, die in Artikel 3 der Richtlinie ausdrücklich, allerdings mit dem Hinweis auf die Energiegewinnung, gestattet ist, nunmehr vollständig verboten werden soll. Die Neuorientierung der Kommission heißt nur, daß unter volkswirtschaftlichen Gesichtspunkten alle aufbereitungsfähigen Altöle zu Zweitraffinaten weiterverarbeitet werden sollen. Die nichtaufarbeitungsfähigen Altöle sollen unter Ausnutzung ihres Energiewertes und bei Beachtung strenger Umweltschutzvorschriften in geeigneten Anlagen verbrannt werden, die eine Vorbehandlung und eine Rauchgaswäsche ermöglichen.

Ein anderes wichtiges Problem, das einen Forschungsaspekt und einen Sammlungsaspekt hat, ist die Herabsetzung der Ölverluste in der Industrie und im Verkehr und ebenso die Teile, die nicht wiederverwertet werden können.

Die Ölverluste auf den Flüssen und Seen durch Schiffe sind beträchtlich. Die Sammlung und Beseitigung dieser Altöle erfordert eine besondere Organisation. Darum kümmert sich in der Bundesrepublik Deutschland der Bilgen-Entwässerungsverband. Diese Organisation ist nicht in allen Mitglied-

staaten ausreichend und befriedigend. Allein der Rhein, um
nur ein Beispiel zu nennen, wird jährlich durch mehr als
10.000 Tonnen Altöle verschmutzt.

Trotz der wirtschaftlichen und technischen Fortschritte,
die bei der Sammlung und insbesondere bei der Behandlung
von Altölen erzielt worden sind, gibt es noch eine ganze
Reihe von Problemen, die zu lösen sind - und von Möglichkeiten, die Behandlung und die Rohstoffausbeute von Altölen
zu verbessern.

Die Gemeinschaft wird prüfen, ob im Rahmen ihrer Forschungs-
und Entwicklungsprogramme, insbesondere auf den Gebieten der
Rohstoff-Forschung und der Umweltforschung, Forschungs-
und Entwicklungsaktionen zur Verbesserung der Altölverarbeitung, der Neutralisierung und der Wiederverwendung der Rückstände aus den Regenerierungsprozessen ergriffen werden
können.

So ist die Kommission insbesondere auch der Auffassung, daß
die Verarbeitungsrückstände aus der Zweitraffination ebenfalls weitgehend genutzt werden sollen.

Die Altölwirtschaft ist weit komplexer als viele meinen :
mit der Sammlung, der Umsetzung der Altölrichtlinie vom
16. Juni 1975, der Förderung der Aufbereitung und der Kontrolle der Verbrennung ist das Altölproblem durchaus
noch nicht gelöst.

Es stellen sich weitere Probleme, zum Beispiel der Definition
und der Abgrenzung von Altöl.

Die Definition der Altölrichtlinie umfaßt unter anderem auch
PCB. Dafür gibt es eine spezielle Gemeinschaftsrichtlinie.
Hier bestehen noch bisher nicht gelöste Abgrenzungsprobleme.

In diesem Zusammenhang gehört auch die Bestimmung der Fremdstoffanteile im Altöl.

Die synthetischen Schmierstoffe werden von der Altöldefinition der Richtlinie ebenfalls erfaßt. Sie nehmen ständig zu und werfen dabei Probleme für die Zweitraffination auf. Die Kommission überlegt, eine spezielle Untersuchung über die synthetischen Schmierstoffe durchzuführen.

Und schließlich gehört zur Zielsetzung einer maximalen Nutzung des wichtigen Zweitrohstoffs "Altöl" auch die Förderung der Verwendung der Zweitraffinate und die Beseitigung von Diskriminierungen, denen sie noch immer gegenüber den Erstraffinaten ausgesetzt sind. Die Kommission prüft die Frage, ob sie ähnlich wie beim Altpapier und den Altpapierprodukten ebenfalls eine Ratsempfehlung ausarbeiten soll, um die vorrangige Verwendung von Zweitraffinaten insbesondere im öffentlichen Sektor vorzuschreiben.

Diese Fragestellung führt dazu zurück, daß dieser Zweite Europäische Altölkongreß einen besonderen Schwerpunkt auf das Altöl als Rohstoff gelegt hat, der zur Verbesserung der Versorgungssituation der Gemeinschaft bei Rohstoffen und Energie herangezogen werden soll. Die Tatsache, daß durch die Wiederverwendung von Altölen, die in der Gemeinschaft anfallen, die Einfuhr von 10 bis 20 Millionen Tonnen Rohöl eingespart werden kann, zeigt die ganze wirtschaftliche Bedeutung der Altöle als zweite Versorgungsquelle für die Herstellung von Schmierstoffen.

Der Beitrag der Altöle für die direkte Energieversorgung der Gemeinschaft ist aus verständlichen Gründen allerdings weit weniger spektakulär als die Versorgung mit Schmierstoffen, die einen sehr wichtigen Rohstoff darstellen für die Industrie der Gemeinschaft und unsere technisierte Gesellschaft. Aus diesem Grunde scheint es uns durchaus vernünftig im Sinne einer rationellen Abfallwirtschafts- und in diesem Falle Altölwirtschaftspolitik, der Rohstoffwirtschaft einen Vorrang vor der Energiewirtschaft einzuräumen.

Die Bedeutung dieses Zweiten Europäischen Altölkongresses ist gerade in seiner Thematik des Recycling für die Kommission Beispiel, Signal und Modell für die zweite Phase der gemeinschaftlichen Abfallwirtschafts- und Ressourcenpolitik, in deren Mittelpunkt die Abfallverwertung stehen wird.

Wir werden die auf diesem Kongreß vermittelten Informationen und die Erfahrungen Ihres Wirtschaftszweiges, der spezialisiert ist in den Bereichen der Sammlung, der Wiedergewinnung, der Wiederaufbereitung, der Wiederverwendung und der Kommerzialisierung von Abfällen und Rückständen als Zweitrohstoffe sorgfältig prüfen und analysieren, und wir hoffen für die nächsten Jahre auf die Fortsetzung und Intensivierung der guten und vertrauensvollen Zusammenarbeit, die unter anderem zur Organisierung der Kongresse von 1976 in Brüssel und dieses Mal in Paris geführt hat.

Ich danke Ihnen für Ihre Aufmerksamkeit.

ANLAGE

Die Beseitigung und Verwertung von Altölen in den EG-Mitgliedstaaten (1977)

LÄNDER	Gesamtverbrauch von Schmierstoffen auf dem Inlandmarkt T.	Rückgewinnbare Altöle (Schätzungen) T.	Gesammelte Altöle T.	Verbrannte Altöle T.	Aufbereitete Altöle T.	Wiederaufbereitungsfähige Altöle (Schätzungen) T.	Direkte Beseitigung T.	Unkontrollierte Beseitigung T.
Belgien	200.000	100.000	15.000	13.000	2.000	–	13.000	–
Dänemark	90.000	45.000	–	–	–	–	–	–
Frankreich	1.020.000	500.000	380.000	40.000	93.000	170.000	60.000	–
Großbritannien	960.000	520.000	180.000	–	50.000	–	–	–
Irland	35.000	17.000	–	–	–	–	–	–
Italien	546.000	270.000	110.000	40.000	70.000	180.000	–	–
Luxemburg	9.000	4.000	–	–	–	–	–	–
Niederlande	229.000	110.000	80.000	80.000	keine	–	–	–
Deutschland	1.108.000	527.000	333.000	30.952	303.000	400.000	184.248	8.800
Insgesamt	4.197.000	2.073.000	1.098.000	203.952	518.000	750.000	257.248	8.800
Aufgerundet	4.200.000	2.100.000	1.100.000	204.000	520.000	750.000	260.000	8.800

DAS GEBRAUCHTÖL-RECYCLING IN DEN VERSCHIEDENEN MITGLIEDSSTAATEN DER EG
THE RECYCLING OF USED OILS IN DIFFERENT MEMBER COUNTRIES OF THE C.E.C.
LE RECYCLAGE DES HUILES USAGEES DANS DIFFERENTS PAYS MEMBRES DE LA C.C.E.
IL RICICLAGGIO DEGLI OLI USATI NEI DIFFERENTI PAESI MEMBRI DELLA C.C.E.

GEBRAUCHTÖL-RECYCLING IN BELGIEN
THE RECYCLING OF USED OILS IN BELGIUM
LE RECYCLAGE DES HUILES USAGEES EN BELGIQUE
IL RICICLO DEGLI OLI USATI IN BELGIO

J. RENGUET
Präsident der Abteilung "Schmierstoffe" in der I.H.M.B. Industrie des Huiles Minérales de Belgique, Brüssel
President of the Lubricants Section of the I.H.M.B. (Industrie des Huiles Minérales de Belgique), Brussels
Président de la section "Lubrifiants" de l'I.H.M.B. (Industrie des Huiles Minérales de Belgique), Bruxelles
Presidente della sezione "Lubrificanti" dell'I.H.M.B. (Industrie des Huiles Minérales de Belgique), Bruxelles

RESUME

La législation belge en matière d'huiles usagées est quasi inexistante :
- une loi du 26 mars 1971 pour la protection des eaux de surface
- un Arrêté Royal du 5 octobre 1975, règlant partiellement les problèmes des huiles usagées, mais juridiquement inapplicable et donc resté sans effet.

Il est normal dans ces conditions que la situation du ramassage des huiles usagées soit anarchique et que la plupart des huiles soient brûlées sans prétraitement, donc de manière polluante.

La régénération ne peut pas supporter économiquement la concurrence d'un brûlage polluant, elle n'a pas de matière première assurée et ne peut donc encore dépasser le cadre des initiatives limitées.

Nous espérons pour bientôt une législation d'application de la Directive communautaire qui impose un ramassage systématique et permette le développement du recyclage.

Mais, nous estimons que la solution de ce vaste problème doit être globale, dépasser le stade local de la destruction des déchêts et s'appuyer sur l'expérience de spécialistes du raffinage des huiles et de la régénération.

ZUSAMMENFASSUNG

In Belgien gibt es praktisch keine Rechtsvorschriften über Gebrauchtöl; es bestehen lediglich
- ein Gesetz vom 26. März 1971 zum Schutz der Oberflächengewässer,
- ein Königlicher Erlass vom 5. Oktober 1975 zur partiellen Regelung der durch Gebrauchtöl bedingten Probleme, der rechtlich jedoch nicht anwendbar ist und daher ohne Wirkung geblieben ist.

Es leuchtet ein, dass sich die Sammlung von Gebrauchtöl unter diesen Umständen regellos vollzieht und dass das meiste Gebrauchtöl ohne Vorbehandlung, also auf umweltbelastende Weise, verbrannt wird.

Die Regenerierung kann im Wettbewerb mit einer umweltverschmutzenden Verbrennung wirtschaftlich nicht standhalten; sie hat keine gesicherte Ausgangsstoffbasis und kann somit noch nicht über den Rahmen der begrenzten Initiativen hinausgehen.

Wir hoffen auf eine baldige Gesetzgebung zur Anwendung der gemeinschaftlichen Richtlinie, die eine systematische Sammlung vorschreibt und die Entwicklung des Recycling ermöglicht.

Wir sind jedoch der Ansicht, dass dieses umfangreiche Problem einer globalen Lösung bedarf, die über das lokale Stadium der Zerstörung der Abfälle hinausgeht und sich auf die Erfahrung der Sachverständigen für die Ölraffination und die Regenerierung stützt.

SUMMARY

Belgian legislation on the subject of used oils is virtually non-existant, the only provisions being :
- a Law of 26 March 1971 on the protection of surface waters;
- a Royal Decree of 5 October 1975, which regulates some aspects of the problem of used oils but is legally inapplicable and therefore ineffective.

In view of this situation, it is understandable that the collection of used oils is disorganized and that most of the oils are burnt without prior treatment, therefore causing pollution.

Regeneration cannot compete economically with polluting combustion and has no reliable supply of raw material; it cannot therefore develop beyond the level of limited schemes.

We hope that legislation will soon be adopted to implement the Community Directive, which renders systematic collection mandatory and makes for the development of recycling.

We consider, however, that the solution to this vast problem must be all-embracing, go beyond the local waste-disposal stage and draw on the experience of oil refining and regeneration specialists.

RIASSUNTO

La legislazione belga in materia di oli usati è quasi inesistente. Abbiamo :
- una legge del 26 marzo 1971 per la protezione delle acque superficiali ;
- un decreto reale del 5 ottobre 1975 che disciplina parzialmente i problemi degli oli usati, ma che non essendo giuridicamente applicabile resta senza effetto.

E' normale che in tali condizioni la raccolta degli oli usati avvenga nel massimo disordine e che la maggior parte di essi sia bruciata senza pretrattamento, con conseguenti effetti inquinanti.

La rigenerazione non può competere economicamente con la combustione inquinante ; inoltre poiché la materia prima non è assicurata non può superare la fase delle iniziative circoscritte.

Entro breve dovrebbe entrare in vigore una normativa d'applicazione della direttiva comunitaria che impone una raccolta sistematica e favorisce lo sviluppo del riciclo.

Riteniamo tuttavia che la soluzione di questo ampio problema debba essere globale, andando oltre la fase locale della distruzione dei rifiuti e facendo perno sull'esperienza degli specialisti della raffinazione degli oli e della rigenerazione.

Il y a 4 ans, lors du Premier Congrès Européen sur les huiles usagées, qui se tenait à Bruxelles, je signalais qu'en Belgique les conditions de ramassage et d'élimination des huiles usagées étaient mal définies et la situation anarchique.

J'ajoutais que, au plan législatif, en application de la Directive du 15 juin 1975, nous n'avions rien, si ce n'est un Arrêté Royal du 30 octobre de la même année ; mais cet Arrêté était juridiquement inapplicable et était donc resté lettre morte.

Nous espérions avant les 24 mois prévus pour l'application de la Directive une législation qui règlerait les problèmes d'huiles usagées, leur collecte, leur élimination ou leur recyclage.

Plus de 4 ans se sont écoulés, beaucoup d'eau et d'huile ont coulé sous les ponts, les cheminées belges ont envoyé des milliers de tonnes de métaux lourds dans l'atmosphère, et nous attendons toujours une législation d'application de la Directive Communautaire. Alors que nos voisins légiféraient et que certains avançaient grandement dans la reconnaissance de la régénération comme le moyen le plus adéquat, nous n'avons pas dépassé le stade des projets de loi.

x
x x

Essayons d'analyser la situation actuelle. Ne me demandez pas trop de chiffres. Si nous savons que la consommation totale d'huile tourne autour de 200.000 tonnes/an avec une proportion de 50% d'huile moteur, nous ignorons presque tout quant aux quantités d'huiles usagées récoltées, brûlées, importées ou exportées. Nous savons uniquement que les quantités régénérées ne dépassent pas les 4.000 tonnes en ce moment.

En matière de législation, nous avons une loi du 26 mars 1971 pour la protection des eaux de surface, nous en avons des arrêtés d'exécution fixant des normes de rejet sectorielles, mais encore aucune législation spécifique en ce qui concerne les huiles usagées.

En effet, l'Arrêté Royal du 5 octobre 1975 réglait :
- le rejet dans les eaux
- la collecte des huiles usagées et créait la notion de collecteur agréé, charge assortie d'une série de contraintes
- les obligations du détenteur d'huiles
- l'importation et l'exportation.

Mais il était muet quant au brûlage et à la pollution de l'air. En ce qui concerne l'élimination des huiles usagées, l'Arrêté ne la règlementait pas.

Texte incomplet donc, mais surtout inapplicable, car dans l'Art. 1, la définition de l'huile usagée, fondement même de l'Arrêté, était liée à une absence de valeur commerciale. Or, une huile usagée a généralement une valeur commerciale, même faible.

x

x   x

Sans barrière, sans cadre même grossier, nous vivons le règne du laisser-aller, du laisser faire. Certains détenteurs d'huiles, surtout d'huiles difficilement brûlables, essayent vainement de s'en débarasser, tandis que d'autres, à la faveur de la crise de l'énergie, se voient offrir des prix de plus en plus élevés pour des huiles usagées courantes. Ces huiles sont souvent brûlées sans aucun contrôle, sans aucun traitement, à part une éventuelle déshydratation, dans de petites installations de chauffage, dans les serres, ou bien elles sont mélangées au fuel.

Pollution donc par les métaux lourds, mais aussi danger. Permettez-moi un exemple pour illustrer ce danger : une usine de régénération a reçu accidentellement une huile usagée apparemment normale, mais dont l'analyse révèlait un point d'inflammabilité à la température ambiante. Je n'ose imaginer la catastrophe qui se serait produite si cette huile avait été cédée à un petit brûleur ou à un serriste.

x

x x

Pour lutter contre cette anarchie, deux organismes, sous forme d'A.S.B.L. ont été créés : "Huiles de Rebut" organisé par la Fédération Pétrolière Belge et "Luborec" organisé par les indépendants du graissage. Ils sont désireux tous les deux de mettre de l'ordre dans la situation que nous connaissons, mais leurs moyens sont faibles.

"Huiles de Rebut" gère un système de ramassage, mais les quantités récoltées sont encore limitées (5.000 tonnes/an sur les 50 à 60.000 potentielles) et ces huiles récoltées sont en général mélangées au fuel sans prétraitement, donc avec toute leur teneur en additifs.

D'autres ramasseurs récoltent exclusivement les bonnes huiles moteurs. Ils sélectionnent, et c'est normal car ils ne bénéficient d'aucun appui ni soutien et ils n'ont pas d'obligation de récolte.

Quant à la régénération proprement dite, celle des huiles moteurs, des projets existent et ils attendent une législation pour se concrétiser. En effet, quel investisseur risquerait la ou les centaines de millions nécessaires à la création d'une usine moderne de régénération, sans matière première assurée. Ceci, en l'absence du moindre cadre législatif, face à un marché dominé par le brûlage polluant. La faillite serait inévitable.

Malgré cela, et en attendant de pouvoir un jour remplir leur double rôle d'éliminateur non polluant de déchets, et de créateurs d'huiles lubrifiantes, deux usines de régénération traitent et régénèrent des huiles, surtout industrielles, soit à façon pour le client, soit pour en refaire des huiles de base. Les quantités traitées sont encore faibles, elles ne dépassent pas les 4.000 tonnes/an et les installations sont loin de tourner à leur rendement optimum, car il y a encore beaucoup d'immobilismes à vaincre.

Mais il y a une volonté de faire quelque chose et dès maintenant, une expérience se crée. L'initiative privée est donc présente, que ce soit au ramassage ou en régénération.

Le Gouvernement belge est conscient du problème et je tiens ici à remercier les fonctionnaires ministériels pour le dialogue constructif qui s'est créé avec les spécialistes, dont les régénérateurs. Mais, le législateur, lui, ne paraît pas encore motivé, parce que notre population ne l'est pas.

Devant l'importance du problème, la nécessité de le résoudre vite et bien, en dehors des pressions, d'où qu'elles viennent, la Belgique doit dépasser le stade des études et des projets de loi, pour aboutir enfin à une saine application de la Directive communautaire. A l'exemple de nos voisins, nous devons trouver une solution nationale, c'est-à-dire qui ne se limite pas au cadre régional d'une destruction de déchêts pour pouvoir créer à une échelle plus vaste une nouvelle matière : l'huile de base qui sortira des usines de régénération.

Je crois important d'ajouter que des spécialistes, des sociétés qui ont l'expérience du ramassage, du raffinage et de la régénération, des lubrifiants et de leur marché, sont les plus aptes à résoudre le problème intelligemment et aux moindres coûts. Nous craignons en effet des initiatives d'inspiration politique et locale, qui ne reposent sur aucune expérience. Des crédits permettent de construire des usines, ils ne suffisent pas pour les gérer, les rentabiliser, ils ne remplacent pas l'expérience.

x

x   x

L'ordre alphabétique m'a permis de faire le premier exposé sur la situation du recyclage dans les différents pays. J'espère avoir pu vous montrer que l'absence de législation entraîne inévitablement la pollution et ne permet pas d'apporter de réponse au problème des huiles usagées. Une réponse, nos voisins directs allemands et français ont pu en apporter.

Vous entendrez dans quelques minutes l'exposé détaillé du recyclage dans ces deux pays.

Je terminerai en espérant que bientôt notre Communauté Européenne toute entière sera dotée d'une législation qui efface les disparités entre nos pays, et qui permette ainsi de résoudre le problème à l'échelle européenne, globalement, sans barrières, pour le plus grand profit de nos économies respectives.

DIE GESETZLICHEN GRUNDLAGEN SOWIE DIE WIRTSCHAFTLICHE SITUATION DER
ZWEITRAFFINERIEN IN DER BUNDESREPUBLIK DEUTSCHLAND
THE LEGAL BASIS AND THE ECONOMIC SITUATION OF RE-REFINING PLANTS IN
THE FEDERAL REPUBLIC OF GERMANY
LES BASES JURIDIQUES ET LA SITUATION ECONOMIQUE DES USINES DE
REGENERATION EN REPUBLIQUE FEDERALE D'ALLEMAGNE
LE BASI GIURIDICHE E LA SITUAZIONE ECONOMICA DELLE RAFFINERIE CHE
RICICLANO GLI OLI USATI NELLA RF DI GERMANIA

Dr. H. KOEHN
Vorsitzender der AMMRA, Hamburg
President of AMMRA, Hamburg
Président de l'AMMRA, Hamburg
Presidente dell'AMMRA, Amburgo

ZUSAMMENFASSUNG

In der Bundesrepublik Deutschland gibt es zur Zeit 10 Raffinerien und Unternehmen, die in der AMMRA der Arbeitsgemeinschaft Mittelständischer Mineralöl-Raffinerien zusammengefaßt sind und Altöl als Einsatzprodukt verwenden. Zur Verbesserung der Raffinationstechnik und für zusätzliche Umweltschutzanlagen wurden in den vergangenen Jahren erhebliche Investitionssummen aufgewendet. Zweitraffinerien sind Unternehmen, die, wie jede andere Raffinerie, auf langfristige Planungen sowohl bei der Beschaffung des Wirtschaftsgutes Altöl, bei

der Verarbeitung und nicht zuletzt bei dem Vertrieb angewiesen
sind. Die deutschen mittelständischen Raffinerien tragen einen
wesentlichen Versorgungsanteil innerhalb der Mineralölwirtschaft.

Das Ende vergangenen Jahres in Kraft getretene neue Altölgesetz, das eine Anpassung an die EG-Richtlinie bringen sollte,
wird noch eine Reihe von Schwierigkeiten, die bis hin zur
existenziellen Bedrohung der Raffinerie-Unternehmen führen,
bringen. Diese Schwierigkeiten werden im einzelnen aufgezeigt
mit der Notwendigkeit, grundsätzlich wettbewerbsgleiche Voraussetzungen zu schaffen. Dazu sollte auch der Gesetzgeber beitragen. Bei den deutschen Raffinerien handelt es sich um
eine leistungsfähige Branche, die willens ist, sich auch in
der Zukunft zu behaupten.

SUMMARY

In the Federal Republic of Germany there are currently ten refineries and companies grouped under AMMRA (Association of small and medium-sized mineral oil refineries) which employ used oils. Considerable investments have been made in recent years on improving refining techniques and on additional environmental protection installations. Re-refining plants are plants which, like any other refinery, depend on long-term planning for the acquisition, processing and finally sale of used oils. German small and medium-sized refineries supply a major part of the mineral oil sector.
The new used oil act, which came into force at the end of last year on the basis of the EEC Directive, will produce a number of problems, even to the extent of threatening the existence of refineries. These problems are illustrated along with the need to create basically equal competitive conditions. This should also be contributed to by legislation. The German refinery sector is an efficient sector which intends to continue to hold its own.

RESUME

Il existe à l'heure actuelle en République Fédérale d'Allemagne dix raffineries et entreprises regroupées au sein de l'AMMRA (Arbeitsgemeinschaft Mittelständischer Mineralöl-Raffinerien, association des entreprises de raffinage indépendantes), et utilisant les huiles usagées comme matière première. Des investissements importants ont été réalisés ces dernières années en vue de l'amélioration de la technique de raffinage, et de l'installation de systèmes complémentaires de protection de l'environnement. L'exploitation des usines de régénération exige, comme pour toutes les autres raffineries, des plans à long terme tant en ce qui concerne la collecte des huiles usagées que le traitement, et même l'écoulement. Les raffineries indépendantes allemandes occupent dans le secteur pétrolier une place importante en ce qui concerne l'approvisionnement.

La nouvelle loi sur les huiles usagées, entrée en vigueur à la fin de l'année dernière, soulèvera encore un certain nombre de difficultés, dont certaines ont jusqu'à présent mis en cause l'existence des entreprises de régénération. Ces difficultés sont signalées, de même que la nécessité d'instaurer des conditions concurrentielles égales. Le législateur a un rôle à jouer à cet effet. Les raffineries allemandes constituent un secteur productif décidé à s'affirmer à l'avenir.

RIASSUNTO

Attualmente nella RF di Germania esistono dieci raffinerie e imprese che partecipano, all'interno dell'AMMRA, al gruppo di lavoro "Raffinerie di medie dimensioni" e che si dedicano al riciclo degli oli usati. Per migliorare le tecniche di raffinazione e gli impianti antinquinamento sono stati fatti negli ultimi anni considerevoli investimenti. Le raffinerie di riciclo sono imprese che, come tutte le altre raffinerie, sono costrette a fare programmi a lunga scadenza sia per quanto riguarda l'acquisto degli oli usati, che per la trasformazione e la vendita. Le raffinerie tedesche di medie dimensioni contribuiscono in modo considerevole all'approvvigionamento del mercato di oli.

La nuova normativa per questo settore, entrata in vigore alla fine dell'anno scorso per adeguare la legge tedesca alle direttive della Comunità, creerà un'ulteriore serie di difficoltà che potrebbero mettere in pericolo l'esistenza dell'industria di raffinazione. Tali difficoltà derivano principalmente dalla necessità di creare condizioni di concorrenza fondamentalmente identiche. In proposito anche il legislatore dovrebbe dare un proprio contributo. In Germania le raffinerie costituiscono un settore efficiente che desidera affermarsi anche in futuro.

In der Bundesrepublik Deutschland gibt es zur Zeit 10 Raffinerien und Unternehmen, die in der AMMRA der Arbeitsgemeinschaft Mittelständischer Mineralöl-Raffinerien zusammengefaßt sind und Altöl als Einsatzprodukt verwenden. Um es gleich vorweg zu sagen, handelt es sich um einen Industriezweig, der heute eine ganz wesentliche Funktion im Rahmen der Mineralölwirtschaft auf der einen Seite und im Umweltschutz auf der anderen Seite erfüllt.

Um dieses zu verdeutlichen, müssen einige Zahlen genannt werden:

Allein in den Jahren 1975 bis 1979 wurden für verbesserte
Anlagen - Raffinationstechnik - rund 29 Mio. DM aufgewendet,
für zusätzliche Umweltschutzanlagen ca. 8 Mio. DM und für
die nächsten 5 Jahre sind für Anlagen weitere Investitionen
von ca. 40 Mio. DM sowie für Umweltschutzprojekte ca. 14 Mio. DM
geplant. Im Vergleich dazu hat eine Mineralölkonzern-Gesellschaft
im Jahre 1979 Mittel für den Umweltschutzin Höhe von 27,28 Mio.
DM aufgewendet. Die AMMRA-Firmen haben für Forschung und
Entwicklung in den letzten 5 Jahren ca. 3 Mio. DM bereitgestellt.
Diese Zahlen wären nicht vollständig, wenn nicht auch die Menge
des gesammelten und verarbeiteten Altöles genannt würde. Es
waren im Jahre 1979 365.802 t, mit einer Ausbeute von
ca. 270.000 t, während das Gesamtschmierstoffvolumen in unserem
Lande ca. 1,2 Mio. t pro Jahr beträgt. Nach Angaben des
Bundesministeriums für Wirtschaft von September 1979 betrug
die Altölmenge insgesamt für 1977 ca. 527.000 t und unter
Hinzurechnung der Fremdstoffanteile beträgt die Gesamtrest-
stoffmenge etwa 2.400.000 t, wovon der Hauptteil auf Wasser
in Wasser/Ölgemischen und Emulsionen zurückzuführen ist.

Raffinerien, auch kleinere, müssen ein funktionierendes
Management haben und sind insbesondere auf langfristige
Planungen angewiesen. Unsere Investitionsvorhaben brauchen
abgesehen von der Planung etc., allein für die Bauprüfverfahren,
die für genehmigungsbedürftige Anlagen erforderlich sind (hierzu
zählen unsere Raffinerien), eine Vorlaufzeit von ca. 1 Jahr.

Eine umfangreiche Logistik und die entsprechende Anzahl
von Tankfahrzeugen ist notwendig für die Rohstoffbeschaffung,
die nach den Ausführungsbestimmungen des Gesetzes und der
Wettbewerbsbedingungen stattfinden soll. Das in seiner Beschaffen-
heit und Zusammensetzung sich stets wandelnde Einsatzprodukt
Altöl muß im Verarbeitungsprozeß zu stets gleichbleibenden
Produkten verarbeitet werden und dazu sind umfangreiche
Kontroll- und Meßverfahren notwendig, um insbesondere beim
Schmieröl den sehr hohen Qualitätsstandard einzuhalten, der
von den Motoren- und Aggregaten-Herstellern gefordert wird.
Das erfordert Innovation und Ingenieurleistung und beinhaltet
gleichzeitig die inzwischen erfolgende umwelt- und energie-

freundliche Nutzung der bei der Schmierölraffination anfallenden
Nebenprodukte wie der ölhaltigen Bleicherde und den Säureteer
bei der Schwefelsäureraffination. Es fallen ferner Kuppelprodukte
wie Gasöl und Spindelöl-Fraktionen an.

Der Verkauf unterliegt den Marktgegebenheiten mit den, wie
sich in den letzten Jahren gezeigt hat, sich schnell wandelnden
Voraussetzungen. Es hat sich aber gezeigt, daß angesichts
der sich abzeichnenden weltweiten Verknappung der Verfügbarkeit des Rohöles die Zweitraffinerien auch in Krisenzeiten einen
Versorgungsreservoir darstellen und des weiteren ein Wettbewerbsregulativ sind.

Wenn durch unsere Unternehmen zur Zeit 15 - 20 % des gesamten
Schmierstoffverbrauchs wieder in den Mineralölkreislauf zurückgeführt werden können und dies mit der Möglichkeit der Wiederholung
und auch der Mengensteigerung, ist das eine Entlastung des
Rohölbedarfs. Nicht alle Rohöle sind gleich gut für die Schmierölproduktion geeignet, einige Rohöle enthalten überhaupt keine
Schmierölfraktion. Das heißt also, daß die Verfügbarkeit
für diesen qualitativ sehr wichtigen Zweig der Mineralölwirtschaft überproportional abnimmt. Aus 1000 t Altöl können
über 500 t Schmieröl und mehr als 150 t Gasöl gewonnen werden.
Der Rohölbedarf für die gleiche Menge ist mindestens fünfmal so
hoch, das wären also ca. 5000 t.

Diese Tatsachen machen die Notwendigkeit des Recycling von
gebrauchten Ölen deutlich und unterstreichen das große
Leistungspotential der deutschen mittelständischen Raffinerien.
Natürlich kann man Altöl unter Beachtung der Immissionsschutz-Gesetze u.Grenzwerte auch verbrennen (die im Altöl enthaltenen
Additives werden zum Teil emittiert und müssen mit Spezial-Rauchgasfiltern zurückgeholt werden). Das Wirtschaftsgut Altöl
ist zum Verbrennen zu schade.

In der Bundestagsdrucksache 8/3713 vom 27.2.80 heißt es,
daß die Bundesregierung die Ressourcenverknappung, die
sich derzeit vor allem auf dem Energiesektor bereits deutlich
abzeichnet, für eine entscheidende Rahmenbedingung der Umweltpoli-

tik hält. In der Bundestagsdrucksache 7/3455 wird erwähnt, daß die Bundesregierung der Ansicht ist, daß Altöl als Wirtschaftsgut von erheblicher rohstoff- und energiepolitischer, wie auch wettbewerbs- und preispolitischer Bedeutung ist; und volkswirtschaftlich sinnvoll genutzt werden soll.

Zur Zeit sind in der Bundesrepublik Deutschland per 1.7.1980 über 27 Mio. Kraftfahrzeuge registriert, die allein ohne den Industrie-Schmierstoffbedarf bei der Annahme von 5 Ltr. Schmieröl-Inhalt pro Fahrzeug und zweimal Ölwechsel pro Jahr ca. 270.000 t gebrauchtes Öl anfallen lassen würden. Die andere Hälfte kommt aus dem Industrie- und Schiffssektor. Folgende Zahlen machen auch die umweltschutzrelevante Leistung deutlich, die mit der Übernahme und der Sammlung von Altöl in kleineren Mengen verbunden ist. Insgesamt werden jährlich rund 500.000 Einzelpartien übernommen. Hierbei liegen 2/3 aller Übernahmen unter 800 Ltr.

Es sei darauf hingewiesen, daß nach dem Verursacherprinzip für die versteuerten Schmieröle, das ist etwa die Hälfte aller zum Einsatz kommenden Schmieröle, eine Ausgleichsabgabe erhoben wird, die zur Zeit DM 11,-- per 100 kg beträgt, die von allen Mineralölherstellern und Importeuren an das Bundesamt für gewerbliche Wirtschaft abgeführt werden muß. Auch für die Schmieröle aus der Zweitraffination werden erneut Mineralölsteuer und Ausgleichsabgabe fällig. Aus diesem Fonds werden die anderweitig nicht zu deckenden Kosten, die bei der Sammlung und Verarbeitung entstehen, abgedeckt. Natürlich unter Abzug der Erlöse und wie aus der Bundestagsdrucksache 8/1676 hervorgeht, unter Einbeziehung einer angemessenen Rendite von 5 % bis 10 % des Gesamtkapitals im Durchschnitt aller beteiligten Unternehmen.

Angesichts der schon zuvor genannten Investitionsvorhaben von ca. 10 Mio. DM pro Jahr und einem Gesamtkapital um ca. 50 Mio. DM für alle Zweitraffinerien ergibt sich ein Mißverhältnis, welches die Frage der sogenannten angemessenen und ausreichenden Rendite an den Gesetzgeber erneut akut werden läßt und Finanzierungsprobleme aufzeigt.

Diese Frage führt auch direkt zu der derzeitigen Situation
der deutschen Zweitraffinerien, die ja dem Altölgesetz und
den entsprechenden Verordnungen und Ausführungsbestimmungen
unterworfen sind. Hierzu sind folgende Ausführungen zu machen:

Am 23. Dezember 1968 ist das Gesetz zur Sicherung der Altölbeseitigung (Altölgesetz) im Bundestag verabschiedet worden
und ist am 1. Januar 1969 in Kraft getreten. Es regelte die
wirtschaftliche Sicherung der Altölbeseitigung durch ein Sonder-
vermögen des Bundes, mit dem Namen Rückstellungsfonds,
der aus der erwähnten Ausgleichsabgabe der versteuert in
den Gebrauch kommenden Schmieröle gespeist wird. Des weiteren
sind Bestimmungen darin enthalten, wie Altöl definiert wird, in
welchen Bereichen gesammelt werden muß und wie der Verbleib des
Altöles überwacht werden soll. Die Bundesregierung hat alle
3 Jahre dem Bundestag über die Vollzugserfahrungen des
Gesetzes zu berichten. Dies ist geschehen in den Drucksachen des
Deutschen Bundestages:

        6/3312 vom 5. April 1972

        7/3455 vom 9. April 1975

        8/1676 vom 31. März 1978

In diesen Unterlagen sind Einzelheiten und z.T. umfangreiches
Zahlenmaterial enthalten. Dies bezieht sich auf den Altölanfall
in den verschiedenen Verbrauchssparten, auf die Fremdstoffanteile und auf die wettbewerbs- und verbraucherpolitischen
Auswirkungen, Fertigprodukte im Handel betreffend sowie
Auswirkungen des Selbst-Ölwechsels. In der Bundestagsdrucksache 8/1434 wird dem Bundestag der Entwurf zur Änderung
des Altölgesetzes vorgelegt, der sich an die Richtlinie des
Rates der Europäischen Gemeinschaften vom 1. April 1975
anlehnt. In der Drucksache 8/2945 wird schließlich der Bericht
des Wirtschaftsausschusses des Bundestages veröffentlicht
und mit Wirkung vom 11. Dezember 1979 das geänderte Altölgesetz. Trotz der umfangreichen Bemühungen und Erklärungen
der Arbeitsgemeinschaft mittelständischer Mineralöl-Raffinerien
enthält dieses Gesetz die Existenz der Zweitraffinerien bedrohende
Ausführungen. Warum ?:

Während der Verabschiedung des Gesetzes im Bundestag am
15.5.1979 hat der Abgeordnete Wolfram für die SPD erklärt:
Für uns hat die Wiederaufbereitung Vorrang vor der Verbrennung.
Es darf überhaupt keine Zweifel geben: in Zukunft muß das
Prinzip "Recycling vor Verbrennung" volle Gültigkeit haben,
und weiter: "Für uns hat dieses Gesetz auch einen mittel-
standspolitischen Aspekt, uns geht es mit dem Gesetz auch
darum, kleine und mittlere selbständige Unternehmen zu erhalten, Unternehmen, die für die Marktstruktur auf diesem
Gebiet von Bedeutung sind, die nicht verdrängt werden dürfen,
die ihre Wettbewerbsfunktion gegenüber den großen Mineralöl-
Konzernen erfüllen müssen." Sinngemäß hat sich auch der
Abgeordnete Zywietz als Sprecher der FDP-Fraktion geäußert.
Diese Ausführungen finden unsere volle Zustimmung.
Der Vorrang des Recycling findet sich jedoch im gegenwärtig
gültigen Gesetz nicht wieder. Im Gegenteil, wir müssen um
unsere zukünftige Existenz fürchten, denn Entgeltzahlungen
für das Wirtschaftsgut Altöl (das ja einen Wert hat) gehen
in die Kostenrechnung zur Ermittlung der ungedeckten Kosten
nicht ein. Zum besseren Verständnis ist hierzu zu erwähnen daß
die gestiegenen Erlöse bei dem Verkauf von Fertigprodukten in der
Kosten-Ertragsrechnung voll berücksichtigt werden, so daß durch
die Nichtberücksichtigung der gestiegenen Entgeltzahlungen und den
Abzug der Erlöse ein völlig schiefes Bild entsteht. Das wäre
vergleichbar mit der Bilanz einer Mineralölgesellschaft, in welcher
die Rohöl-Einstandskosten eliminiert würden und dann folgerichtig
imaginäre Gewinne erschienen, die dann die Grundlage ergäben für
die Ermittlung der Gewinnsituation. Bei uns aber sind diese Zahlen
die Grundlage für die Ermittlung der ungedeckten Kosten inkl. des
schon zitierten angemessenen Gewinns von 5 bis 10 % des Gesamtkapitals.
Da wegen der in den letzten Jahren erheblich gestiegenen
Energiepreise naturgemäß auch der Wert des Wirtschaftsgutes
Altöl gestiegen ist, sind die Entgeltzahlungen wesentlich
höher als der uns zugestandene Gewinn. Mit anderen Worten,
nach dieser Regelung können wir in der Zukunft nur noch
Verluste erwirtschaften, wenn nicht, wie im Gesetz vorgesehen,
ein Teil der Entgeltzahlungen als Kosten anerkannt wird,

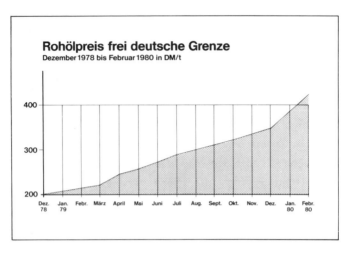

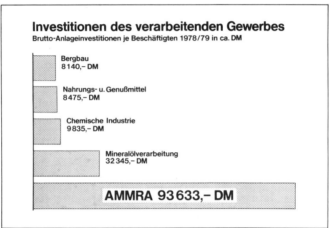

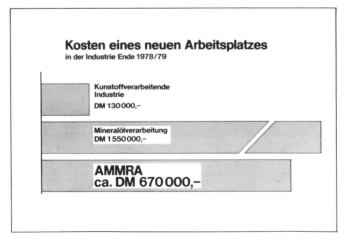

wobei sichergestellt sein muß, daß die ordnungsgemäße Erfüllung
der gesetzlichen Aufgaben nur auf diese Weise erfolgen kann.
Die Bundesbehörden selbst verlangen für das Altöl ein Entgelt,
obwohl der Abgeordnete Zywietz die Überprüfung dieses Verhaltens gefordert hat.

Hier ist der entscheidende Punkt: Wie sind langfristige Planungen der Raffinerien und Investitionen angesichts solcher
Regelungen zu verantworten und wie ist ein vom Gesetzgeber
gewollter Wettbewerb mit dem Wirtschaftsgut Altöl möglich.
Hier können wir nur die verantwortlichen Politiker bitten, für alle
Beteiligten gleiche Wettbewerbsverhältnisse zu schaffen, sowie eine
klare gesetzliche Regelung auch in diesem Punkt. Raffinerie-Anlagen können nicht kurzfristig an- und abgefahren werden, wie
dies beispielsweise bei einem Verbrennungsofen möglich wäre.
Ganz abgesehen von dem nicht vergleichbaren Investitionsaufwand.
Untersuchungen haben übrigens ergeben, daß auch die Energiebilanz beim Recycling des Altöls im Vergleich zur Verbrennung und
im Vergleich zur Verarbeitung von Schmierölen aus Rohöl positiv ist.

Es geht um die Frage, wird Recycling gewollt oder nicht.
In 20 Jahren haben wir das Jahr 2000. Wir müssen heute
wichtige Rohstoffe schonen und die Versorgung mit Schmierstoffen innerhalb der Mineralölwirtschaft auch für die Zukunft
in allen Situationen sicherstellen helfen. Nichts läuft ohne
Öl. Kosten- und Investitionsvergleiche zeigen das hohe Volumen
auf und die Notwendigkeit, sich modernsten technischen Erfordernissen anzupassen.

Zum jetzt gültigen Gesetz selbst und damit auch zur Anpassung
an die Altöl - EG - Richtlinie sind noch zwei Schwierigkeiten
zu erwähnen

A) Um Altöl und Abfall abzugrenzen, soll das Altöl an der
   Anfallstelle nach Fremdstoffanteilen getrennt gelagert werden,
   und zwar Altöl mit beispielsweise 10 % Fremdstoffanteilen
   in den einen Tank und Altöl mit darüberliegenden Anteilen
   in den anderen Tank. Nicht betriebsbedingte Fremdstoffe
   lassen Altöl zum Abfall werden. Wer kontrolliert dies an-

gesichts der Tatsache, daß die Untersuchungen von Proben nur im Laboratorium durchgeführt werden können und bei den vielen Einzelübernahmen einen Kostenfaktor von ca. 20 Mio. DM ausmachen würden.

B) Syntheseöle sind auch nach dem Altölgesetz zu beseitigen, obwohl eine Reihe von synthetisch hergestellten Schmierstoffen gar nicht mit Mineralölen mischbar sind und insbesondere die halogenhaltigen Kohlenwasserstoffe wie Fluor und Chlorkohlenwasserstoffe stark toxische Wirkung haben. Im Umweltbericht 8/3713 der Bundesregierung heißt es beispielsweise:

"Die Toxität und Persistenz der polychlorierten Biphenyle (PCB) und der polychlorierten Terphenyle (PCT), die wegen ihrer besonderen Eigenschaften bereits seit Anfang der 30er Jahre für einen weiten Anwendungsbereich große wirtschaftliche Bedeutung haben, wurde erstmals im Jahre 1966 festgestellt. Die außerordentlich gesundheitsschädliche Wirkung von PCB ist insbesondere gekennzeichnet durch Hautveränderungen, Eingriffe in den Leberstoffwechsel und andere gesundheitschädigende Wirkungen. Nach unserer Auffassung gehören solche giftig-schädlichen Substanzen nicht zu dem Begriff Altöl und müßten gesondert erfaßt und unter besonderen Vorsichtsmaßnahmen vernichtet werden. Verfahrenstechniken zum Recycling solcher oder ähnlicher Substanzen werden zur Zeit nach unserem derzeitigen Wissensstand nicht praktiziert.

Das Recycling von Altöl in der Bundesrepublik ist ein wichtiger Faktor der Mineralölindustrie. Die verfahrenstechnische Entwicklung ist in unserem Lande weit fortgeschritten und dient vielfach auch außerhalb Europas als Maßstab. Es wird weiter daran gearbeitet. Wir wollen mit dazu beitragen, daß die noch vorhandenen Schwierigkeiten innerhalb der Europäischen Gemeinschaft gemeinsam beseitigt werden und hoffen, daß politische Gesichtspunkte sowie die wirtschaftliche Notwendigkeit der Zweitraffination förderlich sein mögen. Neue Gesetzes-Initiativen oder Verordnungen werden daher erforderlich sein -getragen von der gesamten Mineralöl-Industrie der Bundesrepublik Deutschland.

GEBRAUCHTÖL-RECYCLING IN FRANKREICH
THE RECYCLING OF USED OILS IN FRANCE
LE RECYCLAGE DES HUILES USAGEES EN FRANCE
IL RICICLO DEGLI OLI USATI IN FRANCIA

R. KACHLER
Direktor der S.N.F.R.H.G., Paris
Director of S.N.F.R.H.G., Paris
Directeur du S.N.F.R.H.G., Paris
Direttore del S.N.F.R.H.G., Parigi

RESUME

Remontant aux années trente, le recyclage des huiles usagées, rendu obligatoire en France pendant la guerre, n'a jamais cessé de s'y développer jusqu'en 1974.

C'est alors qu'apparaît, en conséquence du renchérissement subit des fuel-oils et de l'encadrement règlementaire de leur consommation, une nouvelle manière d'utiliser ces huiles : le brûlage, qui s'étend rapidement à la faveur d'une règlementation incomplète.

Soucieux, non seulement de protéger les sols et les eaux des rejets sauvages d'huiles usagées, mais encore d'éviter la pollution de l'air due à leur brûlage non contrôlé, les pouvoirs publics français instaurent finalement en 1979 une règlementation nouvelle et complète sur la récupération et l'élimination des huiles usagées.

Appliquant le principe du pollueur-payeur, cette règlementation prescrit la détention soigneuse de ces huiles et en organise la récupération systématique grâce à un zonage du Territoire et à un système d'agrément zonal de ramasseur exclusif, les huiles collectées étant ensuite dirigées obligatoirement vers des centres d'élimination officiellement agréés.

Deux modes d'élimination sont admissibles à l'agrément : la régénération et le brûlage non polluant. - Considérant cependant l'avantage économique du premier, la règlementation oriente préférentiellement les huiles usagées vers la régénération.

Actuellement en cours de mise en place, cette règlementation ne doit entrer en vigueur que le 23 novembre 1980. Ses effets ne pourront donc être mesurés que dans quelques mois. Les régénérateurs français, quant à eux, pensent qu'elle est parfaitement adaptée à son objectif : le recyclage maximum des huiles usagées axé sur l'intérêt économique optimum du pays.

ZUSAMMENFASSUNG

Seit den dreissiger Jahren hat sich das Gebrauchtöl-Recycling, das im Krieg in Frankreich obligatorisch war, stetig bis zum Jahre 1974 entwickelt. Zu diesem Zeitpunkt kommt infolge der Verteuerung des Heizöls und der rechtlichen Regelung seines Verbrauchs eine neue Art der Altöl-Verwendung auf, nämlich die Verbrennung, die sich aufgrund einer unvollständigen Reglementierung schnell ausbreiten kann.
In dem Bestreben, nicht nur den Boden und die Gewässer gegen wilde Ableitungen von Gebrauchtöl zu schützen, sondern auch eine Luftverschmutzung durch unkontrollierte Verbrennung von Altöl zu vermeiden, führen die französischen Behörden 1979 schliesslich eine neue und vollständige Regelung für die Rückgewinnung und die Beseitigung von Gebrauchtöl ein.
Unter Anwendung des Verursacherprinzips schreibt diese Regelung die sorgfältige Aufbewahrung des Gebrauchtöls vor und organisiert die systematische Gebrauchtöl-Rückgewinnung durch eine Einteilung des Staatsgebiets in Zonen und ein System der zonalen Zulassung ausschliesslicher Sammler, wobei das gesammelte Öl dann zu amtlich zugelassenen Beseitigungszentren geleitet werden muss.
Zwei Arten der Beseitigung sind zulässig: die Regenerierung und die umweltfreundliche Verbrennung.-In Anbetracht der wirtschaftlichen Vorteile des ersteren Verfahrens wirkt die Regelung vorzugsweise auf die Regenerierung von Gebrauchtöl hin.
Diese Regelung befindet sich in der Einführung und wird erst am 23. November 1980 in Kraft treten. Ihre Auswirkungen können daher erst in einigen Monaten gemessen werden. Nach Ansicht der französischen Regenerierungsunternehmen ist sie ihrem Ziel, dem auf das optimale wirtschaftliche Interesse des Landes ausgerichteten maximalen Gebrauchtöl-Recycling, vollkommen angepasst.

SUMMARY

The recycling of used oils in France dates back to the thirties; it was made mandatory during the war and enjoyed uninterrupted development until 1974.
Then, as a result of the sudden rise in the price of fuel oils and the official restrictions on the consumption thereof, a new way of disposing of used oils, namely burning, was discovered and developed rapidly owing to loopholes in the regulations.
Anxious not only to protect land and water resources from indiscriminate discharges of used oils, but also to prevent atmospheric pollution due to uncontrolled burning, the French authorities finally adopted in 1979 a new and complete set of regulations on the recovery and disposal of used oils.
These regulations, which apply the "polluter pays" principle, divide the country into sectors within which only one collector is authorized, and thus provide for used oils to be carefully stocked, systematically recovered and compulsorily conveyed to officially approved disposal centres.
Two disposal methods are eligible for official approval: regeneration and pollution-free burning. In view of the economic benefit that can be derived from the former, however, the regulations give preference to regeneration.
These regulations, which are currently being introduced, will not enter into force until 23 November 1980. Their effects will therefore be measurable only in a few months' time. The French regenerating industry, for its part, considers them to be perfectly suited to their aim: the recycling of as much used oil as possible in the optimum economic interest of the country.

RIASSUNTO

Dagli anni 1930 il riciclo degli oli usati, reso obbligatorio in Francia durante la guerra, ha continuato a svilupparsi fino al 1974. E' a questo punto che appare, a seguito del rincaro degli oli combustibili e dell'apparsa di disposizioni che ne disciplinano il consumo, un nuovo modo di utilizzare questi oli la combustione, che si diffonde rapidamente in assenza di una normativa incompleta.
Le autorità pubbliche francesi, preoccupate non soltanto di proteggere il suolo e le acque dagli scarichi abusivi di oli usati, ma anche di evitare l'inquinamento dell'aria causato da una loro combustione incontrollata, hanno fissato nel 1979 una normativa nuova e completa sul recupero e l'eliminazione di tali sostanze.
Applicando il principio del "chi inquina, paga", detta normativa prescrive una detenzione responsabile di tali oli e ne organizza il recupero sistematico mediante una suddivisione in zone del territorio e un sistema di monopolio zonale dell'operazione ; gli oli raccolti devono successivamente essere avviati verso centri d'eliminazione ufficialmente riconosciuti.
Due sono i metodi d'eliminazione consentiti : la rigenerazione e la combustione non inquinante. Tuttavia, tenuto conto dei vantaggi economici della prima soluzione, la normativa si orienta di preferenza verso la rigenerazione degli oli usati.
Detta normativa, attualmente in via di elaborazione, dovrà entrare in vigore il 23 novembre 1980. I suoi effetti potranno pertanto essere valutati soltanto tra qualche mese. Secondo i responsabili francesi della rigenerazione, essa è perfettamente adeguata all'obiettivo che intende raggiungere : il massimo riciclo possibile degli oli usati orientato verso il massimo beneficio economico del paese.

Mesdames, Messieurs,

Si la France ne s'est dotée que depuis peu d'une règlementation d'administration publique sur la récupération et l'élimination des huiles usagées, il faut rappeler que, dans notre pays, le recyclage de ces produits remonte déjà à près d'un demi-siècle ; c'est en effet à partir de 1930 que les premières unités de régénération d'huiles usagées ont été construites en France.

La Régénération française peut donc faire état de quartiers de noblesse qui lui donnent, non seulement une place très honorable dans l'industrie des lubrifiants, mais encore une référence de tout premier ordre dans le domaine général du *recyclage*.

La guerre de 1939 entraîna la destruction très rapide de 70 % de l'outil de raffinage tout neuf né des fameuses lois pétrolières françaises de 1928, la coupure du Territoire de ses sources normales d'approvisionnement pétrolier et de très sévères restrictions imposées au pays ; ces restrictions, on peut en imaginer l'ampleur si l'on considère qu'en 1943, au milieu de la guerre, le contingentement forcé avait réduit la consommation intérieure des produits pétroliers à moins de 6 % de ce qu'elle avait été à la veille de la guerre.

On comprend dès lors que les pouvoirs publics de l'époque aient mis en oeuvre tous les moyens de faire face à une telle situation.

Dans le domaine des lubrifiants, ce furent trois lois de septembre, octobre et novembre 1940 rendant obligatoires la récupération et la régénération des huiles minérales usagées dans des usines spécialement agréées à cet effet. Rappelons pour la petite histoire qu'à l'époque, pour pouvoir acheter 5 kg d'huile neuve, il fallait remettre à son fournisseur, en échange, 2 kg d'huile usagée.

Après la guerre, les sources d'approvisionnement étant à nouveau accessibles et les voies de communications maritimes rétablies, il fallut encore plusieurs années pour reconstruire les raffineries de pétrole et c'est en 1949 seulement que, dans ce pays, les mesures de rationnement des produits pétroliers purent être supprimées.

Une ère nouvelle s'ouvre donc à partir de 1950 : l'industrie pétrolière française a retrouvé son dynamisme d'avant-guerre et, couvrant largement les besoins du pays, devient de plus en plus exportatrice ; l'industrie de la régénération ne disparaît pas pour autant, bien au contraire.

Après avoir été intimement associée aux pétroliers pendant la guerre et l'après-guerre, elle devient la concurrente de ces mêmes pétroliers sur le marché des huiles de base. Certains régénérateurs disparaissent durant que d'autres s'adaptent et soutiennent la concurrence en améliorant leurs procédés.

Dans son ensemble, la régénération française se développe très honorablement puisque, de 1950 à 1973, sa production d'huiles de base quadruple tandis que la consommation intérieure des lubrifiants finis augmente, en gros, de 200 %.

Entre-temps, il y a eu en automne 1956 l'affaire de Suez, la fermeture du canal et la crainte, en France, d'une nouvelle période de pénurie.

Les pouvoirs publics français, se souvenant de l'expérience encore récente des années de guerre, adoptent alors, dans le domaine des lubrifiants, la même solution qu'en 1940 et décident, par un arrêté ministériel du 20 novembre 1956, que *"les huiles minérales de graissage usagées sont intégralement destinées à la régénération, à l'exclusion de tout autre emploi."*

Mais la crise de Suez ne dure pas et, très rapidement, l'arrêté de novembre 1956 cesse d'avoir un intérêt pratique : personne ne se soucie en réalité du sort des huiles usagées. Il s'agit d'un déchet encombrant auquel seuls les régénérateurs s'intéressent et pour la collecte duquel ils ont maintenu en activité la société commune de ramassage créée au début de la guerre. Quant aux huiles usagées non ramassées pour le compte des régénérateurs, elles sont purement et simplement jetées dans la nature.

Surviennent alors deux phénomènes nouveaux :

- tout d'abord, à partir des années soixante, le développement de la préoccupation de l'environnement qui conduit à constater que le rejet des huiles usagées dans le milieu naturel constitue, à lui tout seul, une cause de pollution très importante puisqu'un bon quart de la pollution de toutes les eaux françaises est directement imputable à ce déchet,

- puis la crise pétrolière, déclenchée en octobre 1973 par la nouvelle politique de l'OPEP, qui se traduit par le quadruplement en quelques mois du prix des produits pétroliers et, en France, par la mise en oeuvre progressive d'une politique nouvelle axée sur les économies de devises, et donc d'énergie. Des mesures de contingentement interviennent début juillet 1974 sur le fuel domestique et sont complétées en septembre 1975 par un encadrement des consommations de fuel lourd.

Certains découvrent alors que l'huile usagée peut constituer un produit de remplacement bon marché à du fuel devenu cher et que son utilisation permet aussi de tourner aisément les mesures d'encadrement s'appliquant aux fuels puisque l'accès à l'huile usagée est libre.

Enfin, *pratiquement* libre car, en droit, cette huile est toujours exclusivement destinée à la régénération puisque l'arrêté de novembre 1956 n'a pas été abrogé, ainsi que le souligne du reste une circulaire ministérielle du 14 mai 1976 qui rappelle l'interdiction de brûler les huiles usagées. Mais voilà ! La sanction des infractions à cet arrêté est si bénigne qu'elle est inopérante : il suffit de brûler un baril d'huile usagée pour amortir l'amende en principe encourue et le reste est pur profit.

C'est ainsi qu'en France prend naissance en 1974 et se développe par la suite le brûlage des huiles usagées. Cette utilisation nouvelle provoque une augmentation rapide des prix des huiles usagées car, contrairement aux régénérateurs qui doivent prendre en considération les coûts de traitement liés à la distillation et au reraffinage de *l'huile usagée - matière première*, les utilisateurs de *l'huile usagée - combustible* ont pour seul critère d'utilisation le prix du fuel auquel l'huile usagée doit se substituer, compte tenu, bien sûr, de l'équivalence calorifique existant entre les deux produits (en gros 80 %).

Parallèlement à cette évolution, la préoccupation de l'environnement dont j'évoquais l'essor tout-à-l'heure, s'est développée et s'est finalement traduite par quelques mesures légales et réglementaires : c'est tout d'abord la loi du 15 juillet 1975 relative à l'élimination des déchets qui rend le détenteur d'un déchet polluant responsable de l'élimination *propre* de ce déchet et un décret du 19 août 1977 donnant à l'administration des pouvoirs de contrôle sur les modalités d'élimination des déchets transportés.

En ce qui concerne plus particulièrement les huiles usagées, un décret du 8 mars 1977 interdit leur déversement dans les eaux superficielles ou souterraines ainsi que dans les eaux de mer.

Mais ces quelques textes sont loin de constituer la véritable réglementation nécessaire. Nécessaire *et attendue* depuis que le Conseil des Communautés Européennes, dans une directive du 16 juin 1975, a prescrit aux états membres de réglementer dans un délai de deux ans la détention, la collecte et l'élimination des huiles usagées de manière à protéger les sols, les eaux et l'air.

Finalement, cette réglementation française sur la collecte et l'élimination des huiles usagées voit le jour en 1979 sous la forme de quatre textes : deux décrets des 30 juin et 21 novembre 1979 et deux arrêtés d'application datés également du 21 novembre 1979.

o
o   o

Je ne vous infligerai pas la lecture des textes. Je m'efforcerai simplement d'en indiquer ici les grandes lignes en partant des principes qui les inspirent. Ces principes sont au nombre de trois :

    1° le principe du *pollueur-payeur*
    2° le principe de *l'exhaustivité du ramassage des huiles usagées*
    3° le principe d'une *élimination rationnelle* de ces huiles usagées.

Le principe du *pollueur-payeur* - qui est du reste expressément énoncé dans la Directive de Bruxelles - veut que ce soit le pollueur qui paie pour les nuisances qu'il provoque - et non les autres. Le décret du 30 juin 1979 crée donc une taxe parafiscale qui vient se greffer sur le prix de chaque kilo d'huile neuve achetée. Dès son achat, le consommateur d'huile, automobiliste, transporteur, industriel etc. paie ainsi par avance le coût de récupération de cette huile lorsqu'il viendra à s'en débarrasser après usage. Le produit de cette taxe (40 F/tonne, soit quelque 7 centimes sur un bidon de deux litres) est centralisé dans un compte spécial géré par l'Agence Nationale pour la Récupération et l'Elimination des Déchets (ANRED) et est affecté à des aides à la collecte et à des actions collectives d'information et d'assistance destinées aux producteurs et détenteurs d'huiles usagées. La répartition de ce fonds est décidée par un Comité de Gestion de 10 membres : six représentants des administrations concernées, deux représentants de la profession du ramassage, un représentant des entreprises de traitement et un représentant des utilisateurs d'huiles régénérées.

Le deuxième principe dont s'est inspirée la règlementation procède de l'idée que la protection maximum de l'environnement implique que la totalité des huiles usagées soit collectée. Pour assurer cette *exhaustivité du ramassage*, les mesures suivantes ont été prises.

Tout d'abord, chaque détenteur d'huiles usagées est désormais tenu de conserver ses huiles dans des installations étanches et aménagées de manière à éviter tout mélange avec de l'eau ou un autre produit et ce, jusqu'à la remise de ces huiles, soit à un ramasseur agréé, soit à un éliminateur agréé.

On constate ainsi que, dès leur stockage chez le détenteur, les huiles usagées doivent faire l'objet d'une ségrégation ; cette ségrégation doit permettre, d'une part d'éviter que le produit de la taxe parafiscale payée par le consommateur de lubrifiants pour la récupération de l'huile usagée ne soit détourné de sa destination spécifique en servant également à financer la récupération d'autres produits que ces huiles tels que solvants usés, déchets liquides de toute sorte etc. ; cette ségrégation doit permettre, d'autre part, de conserver aux huiles usagées les caractéristiques qui les rendent propres à être retraitées correctement, tant sur le plan technique que sur le plan économique.

Pour ce qui est de la collecte, le Territoire français a été divisé en zones géographiques correspondant aux départements. Dans chacune de ces zones le ramassage des huiles usagées ne peut être effectué que par les soins d'une personne, physique ou morale, ayant reçu un agrément des pouvoirs publics. Pour chaque zone, un seul agrément est délivré et ce, après un appel d'offre public comportant l'obligation de se plier à un cahier des charges. Cet agrément est accordé pour trois ans par le ministre chargé de l'environnement assisté d'une commission interministérielle spéciale. Sous réserve que les conditions techniques de ramassage proposées par chaque soumissionnaire soient conformes aux exigences du cahier des charges, l'agrément est délivré au candidat le mieux-disant, c'est-à-dire à celui qui offre le prix de collecte le meilleur du point de vue de l'intérêt général.

Le ramasseur agréé a, notamment, pour obligation de collecter *toutes* les huiles usagées disponibles dans la zone qui lui est attribuée en appliquant aux détenteurs d'huiles un barème de prix fixé par les pouvoirs publics.

Il est, d'autre part, tenu de ne remettre les huiles collectées qu'à des éliminateurs eux-mêmes titulaires d'un agrément officiel et le prix auquel il doit céder ces huiles aux éliminateurs agréés doit être obligatoirement celui sur la base duquel l'agrément lui a été délivré ; il faut noter à ce propos que, l'agrément de ramasseur durant trois ans, ce prix fait l'objet d'une indexation officielle.

Le système ainsi mis en place au niveau du ramassage appelle quelques commentaires :

- la procédure de l'appel d'offres permet à la libre concurrence de s'exercer tous les trois ans puisque peut soumissionner librement pour chaque adjudication tout entrepreneur s'estimant capable d'exercer l'activité de collecte dans la zone considérée ;

- l'attribution d'un seul agrément par zone de ramassage permet de situer de façon précise la responsabilité de l'exhaustivité de la collecte ; les pouvoirs publics savent ainsi à qui s'en prendre si le travail est mal fait ; de plus, la formule de l'agrément exclusif permet d'éviter toute surenchère de prix auprès des détenteurs, ce qui est normal puisque la taxe parafiscale d'élimination payée par le consommateur de lubrifiants est destinée à rémunérer la ségrégation et à financer le service de récupération de l'huile usagée - et non à procurer un profit quelconque au détenteur du déchet "huile usagée".

Le troisième principe, celui de *l'élimination rationnelle* des huiles usagées, s'applique dans la réglementation française de la manière suivante :

Comme dit, les ramasseurs agréés ont l'obligation de livrer les huiles usagées collectées aux seuls éliminateurs bénéficiant d'un agrément. Cet agrément, d'une durée de sept ans, est délivré par le ministre chargé de l'environnement, selon un cahier des charges précis.

Pour les candidats éliminateurs, il n'y a ni appel d'offres, ni exclusivité territoriale. Les dossiers sont instruits sur la base de données essentiellement techniques, telles que : capacités de traitement, procédés, modes d'élimination des déchets de traitement, équipements de protection de l'environnement etc.

Aux termes de l'Art. 7 du décret du 21 novembre 1979, les seules modes d'élimination susceptibles d'être agréés sont le recyclage ou la régénération dans des conditions économiques acceptables ou, *à défaut*, l'utilisation industrielle comme combustible.

Le sens à donner à cette disposition est explicité par l'Art. 2 de l'arrêté d'application de même date qui précise en effet que "Les huiles usagées collectées sont *préférentiellement destinées à être éliminées par régénération ou recyclage* dans des installations agréées (...). A défaut, les huiles usagées collectées ne peuvent être éliminées par brûlage que dans des installations agréées au titre de la protection de l'environnement et comportant un dispositif de récupération de chaleur ...".

En d'autres termes, la nouvelle réglementation française n'accepte d'autoriser le brûlage avec récupération de chaleur que des seules huiles usagées non régénérables et dans la mesure seulement où l'installation de brûlage est équipée des dispositifs voulus pour protéger l'environnement.

Il y a donc là une prise de position nette des autorités gouvernementales en faveur de la régénération des huiles usagées.

Pour trouver l'explication de ce choix, il faut se reporter aux déclarations que le ministre français de l'industrie faisait le 11 avril 1979 devant l'assemblée nationale, au cours des débats consacrés aux économies d'énergie : "... Parmi les usages des huiles usées, la régénération présente l'avantage d'économiser de la matière première et permet, en outre, une concurrence salutaire avec les huiles de marque fabriquées par les sociétés de raffinage. Le Gouvernement est donc favorable à cette activité qu'il considère comme *prioritaire* dans ce domaine ...".

Et le Ministre précisait quelques instants après, à propos de la combustion des huiles usagées : "... Cette dernière ne constitue pas une bonne utilisation, d'abord parce qu'elle est un gaspillage, ensuite parce qu'elle peut créer des nuisances préjudiciables à l'environnement".

Telles sont les dispositions réglementaires émanant du pouvoir exécutif français sur l'élimination des huiles usagées. Il est remarquable que le pouvoir législatif ait choisi la même voie puisque, dans la loi du 15 juillet 1980 relative aux économies d'énergie et à l'utilisation de la chaleur, le parlement français a décidé :"Art. 23 - Les seules utilisations des "huiles minérales et synthétiques qui, après usage, ne sont plus aptes à "être utilisées en l'état pour l'emploi auquel elles étaient destinées "comme huiles neuves (...), sont, lorsque la qualité de ces huiles usa-"gées le permet, la régénération et l'utilisation industrielle comme com-"bustible. Cette dernière utilisation ne peut être autorisée que dans des "établissements agréés et *lorsque les besoins des industries de régénéra-"tion ont été préférentiellement satisfaits.*"

La loi française sur les économies d'énergie confirme donc sans ambiguité la position prise par le gouvernement dans sa règlementation sur l'élimination des huiles usagées quant à l'orientation vers la régénération à donner aux huiles usagées régénérables.

o
o o

La règlementation française que je viens de présenter n'est pas encore, il faut le préciser, entrée pleinement en vigueur : elle ne sera totalement applicable que dans quelques semaines, le 23 novembre 1980 exactement.

Il est donc trop tôt pour porter un jugement de valeur sur cette règlementation mais on peut d'ores et déjà en souligner deux aspects intéressants :

- le choix préférentiel, pour l'élimination des huiles usagées, de la solution de la régénération considérée comme répondant le mieux au souci d'économiser de la matière première peut constituer un précédent notable dans une société qui ne peut plus se permettre le gaspillage ;

- la prise en compte du facteur concurrentiel apporté par la régénération sur les marchés des lubrifiants traduit sans doute le souci de ménager l'intérêt du consommateur, mais elle comporte aussi une sorte d'effet "boomerang" qui apparaît logique et équitable. Si l'on considère, en effet, que ce sont les consommateurs de lubrifiants qui financent la collecte des huiles usagées, on constate que la nouvelle règlementation les rembourse d'une certaine manière en organisant sur le marché des huiles neuves une concurrence qui a sur leurs prix d'achat un impact beaucoup plus important que le sacrifice des 40 francs par tonne qui leur est imposé, au titre de la taxe parafiscale.

Pour le reste, nous attendrons que l'arbre ait porté ses fruits avant d'émettre une appréciation définitive sur l'ensemble des dispositions prises.

Aujourd'hui, tous les éléments semblent réunis pour que la nouvelle réglementation française atteigne son triple but : protéger l'environnement, économiser des devises et économiser de la matière première.

Aller plus loin en prétendant, comme le font certains, que le système dont notre pays s'est doté dans ce domaine peut avoir demain valeur d'exemple pour d'autres pays est peut-être hasardeux.

Mais après tout, pourquoi pas ?

Mesdames, Messieurs, je vous remercie.

ZWEITRAFFINATION VON GEBRAUCHTÖLEN IN VERSCHIEDENEN MITGLIEDSTAATEN
DER EG - GROSSBRITANNIEN
THE RECYCLING OF USED OILS IN DIFFERENT MEMBER COUNTRIES OF THE
CEC - GREAT BRITAIN
LE RECYCLAGE DES HUILES USAGEES DANS DIFFERENTS PAYS MEMBRES DE LA
CCE - GRANDE-BRETAGNE
IL RICICLO DEGLI OLI USATI NEI DIVERSI PAESI MEMBRI DELLA
CCE - GRAN BRETAGNA

C. MITCHELL
Chairman der Century Oils Group Ltd., Grossbritannien
Chairman of Century Oils Group Ltd., Great Britain
Chairman of Century Oils Group Ltd., Grande-Bretagne
Chairman of Century Oils Group Ltd., Gran Bretagna

SUMMARY

No basic changes in legislation have occurred during the four years since the last European Congress.

The Clean Air Acts of 1956 and 1968 and the Control of Pollution Act of 1974 give powers to the Department of the Environment and to local authorities to control ground and atmospheric pollution.

There is no national plan for collection or licensing plan for collectors. No effective disposal solution has been found for waste oil generated by the D.I.Y. motorist.

The operation of unrestrained normal market forces has resulted in a fragmented collection industry and a very wide disposal pattern making the job of monitoring and control practically impossible.

The continuation of an unrestricted trading situation gives maximum economic viability to the "in house" burner of waste arisings followed by the small collector with minimal facilities.

The downturn in the economy and the large increase in the number of small burners using waste oil for space heating have resulted in a decrease in amounts available for collection and sharply increased prices.

The re-refining industry cannot now operate viably due to insufficient feedstock at too high prices. Plant capacity has been seriously cut back and outputs reduced to an estimated total rate of 30,000 tons per annum. This important industry seems unlikely to survive without at least some improved implementation of existing regulations.

ZUSAMMENFASSUNG

In den letzten vier Jahren seit dem letzten europäischen Kongress sind die Rechtsvorschriften nicht grundlegend geändert worden. Die Gesetze über die Reinhaltung der Luft von 1956 und 1968 und das Gesetz von 1974 über die Bekämpfung der Verschmutzung ermächtigen das Umweltministerium und die Gemeindebehörden zur Bekämpfung der Boden- und Luftverschmutzung.
   Es gibt keinen landesweiten Plan zur Sammlung von Gebrauchtöl oder einen Plan zur Konzessionserteilung an Sammelstellen. Zur Beseitigung der Gebrauchtöle aus Kraftfahrzeugen (D.I.Y.) ist keine wirksame Lösung gefunden worden.
   Das freie Spiel der Marktkräfte führte zur Entstehung einer zersplitterten Sammelindustrie und einer umfangreichen Beseitigungsstruktur, die die Überwachung und Steuerung praktisch unmöglich macht.
   Die Beibehaltung der restriktionslosen Marktlage begünstigt in erster Linie die wirtschaftliche Lebensfähigkeit von Altölverbrennungsanlagen in den Betrieben und sodann die kleinen Sammelstellen mit minimalen Einrichtungen.
   Die rückläufige Wirtschaftsentwicklung und die starke Zunahme der mit Gebrauchtöl befeuerten Raumheizungsanlagen haben eine Abnahme der zur Einsammlung verfügbaren Altölmengen und einen starken Preisanstieg zur Folge gehabt.
   Den Raffinerien ist wegen der Versorgungsschwierigkeiten und zu hohen Preise ein wirtschaftlicher Betrieb zur Zeit nicht möglich. Die Anlagekapazitäten sind beträchtlich gekürzt worden; die Produktion wird auf jährlich 30.000 t geschätzt. Dieser wichtige Industriezweig scheint ohne zumindest einige Verbesserungen der existierenden Vorschriften nicht lebensfähig.

RESUME

   Aucun changement important n'a été apporté dans la législation au cours des quatre années qui ont suivi le dernier congrès européen. Les lois sur la pollution de l'air (Clean Air Acts) de 1956 et de 1968 et la loi sur la limitation de la pollution (Control of Pollution Act) de 1974 habilitent le Ministère de l'Environnement et les collectivités locales à prendre des mesures pour limiter la pollution du sol et la pollution atmosphérique.

Il n'existe pas de programme national de collecte des huiles usagées ou de programme d'autorisation des entreprises effectuant la collecte. Aucune solution d'élimination efficace n'a été trouvée pour les huiles usagées produites par les automobilistes bricoleurs.

Les forces normales du marché, qui s'exercent en toute liberté, ont entraîné le morcellement de l'industrie de collecte des huiles usagées et la naissance de systèmes très divers d'élimination, ce qui rend les opérations de surveillance et de contrôle pratiquement impossibles.

La poursuite sans restriction des pratiques commerciales actuelles confère une fiabilité économique maximale aux brûleurs "internes" de déchets et ensuite aux petits collecteurs équipés d'installations minimales.

La récession économique et l'augmentation importante du nombre de petits brûleurs utilisant des huiles usagées pour le chauffage des locaux ont entraîné une réduction de quantités disponibles pour la collecte et une hausse brusque des prix.

L'industrie de régénération ne peut fonctionner de façon rentable en raison de l'insuffisance des matières premières et de leur prix trop élevé. La capacité des installations a été fortement réduite et la production ramenée à un taux total estimé de 30.000 tonnes par an. Il semble peu probable que cette industrie importante puisse survivre si l'on n'améliore pas au moins la mise en oeuvre des réglementations existantes.

RIASSUNTO

Non si sono registrati cambiamenti basilari nella legislazione durante i quattro anni trascorsi dall'ultimo congresso europeo. I Clean Air Acts del 1956 e 1968 e il Control of Pollution Act del 1974 conferiscono al Department of the Environment e agli enti locali il potere di controllare l'inquinamento atmosferico e del suolo.

Non esistono programmi nazionali di raccolta o programmi di concessione di licenze per ghiaddetti alla raccolta. Non è stata trovata una soluzione efficace per lo smaltimento di oli usati prodotti dai singoli automobilisti che provvedono da soli alla manutenzione.

La normale azione delle libere forze di mercato ha dato origine ad una industria frammentaria di raccolta e ad uno schema molto vario di smaltimento che rende praticamente impossibile svolgere una funzione di "monitoring" e di controllo.

La continuazione di tale situazione incontrollata rende particolarmente economico bruciare "in loco" gli oli usati, mentre viene in seconde posizione la raccolta degli oli e la loco combustione in impianti minimi.

La crisi economica e il grande aumento del numero di privati che bruciano oli usati per il riscaldamento hanno provocato una diminuzione delle quantità disponibili per la raccolta e un notevole incremento dei prezzi.

L'industria di rigenerazione non può funzionare efficientemente a causa di cariche insufficienti a prezzi troppo elevati. La capacità degli impianti è stata notevolmente limitata e la produzione ridotte ad una quantità stimata totale di 30.000 t all'anno. Questa industria importante non sembra poter sopravvivere senza ricorrere almeno ad una migliore attuazione dei regolamenti esistenti.

In the U.K., no basic changes have occurred in the legislation regarding waste oil elimination since the last European Congress four years ago, which is that a comprehensive framework for controlling pollution from waste oil already exists which would meet the requirements of the 1975 EEC Directive on the Disposal of Waste Oils when fully implemented.

The main control lies in the Deposit of Poisonous Waste Act 1972 and the Control of Pollution Act of 1974.

The Alkali Inspectorate is responsible under the Health and Safety at Work Act 1974 for the control of atmospheric pollution from burning waste oil in scheduled processes, and the Environmental Health Authorities are responsible under the Clean Air Acts 1956 and 1968 and the Control of Pollution Act 1974 for the control of atmospheric pollution from all other premises.

There is a national network for monitoring air pollution.

Waste disposal authorities are responsible under the Control of Pollution Act for ensuring that arisings of waste oil are disposed of in an environmentally accepted way at licensed sites.

Producers of waste oil are required to keep a record of their disposal which they can be required to produce at any time.

Water Authorities are responsible under the Control of Pollution Act and previous legislation for the control of all discharges and entries of waste oil.

The Health & Safety Executive are responsible for the safe handling of waste oil within the workplace.

The Department of the Environment has published a technical memorandum on mineral oil wastes.

The control of environmental pollution in the U.K. is said to be based on the principle of the best practicable means rather than a uniform national standard.

In practical terms, the exercising of such control would depend on evidence being produced that the environmental regulatory agencies were failing to meet their responsibilities.

Since there is no requirement for persons or organisations to notify that they are burning waste oil, the only likelihood of any form of monitoring taking place would be in the event of excessive smoking being observed from an installation, or an obviously excessive particle emission causing complaints from neighbours. There are no levels of acceptable pollution prescribed other than dust, grit and sulphur dioxide.

Where processes are being carried out where a pollutant hazard is known, such processes are required to be licensed and can be monitored. Where waste oil is being burned, the user is unlikely to know its contents or to be aware that a pollution risk exists, nor will the local authorities know of its existence. Therefore it is equally unlikely that any significant monitoring of pollution from installations burning waste oil exists anywhere in the U.K.

Although producers of waste oil are required to keep records of their disposals, there is no evidence currently of such records being regularly monitored other than that waste going to a licensed disposal facility.

On the collection side of waste oil arisings, there is a wide variation in the types of organisation involved in this activity.

A significant percentage of waste oil arisings is collected "in house" by the companies and organisations creating such waste. There are no figures available to help estimate this quantity, but the proliferation of small burner systems used within the U.K. would seem to indicate that the quantity is very substantial.

For the remainder, there is no overall national plan for collection or any form of licensing for collectors.

As a result there exists a very fragmented collection industry ranging from large fleet operations with national coverage and central collection points capable of comprehensive treatment of the incoming waste, to single vehicle operations with no treatment or waste disposal facilities.

In practice, any adult person can purchase and operate a collection vehicle without having to have any storage accommodation or facilities to treat separated waste water.

No effective solution has yet been found to recover the waste oils generated by D.I.Y. motorists.

Some major oil companies have set up schemes designed to offer incentives to the motorist to return used oil by offering credit on such returns against the purchase of new engine oil.

Also, a few local authorities and independent organisations have set up disposal points to try to encourage the D.I.Y. motorist to properly dispose of his arisings.

None of these schemes has had other than very limited success, and there appears no likelihood at the moment that any further arrangements will be made unless local authorities get together with collectors to try to find a mutually acceptable way of fulfilling this need.

On the disposal side, the "in house" burner continues to operate without any effective control on emissions from his installations, and, in nearly all cases, without any form of treatment of the waste prior to burning it.

The collectors operate with a very wide disposal pattern. Some arisings go to re-refiners for full recycling treatment, although the volumes disposed of in this way have been rapidly diminishing.

Some simply resell the waste as fuel without testing or treatment. Others take the arisings to a collection point where it may receive some treatment, such as excessive water removal, filtration, centrifuging or, in some cases, more substantial treatment.

Generally, the minimum possible treatment is effected compatible with the end use of the material. This generally means the minimum treatment necessary to physically allow the waste to be burnt on a particular installation.

There is usually no form of testing carried out on the waste, and there are currently no restrictions whatsoever on the type of installations permitted.

Badly contaminated waste oil and mixed wastes containing waste oil are collected by waste disposal companies for incinerating or dumping.

The D.I.Y. motorist continues to dispose of his arisings, apparently without detection, and presumably does so in a variety of ways such as burning on a bonfire or dumping in the ground or drains. The build up of pollution from such disposals must be significant since it is estimated that probably some 60,000,000 litres per year are now disposed of in this way.

Since collectors are neither licensed nor controlled, they operate in a free market situation.

This means that they can compete together for the purchase of arisings, and that the prices paid vary according to the competition. In heavily industrialised areas this means that the waste oil prices can approach the prices being paid for fuel oil, and in remote country areas there might be some difficulty in arranging for collection.

The steadily increasing number of small burner installations using waste oil for space heating and other "in house" use, together with the downturn in the economy, have resulted in further increased competition by the collectors, and further pressure on them to minimise treatment and collection expenses to remain viable operations.

In the U.K. market where normal market forces are allowed to operate, maximum viability is achieved by the "in house" burner who has no significant associated costs and compares the value of the waste to that of fuel oil.

Next in viability comes the small local collector with minimum facilities, and so on up to the least viable, being the re-refiner with extremely high proportional costs and heavy financing requirements.

On the other hand, the disposal price of the re-refiners end products is considerably higher than that when the end use is fuel. In spite of this fact, the re-refining industry cannot now operate viably due to insufficient feedstock at too high prices.

Over the past few years there has been a sharp decline in the industry, and re-refining capacity has been seriously cut back. One 50,000 tons/annum capacity plant has been discontinued, and total U.K. rate of output reduced to approximately 30,000 tons/annum. Under present conditions it seems that a further decline will inevitably take place in the current year. This important industry that has served so well in times of crisis seems unlikely to survive without at least some improved implementation of existing regulations.

It is interesting to note that the differing regulations covering the burning of waste oil in the different countries of the European Community allows burner equipment to be used in some countries which is prohibited in others on environmental grounds.

These anomalies can only lead one to conclude that either the legislation now existing in some countries is unnecessary, or that the environment in others is not being adequately protected at the risk of danger to the health of the population.

Also, since such regulations and standards of enforcement greatly affect the price of waste oil arisings, it is clear that the ability of the seller of recycled oil to compete successfully, whether the end use be as fuel or as lubricant, will vary from country to country within the EEC, from which it would appear that a state of unfair competition exists.

Until there is some uniform legislation and enforcement of such legislation regarding the conditions under which waste oil may be burnt as fuel and what degree of treatment or product specification limits are to be met for such fuel, there appears to be little chance of achieving a stable industry in the U.K., or of suitable levels of investment being made to ensure adequate environmental protection.

RECYCLING IN IRLAND
RECYCLING IN IRELAND
LE RECYCLAGE EN IRLANDE
RICICLO IN IRLANDA

F.M.J. DUFFY
Präsident der Association of the Irish Independent Lubricant
Industry, Dublin
President of the Association of the Irish Independent
Lubricant Industry, Dublin
Président de l'Association of the Irish Independent Lubricant
Industry, Dublin
Presidente dell'Association of the Irish Independent Lubricant
Industry, Dublino

SUMMARY

The initial development and growth of the Irish oil recycling industry only took place in the 1970's and has very successfully established a country wide collection service for even quite small collections.

The table shows the market position:

Total sales	58 000 tons
Less agriculture	15 000 tons
Less Home Use	6 000 tons
Balance	37 000 tons

```
Total Recycling potential
    (40% x 37 000)              15 000 tons
Presently recycled
    (estimate)                  10 000 tons
Presently burned in
    garage heaters               1 500 tons
Balance dumped                   3 500 tons
```

The paper argues that the Agriculture Market is impossible to collect but that the Home Use or Do-It-Yourself Market must be collected.

Future recycling potential is shown by this table.

```
1. Now being dumped              3 500 tons
2. Now being burned in
   garage heaters                1 500 tons
3. D.I.Y. market
   (40% x 6000 tons)             2 400

   Total potential               7 400
```

The paper states that most legislation refers to toxic waste and is designed for conservation of the environment and not conservation of energy.

In Ireland such legislation is hardly enforced due to the apathy of local councils.

The Irish industry is now finishing the construction of two recycling plants, one in Northern Ireland and one in the Republic of Ireland. The process is described in the paper.

They are anxious to ensure the supply of raw material and ask the following:

1. That local councils will adequately control the disposal of used oil.
2. That burning of untreated used oil be made illegal.
3. That the D.I.Y. market be adequately collected.
4. That marine authorities will adequately control disposal of oil from shipping and fishing fleets.

5. That central government in each country would programme a preference for the purchase of recycled lubricant instead of virgin lubricant.

ZUSAMMENFASSUNG

Die irische Gebrauchtöl-Recycling-Industrie ist erst in den 70er Jahren entstanden; sie hat mit grossem Erfolg einen das ganze Land erfassenden Altöl-Sammeldienst geschaffen, der sich auf ziemlich kleine Mengen erstreckt.
Die nachstehende Tabelle zeigt die Marktposition:

Gesamtverkäufe	58.000 Tonnen
abzüglich Landwirtschaft	15.000 Tonnen
abzüglich Eigenverbrauch	6.000 Tonnen
Differenz	37.000 Tonnen
Recycling-Potential insgesamt (40% x 37.000)	15.000 Tonnen
Gegenwärtiges Recycling (Schätzung)	10.000 Tonnen
Gegenwärtig in Garagenheizungen verbrannt	1.500 Tonnen
Beseitigte Differenz	3.500 Tonnen

In dem Bericht heisst es, dass es unmöglich sei, die in der Landwirtschaft anfallenden Gebrauchtölmengen zu sammeln, dass aber die Mengen aus dem Eigenverbrauch oder Do-It-Yourself-Verbrauch gesammelt werden müssen.
Das künftige Recycling-Potential wird wie folgt angegeben:

1. Gegenwärtig abgeleitet	3.500 Tonnen
2. Gegenwärtig in Garagenheizungen verbrannt	1.500 Tonnen
3. Do-It-Yourself-Markt (40% x 6.000 Tonnen)	2.400 Tonnen
Gesamtpotential	7.400 Tonnen

Dem Bericht zufolge betreffen die Rechtsvorschriften überwiegend giftige Abfälle und zielen auf die Schonung der Umwelt, nicht aber auf Energieeinsparung ab.
Bei der Apathie der Gebietskörperschaften werden diese Rechtsvorschriften in Irland kaum durchgesetzt.
Die irische Industrie stellt zur Zeit den Bau zweier Recycling-Anlagen, einer in Nordirland und einer in der Republik Irland, fertig. Der Prozess ist in dem Bericht beschrieben.
Die irische Industrie ist darauf bedacht, die Rohstoffversorgung zu sichern und sie stellt folgende Forderungen:
1. Die Gebietskörperschaften sollen die Gebrauchtöl-Beseitigung angemessen kontrollieren.
2. Das Verbrennen von nichtbehandeltem Gebrauchtöl soll gesetzlich verboten werden.
3. Die beim Do-It-Yourself anfallenden Gebrauchtölmengen sollen in ausreichendem Umfang gesammelt werden.
4. Die Meeresbehörden sollen die Ölverklappung von Schiffen und Fischereiflotten aus angemessen kontrollieren.
5. Die Zentralregierung in jedem Land sollte dem Kauf von Schmierstoff aus dem Recycling gegenüber frischen Schmierstoffen den Vorzug geben.

RIASSUNTO

Lo sviluppo iniziale e l'espansione dell'industria irlandese del riciclo degli oli usati risole solo agli anni 1970 ma ha consentito di introdurre con molto successo un servizio di raccolta nazionale anche per piccolissime quantità.
La tabella indica la situazione del mercato :

Vendite complessive	58.000 t
Meno agricoltura	15.000 t
Meno usi domestici	6.000 t
Residuo	37.000 t
Totale potenziale di riciclo (40% x 37.000)	15.000 t
Riciclo svolto attualmente (valutazione)	10.000 t
Quantità bruciata nelle autorimesse	1.500 t
Quantità scaricata tra i rifiuti	3.500 t

Nel documento si afferma che non è possibile la raccolta nel settore agricolo mentre invece deve avvenire in quello domestico o del "Far da sé".
Il futuro potenziale di riciclo è indicato nella tabella seguente.

1.	Scaricato attualmente tra i rifiuti	3.500 t
2.	Bruciato nelle autorimesse	1.500 t
3.	"Far da sé" (uso domestico) (40% x 6.000 t)	2.400 t
	Potenziale totale	7.400 t

Nel documento si sostiene che le legislazioni si riferiscono in genere ai residui tossici ed hanno come obiettivo la protezione dell'ambiente e non il risparmio dell'energia.
In Irlanda, tale legislazione non viene quasi mai fatta rispettare a causa dell'apatia dei consigli locali.
L'industria irlandese sta ora terminando la costruzione di due impianti di riciclo, uno nell'Irlanda del nord ed uno nella Repubblica d'Irlanda. Il processo viene descritto nel documento.
Tale industria intende assicurarsi l'approvvigionamento di materie prime e chiede pertanto i seguenti provvedimenti :
1. Che i consigli controllino adeguatamente l'eliminazione di oli usati.
2. Che sia dichiarata illegale la combustione di oli usati non trattati.
3. Che sia predisposta un'adeguata raccolta nel settore domestico.
4. Che le autorità marittime controllino adeguatamente l'eliminazione di oli provenienti dalle navi da carico e da pesca.
5. Che il governo di ogni paese preveda misure preferenziali a favore dell'acquisto di oli lubrificanti riciclati invece di oli lubrificanti nuovi.

RESUME

L'industrie irlandaise du recyclage des huiles usagées n'a commencé à se développer et à croître que pendant les années 70; elle a obtenu des résultats excellents dans l'organisation d'un service de collecte au niveau national, même pour les quantités très faibles.

Le tableau ci-dessous donne les caractéristiques du marché :

Ventes totales	58.000 t
Moins agriculture	15.000 t
Moins utilisation domestique	6.000 t
Différence	37.000 t
Possibilités de recyclage totales (40% x 37.000)	15.000 t
Quantité actuellement recyclée (estimation)	10.000 t
Quantité brûlée dans des appareils de chauffage de garages	1.500 t
Quantité déversée	3.500 t

L'auteur précise qu'il est impossible de procéder à la collecte sur le marché de l'agriculture, mais que cette collecte doit être effectuée sur le marché de l'utilisation domestique ou du "do it yourself".

Les possibilités futures de recyclage sont indiquées dans le tableau ci-dessous.

1. Quantité actuellement déversée	3.500 t
2. Quantité actuellement brûlée dans les appareils de chauffage des garages	1.500 t
3. Marché du do it yourself (40% x 6.000 t)	2.400 t
Total	7.400 t

La plupart des dispositions législatives concernent les déchets toxiques et ont pour objectif la protection de l'environnement, et non la conservation de l'énergie.

En Irlande, cette législation est rarement appliquée, en raison de l'inertie des autorités locales.

L'industrie irlandaise termine actuellement la construction de deux installations de recyclage, l'une en Irlande du Nord et l'autre en république d'Irlande. Le procédé utilisé est décrit.

Le problème principal est d'assurer l'approvisionnement de ces installations en matières premières, et les mesures suivantes sont demandées:

1. Contrôle approprié de l'élimination des huiles usagées par les autorités locales.
2. Promulgation de dispositions au terme desquelles l'utilisation des huiles usagées comme combustible est déclarée illégale.
3. Organisation d'une collecte appropriée sur le marché du "do it yourself".
4. Contrôle approprié par les autorités maritimes de l'élimination des huiles par les navires de transport et de pêche.
5. Mesures du gouvernement de chaque pays prévoyant l'acquisition de lubrifiants recyclés de préférence aux lubrifiants non recyclés.

Because our group of companies and our association operate throughout Northern Ireland and the Republic of Ireland and because the total supply of used oils is treated by our Industry as coming from one source, I intend in this paper to include both economies.

During the 1950's and 1960's there was virtually no oil recycling industry in Ireland.

With an abundance of low cost oils efficiently distributed by the major oil companies, the development of all other energy sources was inhibited.

Ireland had a small industrial base and being quite a large country in relation to its population, all automotive and manufacturing industry was widely scattered throughout a largely rural population with small urban centres.

During the 1960's an important change took place in the Irish economy. A strong government introduced a dynamic industrialisation policy assisted by tax holidays and capital grants. Overnight Ireland became an attractive location for industry and an important industrial sector began to develop.

The modern Irish oil recycling industry dates from the early 1970's, after ten years of industrial growth. A serious total collection service for all used oil was inaugerated. Practices which had proven successful in the virgin oil industry were adopted. One of the most successful of these practices was "Planned Collection" (the reverse of "Planned Delivery") whereby up to 5000 firms throughout the country were programmed for orderly collection on an agreed schedule. While this system allows for a measure of emergency calls, it removes the inefficiency arising from a haphazard "ad hoc" demand.

Given the diverse nature of the Irish economy this type of collection was the key to a highly successful collection industry. Large 25 000 litre tankers are routed on 2, 3 and even 4 day runs into the rural areas of the western seeboard while compact 10 000 litre tankers are used for city work.

The industry welcomes each and every type of oil except water based emulsion oils. Payment is made dependent on quality and contamination. All quantities down to 200 litres can be efficiently collected in almost every part of Ireland.

Companies Active in Recycling

The companies collecting oil in Ireland are as follows:-

Atlas Oil Laboratories Ltd.,	N. Ireland and Dublin
Waste Recyclers (N.I.) Ltd.,	Belfast
Irish Waste Recyclers Ltd.	Belfast
Northern Lubricants	Belfast
A. & A. Waste Oil Co.	Dublin
Allied Oils Ltd.	Dundalk

Recycling Market Size

The total sale of lubricating oils in Ireland is approximately 58 000 tons. The potential for recycling is shown by the following table.

	Tons
Total Sales	58 000
Less Agricultural market (Note 1)	15 000
Less D.I.Y. market (Note 2)	6 000
Balance	37 000
Total Recycling Potential	15 000
(estimated at 40% of balance)	
Presently recycled (estimated)	10 000
Presently burned in garage heaters	1 500
Balanced dumped	3 500

Note 1 :   The agricultural sector represents a significant part of the total market. It has a diverse character having many small outlets. The farmers in most cases re-use the waste oil for various purposes including weed killing, burning scrub, as a rust preventative etc. It is not felt that this market can be efficiently collected.

Note 2 :   The self-service of cars from private homes and small unregistered businesses (known as the Do-It-Yourself or D.I.Y. market) has become a very important sector. Most of this is currently being dumped. It is felt that much of this could be recovered.

Future Expansion Possibilities

Targets

1. Balance now being dumped   3500 tons
2. Now being burned in garage heaters 1500 tons
3. D.I.Y. market   2400 tons
   (estimated at 40% of 6000 tons)
Total possibilities   7400 tons

The D.I.Y. market has been shown separately from general industry because it is an identifiable sector which must be treated in an individual manner if it is to be collected satisfactorily.

Legislation

With a growing awareness of the need to protect the ecological balance several laws have been introduced which refer to water pollution by toxic waste.

Northern Ireland :

The introduction of the Pollution Control and Local Government (N.I.) Order 1978 corresponds to the Control of Pollution Act, 1974 in Great Britain. This order came into Force on 2nd June 1980 and provides that waste may only be disposed of at plants which have been licenced by the District Councils. Councils are also given new powers to deal with reclamation of waste.

Republic of Ireland :
The following acts apply:
Local Government (Planning and Development Act) 1963
Local Government (Planning and Development Act) 1976
Local Government Water Pollution Act 1977

In addition the Minister for the Environment recently introduced an order which gave effect to the E.E.C. Directive on Waste (No. 75/442/EEC) of 15th July, 1975.

However, while it may be seen that there is adequate protection for the environment these laws relate to waste generally.

There are no laws which relate to the control and disposal of used oils specifically. Furthermore a prosecution under these laws will only be brought in the event of serious pollution and little control is exercised over the hundreds of small polluters.

The District Councils and County Councils have now a responsibility to list and exercise control over all producers of used oil. There has been almost no evidence of their interest in this area. There is ample evidence of a broad gulf between central government which passes such laws and the local authorities who are supposed to impliment them but do not do so.

Legislative Requirements and Implimentation
The Irish recycling industry recommends that steps be taken to ensure as follows:
1. That every firm purchasing more than 250 litres of lubricant per year be obliged to keep a record of its disposal and that the names of such firms be communicated to each local council.

co-operating in this venture.

For the last 6 years the Irish recycling industry has geared itself to develop a lubricant refining facility which would provide the optimum combination of process flexibility and quality performance having regard to the small size and diverse nature of the Irish industrial and automotive market. Most of the more widely used re-refining methods relate high capital costs to large throughputs on the basis of continuous processes refining mixtures containing all types of used lubricant. Our studies showed that scaling down of size did not produce a straight line reduction in capital cost. Furthermore the materials used were either not available locally and therefore expensive or else environmentally unacceptable. Many of these processes were studied in depth and options of design, franchise and joint venture were explored. The requirements of the small Irish industry were as follows: small size, modest capital cost, economic viability, ease of operation, flexibility in terms of process control and throughput. These questions led to the choice of a "Batch" type design which includes the following processes:

        Initial segregation and analysis
        Dehydration
        Treatment with chemicals
        High temperature centrifugation
        Vacuum Distillation
        Pressure Filtration
        Adsorbtion
        Quality control including spectrographic and
          infra-red analysis
        Blending to specification.

One of these plants is currently being commissioned and the other will commence production next month, November 1980.

Our members believe that they have developed a process which would be suitable for many of the worlds smaller economies and we intend to welcome overseas visitors whenever our plants are operating viably.

Action Required

1. Our industry has very strong reservations about the proliferation of small heaters and burners which are highly dangerous to the users and toxic to the atmosphere. Their use appears to be growing in numbers.

These heaters eliminate by combustion a valuable indigenous raw material which could otherwise be recycled many times. We believe such combustion should be made illegal and their used lubricant sent for re-refining or laundering.

2. The loss of oil into the ecology from the D.I.Y. market must somehow be halted. We would like this Congress to examine what can be done to achieve successful collection of this waste material.

3. Each of our governments are very large users of lubricant. We believe that governments are now prepared to specify a preference for re-refined product. The effect of this buying force in the lubricant market would harden the economics of re-refining to a point where we could offer sufficient payment for used material to offset the alternatives of burning and also to help overcome the apathy of the dumpers.

Given the present emphasis by all our governments on the conservation of energy and also upon the conservation of the environment the time is fully ripe for our industry to press for government programmes to purchase re-cycled lubricants in preference to virgin lubricants. This we believe will give our industry a firm basis upon which to build and expand into the third millennium.

2. That each local council be obliged to appoint an oil control officer who must control the disposal of used oil.

3. That the burning of untreated used oil in local heaters be made illegal.

4. That all firms selling lubricant especially to the D.I.Y. market must arrange a convenient accessible facility to which the client can return his oil, and that this information be conveyed on the container of used oil.

5. That all fishing fleets and merchant ships be closely controlled as regards the disposal of their used oil.

Government Policy

The attitude of our central governments to recycling is to encourage it.

1. The development of the industry towards lubricant manufacture has been positively promoted by government agencies.

2. Large grants have been paid to assist with the development of suitable plants.

3. There are indications that a preference will be shown by government towards the use of re-refined lubricant by state and semi-state organisations. As these organisations are very large lubricant users, such a policy would be most helpful to the infant lubricant refining industry.

Present State of the Art

Initially the Irish recycling industry relied upon fuel oil production for its market outlets and this continues to this day.

However two plants are nearing completion to launder and refine lubricant back to a base oil suitable for re-blending as a lubricant. One plant is located in Northern Ireland, the other at Port Laoise in the Republic of Ireland. Total capital investment is approximately £Ir 1.5 M. Several different firms are

GEBRAUCHTÖL-RECYCLING IN ITALIEN
THE RECYCLING OF USED OILS IN ITALY
LE RECYCLAGE DES HUILES USAGEES EN ITALIE
IL RICICLAGGIO DEGLI OLII USATI IN ITALIA

Dr. R. SCHIEPPATI
Präsident der Gruppo Aziende Indipendenti Lubrificanti
(ASCHIMICI), Milano
President of Gruppo Aziende Indipendenti Lubrificanti
(ASCHIMICI), Milan
Président du Gruppo Aziende Indipendenti Lubrificanti
(ASCHIMICI), Milano
Presidente del Gruppo Aziende Indipendenti Lubrificanti
(ASCHIMICI), Milano

RIASSUNTO

La rigenerazione degli olii lubrificanti usati ha in Italia una grande tradizione.
I primi impianti risalgono infatti al 1930 e, anche sotto l'aspetto legislativo, norme specifiche regolano la raccolta e la rigenerazione degli olii usati fin dal 1940 con la legge n. 671 del 1940 ed il Decreto di attuazione 884 del 1941 che fissano l'obbligatorietà della raccolta e la destinazione prioritaria alla rigenerazione dell'olio usato.
Queste disposizioni sono dettate però solo da interessi economici mentre il problema ecologico sollevato dalla Direttiva CEE del 1975 rimane igno-

rato ed è affidato alle leggi generali che tutelano le acque e l'aria dall'inquinamento.

La legge n. 1852 del 1962 prevede, inoltre, per i prodotti ottenuti dalla rigenerazione di olii usati uno sgravio del 75% dell'Imposta di Fabbricazione.

Tutto ciò ha permesso il sorgere di un'industria rigeneratrice italiana che è attualmente forte di ben 12 impianti per complessive 212.400 tonnellate annue di lavorazione.

Su un consumo nazionale di 630.000 tonnellate annue di prodotto si stima in 220.000 tonnellate annue l'olio che può essere raccolto mentre attualmente il quantitativo recuperato ed inviato quasi esclusivamente alla rigenerazione, non risultando dati sull'eliminazione degli olii usati attraverso altre procedure quali la combustione e l'incenerimento, si aggira sulle 110.000 tonnellate annue.

Per risolvere i problemi connessi alla raccolta dell'olio usato ed assicurarne il recupero energetico nel pieno rispetto delle norme ecologiche è necessaria una nuova legge organica. Attualmente un disegno di legge si trova all'esame delle competenti autorità e potrà essere tradotto in legge non appena il Governo sarà stato investito dei necessari poteri per attuare la Direttiva Comunitaria del 1975.

Il testo della nuova disposizione di legge prevede l'obbligatorietà della raccolta affidata ad un Consorzio da costituirsi fra le Aziende di riraffinazione e le Aziende produttrici di olii lubrificanti finiti e fissa i criteri per la raccolta del prodotto, la sua destinazione nonchè per l'esercizio degli impianti di rigenerazione.

ZUSAMMENFASSUNG

Die Zweitraffination von Gebrauchtöl hat in Italien eine lange Tradition.
Die ersten Anlagen gehen auf das Jahr 1930 zurück, und was den gesetzgeberischen Aspekt anlangt, so regeln spezifische Normen die Sammlung und Zweitraffination von Gebrauchtöl seit 1940 im Rahmen des Gesetzes N° 671 von 1940 und der Durchführungsverordnung 884 von 1941, nach denen die Sammlung von Altöl und seine vorrangige Bestimmung zur Zweitraffination vorgeschrieben sind.
Diese Bestimmungen sind jedoch nur von wirtschaftlichen Interessen dik-

tiert, während das in der EG-Richtlinie von 1975 angesprochene ökologische Problem nicht berücksichtigt ist und den allgemeinen Gesetzen zum Schutz der Gewässer und der Luft gegen Verschmutzung überlassen bleibt.

Das Gesetz N° 1852 von 1962 sieht unter anderem für die bei der Zweitraffination von Gebrauchtöl erlangten Produkte einen Nachlass von 75 % der Herstellungssteuer vor.

Alles dieses hat die Entstehung einer italienischen Zweitraffinationsindustrie ermöglicht, die gegenwärtig 12 Anlagen mit einer Verarbeitungskapazität von insgesamt 212.400 jato umfasst.

Von dem 630.000 jato betragenden Inlandsverbrauch können schätzungsweise 220.000 jato gesammelt werden, während die Menge, die rückgewonnen und nahezu ausschliesslich der Zweitraffination zugeführt wird, gegenwärtig bei 110.000 jato liegt; über die Gebrauchtölbeseitigung auf andere Weise, z.B. durch Verbrennung und Veraschung, sind keine Angaben verfügbar.

Es ist ein neues Organisationsgesetz notwendig, um die mit der Gebrauchtölsammlung verbundenen Probleme zu lösen und die Energierückgewinnung unter voller Beachtung der ökologischen Normen zu gewährleisten. Eine Gesetzesvorlage wird zur Zeit von den zuständigen Behörden geprüft; sie kann als Gesetz verabschiedet werden, sobald die Regierung die notwendigen Befugnisse zur Durchführung der Gemeinschaftsrichtlinie von 1975 erhalten hat.

Die neue Rechtsvorschrift sieht vor, dass die Sammlung einem von den Raffinerien und den Schmierölherstellern zu bildenden Konsortium zu übertragen ist; sie legt die Kriterien für die Sammlung des Produkts, für seine Bestimmung und auch für den Betrieb der Zweitraffinationsanlagen fest.

SUMMARY

The regeneration of used oils has a long tradition in Italy.

The first plants date back to 1930, and even in terms of legislation there have been specific regulations governing the collection and regeneration of used oils since as early as 1940 in the form of law N° 671 of 1940 and Implementing Decree N° 884 of 1941, which makes the collection of used oils obligatory and their regeneration a matter of priority.

These provisions, however, are governed solely by economic interests whereas the ecological problem raised by the EEC directive of 1975 is not covered, this being left to general laws governing water and air pollution.

Law N° 1852 of 1962 also provides for 75 % tax relief on goods manufactured from regenerated used oils.

This has led to the emergence of an Italian regenerating industry which has 12 plants and a total output of 212.400 tonnes per year.

An estimated 220.000 tonnes per year can be collected from the national consumption of 630.000 tonnes, whereas the quantity recovered and sent almost exclusively for regeneration, excluding figures on the disposal of used oils via other processes, such as combustion and incineration, is currently around 110.000 tonnes per year.

A new organic law is needed if the problems connected with collecting used oils are to be solved and energy is to be recovered from them while fully observing ecological standards. A bill is currently being examined by the competent authorities, and this could become law as soon as the government has been invested with the powers needed to implement the Community directive of 1975.

The bill stipulates the collection of used oils by a consortium consisting of refining companies and manufacturers of lubricating oils, and lays down the criteria for the collection of the oils, their sending to regenerating plants and the operation of the latter.

RESUME

La régénération des huiles lubrifiantes usagées a une longue tradition en Italie.

Les premières installations ont été construites en effet en 1930, et du point de vue législatif, des normes spécifiques réglementent la récupération et la régénération des huiles usagées depuis 1940, avec la loi n°671 de 1940 et le décret d'application 884 de 1941, qui rendent obligatoire la récupération des huiles usagées et leur affectation prioritaire à la régénération.

Ces dispositions ne sont toutefois dictées que par des considérations économiques, alors que le problème écologique soulevé par la directive CEE de 1975 reste ignoré, et relève des lois générales relatives à la lutte contre la pollution atmosphérique et la pollution des eaux.

La loi n°1852 de 1962 prévoit en outre, pour les produits de la régénération des huiles usagées, un dégrèvement de 75% de l'impôt de fabrication.

Ces diverses mesures ont permis la création en Italie d'une industrie de la régénération comprenant actuellement 12 installations traitant au total 212.400 tonnes par an.

Sur une consommation annuelle de 630.000 tonnes de produit, on estime à 220.000 tonnes la quantité annuelle d'huile qui peut être récupérée, alors que la quantité récupérée et affectée presque exclusivement à la régénération peut être estimée, en l'absence de données sur l'élimination des huiles usagées par d'autres procédés, tels que la combustion et l'incinération, à environ 110.000 tonnes par an.

Une nouvelle loi organique est nécessaire pour résoudre les problèmes posés par la récupération des huiles usagées et pour assurer leur récupération énergétique en respectant entièrement les normes écologiques.
Un projet de loi est actuellement examiné par les autorités compétentes, et pourra être adopté dès que le gouvernement aura été investi des pouvoirs nécessaires pour appliquer la directive communautaire de 1975.

Le texte du nouveau projet de loi rend obligatoire la récupération, qui sera confiée à un consortium formé par des entreprises de régénération et des entreprises de production de lubrifiants, et fixe les critères appliqués à la récupération du produit, à son affectation et à l'exploitation des installations de régénération.

La rigenerazione degli olii lubrificanti usati ha in Italia una grande tradizione. Le prime sperimentazioni risalgono agli anni intorno al 1930 e prima del secondo conflitto mondiale erano già sorti i primi impianti di rigenerazione.

Anche sotto l'aspetto legislativo l'Italia si pone tra i primi Paesi che hanno sentito la necessità di regolamentare l'attività della raccolta e della lavorazione degli olii lubrificanti usati.

La Legge 671 del 29.4.1940 sull'obbligatorietà della raccolta e della rigenerazione degli olii usati, il Decreto 844 del 22.5.1941 contenente le norme per l'attuazione della precedente nonchè l'art. 12 della legge 1852 del 31.12.1962 sul regime fiscale dei prodotti petroliferi rigenerati e recuperati, costituiscono un insieme di disposizioni che hanno consentito il sorgere e lo svilupparsi di questa attività in Italia.

Negli anni che vanno dal 1960 al 1970 venivano realizzati ben 22 impianti per complessive 257.660 tonnellate annue di capacità produttiva.

Attualmente la disciplina degli olii usati risulta ancora regolata da queste norme specifiche affiancate, per quanto concerne la protezione dell'ambiente, dai Testi Unici delle leggi sanitarie e sulla pesca nonchè dalle più recenti leggi n. 319 del 10.5.1976 e n. 650 del 24.12.1979 sulla tutela delle acque dall'inquinamento e n. 615 del 13.7.1966 sulla tutela dell'inquinamento atmosferico.

I testi attinenti specificamente la materia che costituiscono il nucleo fondamentale della normativa sono però ispirati a problemi di autarchia che caratterizzarono l'economia italiana negli anni precedenti la Seconda Guerra Mondiale. Si vuole il riutilizzo degli olii minerali usati attraverso la rigenerazione mentre il problema ecologico resta del tutto ignorato. E' pur vero che, in pratica, l'esecuzione della legge comporta un freno allo spandimento degli olii e, quindi, all'inquinamento, ma questo è un fatto accidentale ancorchè benefico, certo non contemplato dal legislatore del tempo.

La legge del 1940 costituisce, tuttavia, un esempio importante di regolamentazione della raccolta e della rigenerazione in quanto recepisce già alcuni principi che verranno poi ripresi dalla Direttiva CEE del 16.6.1975 sull'eliminazione degli olii usati.

La legge del 1940 pone l'obbligo di tenere separati gli olii usati secondo le categorie, ne fa una classificazione secondo le caratteristiche d'impiego, obbliga alla conservazione i detentori di olii usati, destina prioritariamente alla rigenerazione gli olii raccolti, prevede l'autorizzazione ministeriale all'esercizio degli impianti di rigenerazione e fissa tutti gli obblighi, i controlli e gli adempimenti dovuti sia da parte dei consumatori di lubrificanti che delle aziende rigeneratrici sanzionando penalmente il divieto di distruzione, dispersione degli olii usati e, comunque, di qualsiasi operazione che ne impedisca la rigenerazione.

La legge dispone anche le condizioni per la rigenerazione in proprio degli olii usati da parte dei detentori (in tal caso il prodotto non è alienabile e deve essere dagli stessi riutilizzato), per la vendita dell'olio usato agli stabilimenti di rigenerazione o per la sua lavorazione "per conto" dietro compenso.

Oggigiorno tutta l'industria della rigenerazione si trova a dover operare, sebbene mai contro, talvolta al di fuori del descritto sistema soprattutto per quanto concerne le modalità della raccolta e l'obbligo di destinare tutto il prodotto alla rigenerazione essendo alcune norme superate dal venir meno dello scopo autarchico che le aveva ispirate. Tuttavia le Amministrazioni Statali, gli Enti Pubblici e l'Esercito continuano ad attenersi, per quanto concerne la conservazione, segregazione e vendita degli olii usati, alle norme legislative previste.

Per far fronte alle notevoli spese della raccolta e mantenere il prezzo di acquisto del prodotto al livello dell'olio combustibile, onde evitare che il prodotto venga destinato alla combustione, l'industria della rigenerazione necessita di un aiuto esterno oggi fornitole dall'art. 12 della menzionata legge fiscale del 1962 che riduce del 75% l'imposta di fabbricazione vigente sugli olii lubrificanti rigenerati purchè gli impianti siano autorizzati dal Ministero delle Finanze. Questa agevolazione contrasta, però, con le norme del trattato di Roma e, pertanto, in base ad una recente sentenza, l'Italia si trova costretta ad eliminarla o ad estenderla anche agli olii rigenerati provenienti da altri Paesi della Comunità per armonizzare il trattamento fiscale.

La soluzione di questi ed altri problemi connessi con i nuovi assillanti problemi ecologici posti in evidenza dalla Direttiva CEE del 16.6.1975 può trovarsi solo in una nuova normativa specifica che persegua gli scopi previsti dalla Direttiva stessa: impedire che gli olii minerali una volta utilizzati vengano abbandonati o dispersi con conseguente inquinamento del suolo attraverso la loro raccolta, rigenerazione o combustione.

Le coincidenze tra la Direttiva CEE e l'attuale legislazione italiana sono solo occasionali e, pertanto, rare. Per porre rimedio all'attuale stato di cose è stato predisposto un disegno di legge sulla raccolta ed il riciclaggio degli olii lubrificanti usati che attende la delega al governo dei necessari poteri per trovare poi attuazione.

Purtroppo restano tuttora incerti i tempi, prevedibilmente lunghi, entro i quali la nuova legge potrà divenire norma vigente con le conseguenze di carattere ecologico che comporta la dispersione degli olii non rigenerabili, e quindi non completamente raccolti, nonchè la spietata concorrenza della combustione che, oltre a provocare un grave inquinamento dell'aria, sottrae un prodotto prezioso ai fini del recupero energetico e del risparmio valutario.

La nuova disciplina per la raccolta e rigenerazione degli olii usati sancisce l'obbligo per chiunque ottenga olii usati da cicli di produzione o impieghi, di tenerli, conservarli e consegnarli agli operatori che ne effettuano la raccolta, adottando le precauzioni necessarie per evitare l'immissione di sostanze estranee negli stessi.

Ad un Consorzio da costituirsi tra le Aziende di riraffinazione e le Aziende produttrici di olii lubrificanti finiti, dovrebbe essere affidato il compito di organizzare la raccolta degli olii usati su tutto il territorio nazionale e di decidere la diversa destinazione da dare al prodotto qualora le sue caratteristiche qualitative ne rendano impossibile il riutilizzo.

La legge inoltre:

1) introduce, in considerazione del costo della raccolta, una sovvenzione finanziata mediante apposita tassa sugli olii nuovi in applicazione del principio per cui chi inquina paga;

2) statuisce il pagamento dell'olio usato onde incentivare l'interesse dei produttori a conservarlo;

3) fissa i criteri per la determinazione del prezzo di cessione alle aziende di rigenerazione;

4) stabilisce i criteri per la concessione delle autorizzazioni all'esercizio degli impianti al fine di un adeguamento degli stessi al progresso tecnologico;

5) colpisce i contravventori con adeguate sanzioni penali.

All'assenza di un efficace sistema legislativo fa però contrasto in Italia la vitalità dell'industria rigeneratrice.

Sono 12 le aziende che operano attualmente: 9 impianti sono dislocati nel Nord – di cui 6 nella sola Lombardia –, 2 al Centro ed 1 nel Meridione. Le licenze di lavorazione concesse ammontano complessivamente a tonnellate annue 212.400. (Tab. I)

Sono state applicate nuove tecnologie quali la raffinazione degli olii esausti mediante propano ed è in costruzione un impianto nell'Italia centrale che applica il processo SNAM Progetti "Propano + Idrogenazione".

Risolto il problema della regolamentazione del settore, l'industria nazionale della rigenerazione potrà affrontare opere di potenziamento degli impianti per trattare quantità di olio usato superiori a quelle attualmente raccolte.

Gli investimenti nel settore andranno preminentemente indirizzati ad organizzare e potenziare il sistema di raccolta che dovrà funzionare sotto il controllo delle industrie di rigenerazione: non tutti gli olii, infatti, possono essere indiscriminatamente sottoposti a trattamento di riraffinazione e sono necessarie conoscenze tecniche e mezzi tecnologici adeguati per distinguere gli olii lubrificanti riraffinabili da quelli destinabili alla combustione con recupero di calore o all'incenerimento.

Oggi in Italia il consumo di olii lubrificanti si aggira intorno alle 630.000 tonnellate annue e si stima che il quantitativo di olio usato prodotto e recuperabile si aggiri intorno alle 220.000 tonnellate annue. (Tab. II)

Attualmente, nonostante le disposizioni di legge e gli sforzi dell'industria della rigenerazione e delle imprese di raccolta appositamente create per

TAB. I

AZIENDE DI RIRAFFINAZIONE OPERANTI ATTUALMENTE IN ITALIA ADERENTI AL "G.A.I.L. - Gruppo Aziende Indipendenti Lubrificanti -" DELLA "ASSOCIAZIONE NAZIONALE DELL'INDUSTRIA CHIMICA".

NORD

ICEP	DRESANO	(MI)	20.000 T/a.
OMA	RIVALTA TORINESE	(TO)	7.000 T/a.
RIVOL	SPESSA PO	(PV)	15.000 T/a.
RONDINE	PERO	(MI)	12.000 T/a.
ROMA	MILANO		21.400 T/a.
SIRO	CORBETTA	(MI)	9.000 T/a.
VISCOLUBE	PIEVE FISSIRAGA	(MI)	30.000 T/a.

CENTRO

CLIPPER OIL IT.	CECCANO	(FR)	35.000 T/a.

SUD

RAMOIL	CASALNUOVO DI NAPOLI		15.000 T/a.

TAB. II

Quantità degli olii	Consumo (t./a.)	Olio usato recuperabile (t./a.)
Olio motore	378.000	132.000
Trasformatori	30.000	26.000
Turbine	12.000	10.000
Idraulico	46.000	27.000
Taglio metalli solubile	29.000	17.000
Taglio metalli non solubile è tempera	35.000	8.000
Spindle, grassi, cilindri, olii neri sformatura, protettivi vari	100.000	-
TOTALE	630.000	220.000

raggiungere non solo le grandi partite provenienti da importanti consumatori di lubrificanti ma anche i modesti quantitativi reperibili presso le miriadi di piccoli utilizzatori sparsi su tutto il territorio nazionale, si riesce a raccogliere solo il 60% degli olii usati prodotti, pari a circa il 20% degli olii nuovi immessi al consumo.

Sono 130.000 le tonnellate di olio usato provenienti annualmente dal settore automobilistico e dalle industrie. Di queste una parte viene rigenerata mentre gli olii chiari provenienti da trasformatori, olii turbine, olii idraulici, trovano spesso un riutilizzo in settori di lubrificazione secondaria. Non risultano, invece, dati sull'eventuale eliminazione degli olii usati attraverso altre procedure come quelle per combustione o incenerimento.

Soltanto la piena sensibilizzazione di tutti coloro che, direttamente o indirettamente, hanno a che fare con l'olio usato, sorretti da adeguati strumenti legislativi e incentivati nella fase più delicata della raccolta, potrà consentire la realizzazione di un'efficace rete di recupero e di adeguare alle più moderne tecnologie gli impianti esistenti al fine di un miglioramento anche qualitativo della produzione e solo così lo spreco di materia prima potrà essere decisamente ridotto e si otterrà un'importante vittoria per la battaglia della tutela dell'ambiente.

DIE PREISBILDUNG FÜR GEBRAUCHTÖL IN EUROPA
THE ESTABLISHMENT OF PRICES FOR USED OILS IN EUROPE
LA FORMATION DES PRIX DES HUILES USAGEES EN EUROPE
LA FORMAZIONE DEI PREZZI DEGLI OLI USATI IN EUROPA

P. GRARD
Ehrenpräsident der I.H.M.B. Industrie des Huiles Minérales de
Belgique, Brüssel, Vizepräsident der U.E.I.L.
Honorary President of the I.H.M.B. (Industrie des Huiles Minérales de
Belgique), Brussels, Vice-President of E.U.I.L.
Président Honoraire de l'I.H.M.B. (Industrie des Huiles Minérales de
Belgique), Bruxelles, Vice-Président de l'U.E.I.L.
Presidente Onorario dell'I.H.M.B. (Industrie des Huiles Minérales de
Belgique), Bruxelles, Vice Presidente della U.E.I.L.

RESUME

La formation du prix des Huiles Usagées est essentiellement fonction du degré de volonté des gouvernements

- de protéger leur environnement,

- d'économiser les matières premières.

En effet, tout détenteur a la possibilité d'utiliser lui-même ses huiles comme combustible ou de les revendre à un collecteur ou à un éliminateur.

Si la législation sur l'utilisation des Huiles Usagées comme combustible n'est pas rigoureuse, le détenteur de ces huiles sera tenté de les utiliser lui-même comme combustible ou de les valoriser à un prix qui ne sera pas très inférieur à celui du combustible généralement utilisé, sans se préoccuper des conséquences désastreuses pour l'environnement.

Dans un régime que l'on peut donc qualifier de "laisser-aller", le prix des Huiles Usagées dépendra de celui du fuel ; dans un cadre de réglementation limitant son utilisation, le prix de l'huile usagées sera également fonction de celui du fuel, mais diminué des contraintes que feront peser sur sa valeur d'utilisation les réglementations mises en place.

On peut donc affirmer que, s'il existe une législation réglementant l'utilisation des Huiles Usagées, celle-ci a une influence déterminante sur leur prix de reprise.

ZUSAMMENFASSUNG

Die Preisbildung bei Gebrauchtöl hängt in erster Linie von dem Grad des Willens der Regierungen ab,
-ihre Umwelt zu schützen,
-Rohstoffe einzusparen.
Jeder, der über Gebrauchtöl verfügt, kann es selbst verbrennen oder an ein Sammel- oder Beseitigungsunternehmen verkaufen.
Ohne strenge Rechtsvorschriften für die Verwendung von Gebrauchtöl als Brennstoff werden diejenigen, die über solches Öl verfügen, versucht sein, es selbst als Brennstoff zu verwenden oder es zu einem Preis zu verwerten, der nicht weit unter dem Preis für allgemein benutzten Brennstoff liegt, ohne sich über die katastrophalen Folgen für die Umwelt Gedanken zu machen.
In einem System, das der Willkür Raum lässt, wird der Gebrauchtöl-Preis somit vom Heizölpreis abhängen; im Rahmen einer Regelung, die die Verwendung von Gebrauchtöl einschränkt, wird der Preis ebenfalls vom Heizölpreis abhängen, aber in einem entsprechend den Zwängen verringertem Masse, mit denen die Rechtsvorschriften seinen Gebrauchswert belasten.
Ist mithin die Verwendung von Gebrauchtöl durch Rechtsvorschriften geregelt, so haben diese einen entscheidenden Einfluss auf die Preisbildung.

SUMMARY

The establishment of prices for used oils depends basically on how much governments are willing
- to protect their environment, and
- to **economize** on raw materials.
Each holder has the choice of using the oils himself as fuel or of selling them for collecting or disposal purposes.
If legislation on the use of used oils as fuel is not strict, holders of these oils will be tempted to use them themselves as fuel or to increase the price of such oils to little less than that of normal fuel without worrying about the disastrous consequences for the environment.
In this sector with its "devil may care" attitude the price of used oils will depend on the price of fuel oil. Given rules limiting its use, the price of used oil will also depend on the price of fuel oil, but it will

be lessened by the restrictions imposed on it by regulations.
It is therefore clear that legislation determining the use of used oils will have a major influence on their selling price.

RIASSUNTO

La formazione del prezzo degli oli usati dipende soprattutto dal grado di volontà dei governi
- di proteggere l'ambiente,
- di risparmiare le materie prime.

Infatti, chiunque ha la possibilità di utilizzare direttamente gli oli in suo possesso come combustibile oppure di rivenderli a un'impresa di raccolta o di eliminazione.

Se la legislazione sull'impiego degli oli usati come combustibile non è rigorosa, chi dispone di tali oli sarà tentato di utilizzarli direttamente come combustibile o di valorizzarli ad un prezzo che non sarà molto inferiore a quello del combustibile generalmente utilizzato, senza preoccuparsi delle conseguenze disastrose per l'ambiente.

In un regime che si può pertanto definire di "laisser-aller", il prezzo degli oli usati dipendera da quello del gasolio ; nel quadro di una regolamentazione che ne limita l'impiego, il prezzo degli oli usati dipenderà quindi da quello del gasolio più gli oneri che graveranno sul suo valore in seguito alle norme in vigore.

Si può pertanto affermare che una normativa che definisca l'impiego degli oli usati potrà influire in modo determinante sul loro prezzo di acquisto.

Certains produits, après l'emploi auquel ils étaient destinés primitivement, peuvent avoir une utilisation secondaire et, par voie de conséquence, conserver une certaine valeur. C'est le cas de l'huile usagée.

L'huile usagée peut, en effet, être réutilisée, soit comme combustible, soit comme matière première pour la fabrication d'huile de base. Dans ces deux cas, la valeur des huiles usagées s'est de tout temps et partout toujours établie en fonction des volumes unitaires disponibles. En effet, compte tenu des coûts de collecte, le prix des petites quantités s'établit à une valeur nulle, voire négative, les quantités importantes peuvent, elles, au contraire, avoir une valeur relativement élevée. Ce qu'il est -ntéressant de noter, c'est que l'écart de prix entre ces deux extrêmes évolue essentiellement en fonction de la volonté politique des gouvernements de protéger partiellement ou totalement l'environnement, voire depuis 1973 d'économiser les matières premières.

C'est ainsi qu'a priori l'utilisation de l'huile usagée comme combustible paraît économiquement la plus séduisante. En effet, utilisée comme telle sur place, on évite tout frais de collecte et son utilisateur y trouve un profit évident, le p. c. i. de l'huile usagée étant de l'ordre de 8.500 millithermies/Kg, p. c. i. qui est à rapprocher du Fuel Oil Domestique qui est de 10.300 millithermies/Kg. En rapprochant ces p. c. i. respectifs, on peut déterminer une valeur de réemploi de l'huile usagée par son détenteur, représentant environ 80 % de la valeur du combustible utilisé habituellement. Toutefois, l'utilisation comme combustible de l'huile usagée IN SITU suppose d'abord que son détenteur en possède des quantités suffisantes pour justifier l'achat d'une installation poly-combustible, ce qui n'est pas évident pour les petits garages, les petites stations service, les petits transporteurs, etc... qui n'ont alors que deux possibilités : les jeter ou les remettre à un collecteur spécialisé.

Tant qu'aucune réglementation n'existe, le collecteur spécialisé doit évidemment dégager une marge suffisante entre son prix d'achat de l'huile et le prix de revente de celle-ci lui permettant tout à la fois de couvrir ses frais de collecte et de générer un profit. Or, l'expérience démontre que la collecte de petites quantités ne permet pas de dégager une marge suffisante pour le collecteur à moins qu'il ne fasse suyporter au détenteur une partie du coût de la collecte, ce qui conduit l'huile usagée à n'avoir dans ce cas qu'une valeur négative ; la tentation devient alors grande pour son détenteur de la jeter plutôt que d'avoir à payer pour son enlèvement. C'est pour ces raisons que, dans les pays où il n'y a pas de réglementation concernant l'élimination des huiles usagées, l'on constate une valeur élevée de celle-ci lorsqu'elle est détenue en quantités unitaires importantes et parallèlement une pollution élevée des eaux par l'huile usagée créée par le flux des petites quantités non collectées. C'est ainsi qu'il est apparu que la seule loi de l'offre et de la demande pour la formation du prix des huiles usagées conduisait inévitablement à un ramassage écrémant et, par voie de conséquence, à une pollution importante des eaux. C'est ce qui a amené les gouvernements soucieux de protéger l'environnement à charger leurs administrations responsables d'analyser ces questions dont les réponses entraînent les décisions qui conduisent à une formation du prix des huiles usagées très différente de celle que l'on obtient dans un régime que l'on pourrait qualifier de "laisser-aller".

En effet, vue sous le plan d'une collectivité nationale, une analyse simple amène à se poser les questions suivantes et à en tirer immédiatement les conclusions évidentes :

- peut-on jeter sans inconvénients pour l'environnement les huiles usagées ? Non, parce que les huiles usagées présentent un réel danger dans le cadre de la protection des eaux. En effet, chargée d'additifs, ces huiles ne sont pas biodégradables, se déposent éventuellement sur les berges des rivières, empêchent la reproduction du poisson et de la flore ; elles peuvent, par surcroît, si elles arrivent en quantités importantes dans une station d'épuration d'eau, bloquer le processus de fermentation des boues et paralyser de manière définitive la station. Si, au lieu que ce soient les eaux de surface qui soient touchées, il s'agit d'une nappe phréatique, celle-ci peut être définitivement polluée. La conclusion est donc évidente : protéger l'environnement signifie contrôler la collecte et la rendre exhaustive.

- peut-on utiliser les huiles usagées sans inconvénient comme combustible ? Oui, à condition, soit de traiter celles-ci avant leur utilisation pour en supprimer tous les métaux : zinc, phosphate, plomb, etc... contenus dans ces huiles, très dangereux en cas d'évacuation dans l'atmosphère, soit en traitant les fumées de leur combustion à la sortie des installations thermiques.
- peut-on, sans inconvénient pour l'environnement, utiliser les huiles usagées en l'état comme matière première pour la fabrication de lubrifiants ? Oui, à condition que le traitement de reraffinage soit conduit, comme toute production industrielle de déchets, avec tous les équipements voulus pour éviter les nuisances.

La conclusion est donc évidente : protéger l'environnement signifie qu'il faut contrôler l'élimination des huiles usagées.

Ces différentes contraintes d'environnement touchant à la collecte et à l'utilisation des huiles usagées ont évidemment pour effet de diminuer la valeur pondérée des huiles usagées. En effet, si l'on désire obtenir une collecte exhaustive, il faut d'abord s'assurer que tout sera collecté.

Comme il n'est pas possible de placer un policier derrière chaque détenteur, la seule solution économique consiste à ne rien demander aux petits détenteurs pour l'enlèvement de leurs huiles usagées, tout en permettant au collecteur de le faire de manière rentable ; ceci suppose qu'il puisse, pour supporter l'effort économique auquel cela correspond, effectuer une péréquation des coûts, ce qui l'oblige à ne plus reprendre aux gros détenteurs, à un prix comparable au prix de reprise appliqué en régime de "laisser-aller". L'on constate donc qu'une collecte exhaustive inscrite dans un cadre réglementaire a pour effet de diminuer l'écart de prix de reprise entre les petites quantités unitaires et les grandes.

Mais si, également, dans le souci de protéger l'environnement, l'utilisation des huiles comme combustible ou comme matière première pour le reraffinage est tenu à certaines contraintes d'environnement, ces contraintes viennent tout naturellement peser sur les prix de reprise que l'on peut offrir pour ces huiles.

Ces contraintes générales peuvent se traduire par la formule suivante :
- valeur fuel p. c. i.   10 300 millithermies par kilo
- valeur huile usagée p. c. i.   8 500 millithermies = 17,5 % de moins
moins coût de collecte exhaustive
moins contraintes environnement pour élimination = valeur réelle huile usagée.

A ces préoccupations d'environnement des gouvernements, peuvent s'en ajouter d'autres pour certains d'entre eux depuis les événements de 1973, à savoir le souci d'économiser les matières premières. Les responsables des administrations concernées sont alors amenés à se poser la question suivante : quelle est la meilleure destination des huiles usagées ?

Si un choix est alors fait, il peut tout naturellement, en diminuant le nombre des repreneurs possibles de l'huile usagée, réduire les effets de la loi de l'offre et de la demande, ce qui peut tendre évidemment à réduire encore le prix de reprise possible de ces produits.

Toutefois, il est intéressant de noter que ces fluctuations possibles des prix de reprise des huiles usagées sont sans conséquences graves pour leurs détenteurs et c'est ce qu'il faut signaler. En effet, l'on constate qu'au regard de leurs activités, garages, stations service, transporteurs, industriels ou agriculteurs ne voient jamais leurs comptes d'exploitation déséquilibrés du fait du prix de cession plus ou moins élevé des huiles usagées qu'ils détiennent.

Par contre, si l'on se situe au niveau des collectivités nationales, il est important de considérer qu'il est d'intérêt national que ces huiles soient parfaitement collectées car dans le cas contraire, elles peuvent être à l'origine de pollutions graves dont le coût est toujours élevé pour la collectivité. Outre qu'elle permet d'éviter ces pollutions, la récupération de ces huiles engendre des économies de matières premières qui sont loin d'être négligeables dans le contexte que nous connaissons aujourd'hui et qui, sans nul doute, durera encore longtemps.

DAS WIRTSCHAFTLICHE ERFORDERNIS EINER PRODUKTFLEXIBILITÄT IN DER
ZWEITRAFFINATIONSINDUSTRIE
THE ECONOMIC NECESSITY FOR QUALITATIVE FLEXIBILITY OF PRODUCTION IN
THE REGENERATING INDUSTRY
NECESSITE ECONOMIQUE D'UNE FLEXIBILITE QUALITATIVE DE PRODUCTION DANS
L'INDUSTRIE DE LA REGENERATION
LE ESIGENZE ECONOMICHE DI UNA FLESSIBILITA DI PRODUZIONE NEL SETTORE
DELLA RIGENERAZIONE

R. HAVEMANN
Stellvertretender Vorsitzender der AMMRA, Hamburg
Vice President of AMMRA, Hamburg
Vice-Président de L'AMMRA, Hamburg
Vice Presidente dell'AMMRA, Amburgo

ZUSAMMENFASSUNG

Eine Produktflexibilität der Zweitraffinationsindustrie in der Bundesrepublik Deutschland ist aus folgenden Gründen erforderlich:

1. Nationale Gesetzgebung
2. Altölsammlung und Altölqualität
3. Reststoffbeseitigung
4. Marktstrategie

Aufgrund der Gesetzgebung in der Bundesrepublik mit der Definition Altöl und der Art der Altölsammlung ist eine ausschließliche Aufarbeitung von Altölen zu Schmierölen nicht möglich. Knapp 70 % aller Altölübernahmen liegen unter 800 Ltr.

Altöl ist also von unterschiedlicher Beschaffenheit und eine qualitative

Einstufung ist erforderlich, um zu vertretbaren Kosten wettbewerbsfähige Produkte herzustellen. Die Entsorgung der bei der Schmierölproduktion anfallenden Reststoffe ist zur Zeit gesichert, aber sehr kostenintensiv.

Zeitweise sich ergebende Probleme bei der Reststoffbeseitigung und bei der Vermarktung von Schmierölen aufgrund eines stagnierenden Marktes und Überschußmengen haben aufgeschlossene Unternehmer bewogen, aus Altöl weitere Produkte wie Fluxöl, Reduktionsöl und Arten von Gasöl und Schweröl herzustellen.

Die Zweitraffineure in der Bundesrepublik haben den Gesetzesauftrag bislang erfolgreich erfüllt. Für diese Unternehmen ist Altöl ein bodenständiger Rohstoff, aus dem es gilt, hochwertige Produkte herzustellen, die den internationalen und nationalen Normen entsprechen.

Daß Gebrauchtes nicht verbraucht ist, das beweisen diese Unternehmen seit Jahrzehnten jeden Tag.

SUMMARY

Flexibility of production in the Federal Republic of Germany's regenerating industry is necessary for the following reasons:
1. the provisions of domestic law;
2. the conditions of collection and the quality of used oils;
3. the disposal of residues;
4. market strategy.

On account of the provisions of German law governing the definition of used oil and because of the way in which it is collected, regeneration techniques cannot be applied solely to the production of lubricating oils. Just under 70% of all used-oils consignments are of less than 800 litres.

Used oil is of varying composition and a qualitative classification is therefore required if competitive products are to be manufactured at reasonable cost. There is no difficulty at present in disposing of the residues from lubricating oil production, but the costs are high.

Occasional problems arising in the disposal of residues and the marketing of lubricating oils as a result of a stagnating market and surplus quantities have prompted enterprising businessmen to manufacture other products from used oils, such as flux oil, reducing oil and certain types of gas oil and heavy fuel oil.

The regenerating industry in the Federal Republic has so far always fulfilled its obligations under the law. These firms regard used oil an indigenous raw material from which to manufacture high-quality products that conform to domestic and international standards.

For several decades these firms have been demonstrating daily that "used" does not mean "exhausted".

RESUME

La flexibilité de la production dans l'industrie de la régénération s'impose en République Fédérale d'Allemagne en raison des facteurs suivants:

1. La législation nationale.
2. La collecte et la qualité des huiles usagées.
3. L'élimination des déchets.
4. La stratégie du marché.

La législation de la République Fédérale ne permet pas, de par la définition des huiles usagées et les méthodes de collecte de ces produits, de transformer ces huiles en huiles de graissage. Près de 70% de toutes les huiles usagées recueillies représentent moins de 800 l.

Les huiles usagées se présentent donc sous des formes diverses et une classification qualitative s'impose si l'on veut fabriquer des produits concurrentiels à des coûts raisonnables. L'élimination des déchets résultant de la production des huiles de graissage est actuellement assurée mais très coûteuse.

Les problèmes que pose parfois l'élimination des déchets sont liés à la commercialisation des huiles de graissage, à cause de la stagnation du marché, et à l'existence de surplus qui ont amené des chefs d'entreprise dynamiques à produire à partir des huiles usagées d'autres produits comme les huiles de fluxage, les résidus de distillation atmosphérique et certains types de gasoil et d'huiles lourdes.

Les entreprises de régénération de la République Fédérale se sont jusqu'à présent conformées avec succès aux prescriptions de la loi. Elles considèrent que les huiles usagées sont une matière première indigène à partir de laquelle il convient de fabriquer des produits de haute qualité conformes aux normes internationales et nationales.

Ces entreprises démontrent chaque jour depuis des dizaines d'années qu'"usagé" ne veut pas nécessairement dire "épuisé".

RIASSUNTO

Nella Repubblica federale di Germania è necessario che vi sia una flessibilità di produzione nel settore della rigenerazione per i seguenti motivi :
1. Legislazione nazionale.
2. Raccolta e qualità degli oli usati.
3. Eliminazione dei residui.
4. Strategia di mercato.

In base alla legislazione della Repubblica federale tedesca e alla definizione di oli usati e di raccolta degli oli usati, non è possibile produrre con tali oli esclusivamente oli lubrificanti. Quasi il 70% delle consegne di oli usati riguardano quantità al di sotto dogli 800 litri.

Gli oli usati sono di composizione differente ed è necessario classificarli dal punto di vista qualitativo per poter fabbricare dei prodotti concorrenziali a costi ragionevoli. L'eliminazione dei rifiuti che derivano dalla produzione di oli lubrificanti è attualmente possibile ma molto costosa.

Poiché l'eliminazione dei residui e la vendita degli oli lubrificanti può talvolta essere problematica a causa di una stagnazione del mercato oppure per la presenza di eccedenze, alcuni imprenditori particolarmente dinamici hanno deciso di trarre dagli oli usati prodotti diversi come olio di flusso, olio di riduzione e altri tipi di gasolio e di olio pesante.

I produttori di oli rigenerati nella Repubblica federale hanno fino ad ora adempiuto agli obblighi previsti dalla legge. Per tali imprenditori gli oli usati costituiscono una materia prima disponibile dalla quale si possono trarre prodotti di alto livello che corrispondono alle norme internazionali e nazionali.

Che l'usato non sia "consumato" lo dimostrano queste imprese da decenni, giorno per giorno.

Es ist mir ein besonderes Anliegen, im Rahmen dieses 2. Europäischen
Altölkongresses über die Notwendigkeit der Produktflexibilität von Zweitraffinerien zu sprechen, d.h. über die Gründe, aus denen heraus Zweitraffinerien auch andere Produkte als Schmieröle herstellen müssen.

Das Thema läßt zwar in erster Linie nur auf das wirtschaftliche Erfordernis schließen, dieses ist ohne Zweifel auch vorhanden, aber letztlich
nicht allein ausschlaggebend. Ich kann hier natürlich nur speziell zur
Situation in der Bundesrepublik Deutschland Stellung nehmen, bin mir aber
eigentlich aufgrund der vielen Gespräche mit Kollegen aus anderen Ländern
einig, daß diese in der einen oder anderen Art und Weise auch dort zutreffen.

Welche Gründe haben nun zur Produktflexibilität der Zweitraffinationsindustrie geführt:

1. Die jeweilige nationale Gesetzgebung
2. Die Altölsammlung und die Altölqualität
3. Die Reststoffbeseitigung
4. Die Marktstrategie

1. Die jeweilige nationale Gesetzgebung

   Nach dem Altölgesetz der Bundesrepublik sind Altöle "gebrauchte, halbflüssige und flüssige Stoffe, die ganz oder teilweise aus Mineralöl
   oder synthetischem Öl bestehen, einschließlich ölhaltiger Rückstände
   aus Behältern, Emulsionen und Ölwassergemische mit mindestens 4 v.H.
   Ölgehalt".

   Altöle nach dieser Definition dürfen Fremdstoffe nur bis maximal 10 %
   enthalten, die aus gebrauchs- und betriebsbedingten Gründen hineingelangt sind.

   Allein das Gesetz mit dieser Definition Altöl schließt letztlich aus,
   daß die Zweitraffinerien ausschließlich Schmieröle herstellen können,
   denn Mineralöl ist nicht automatisch Schmieröl. Mineralöl kann auch
   Spezial- und Testbenzin, Vergaserkraftstoff, Gasöl, Heizöl sein und
   Mineralöl, welches nie Schmieröl war, kann auch von keinem Zweitraffineur, gleich durch welchen Prozeß, zu Schmieröl umgewandelt
   werden.

2. Die Altölsammlung und die Altölqualität

Lassen Sie mich nun über die Altölsammlung sprechen. Die Verpflichtung nach dem Altölgesetz führt zu 500.000 Einzelübernahmen an den Anfallstellen von Altöl der Branche insgesamt, wobei knapp 70 % aller Mengen unter 800 Ltr. liegen, d.h. 5 Barrel im Durchschnitt.

Die Rohstoffqualität, d.h. die Beschaffenheit des Altöles ist unterschiedlich. Wie bereits ausgeführt, muß es mühsam in Kleinpartien zusammengetragen werden, damit es in entsprechender Menge als Rohstoff für die Aufarbeitung zur Verfügung steht. Es bedarf eines erheblichen Laboraufwandes, die eingehenden Partien zu analysieren und qualitativ einzustufen, um letztlich zu vertretbaren Kosten wettbewerbsfähige Produkte herzustellen.

Man muß ganz deutlich sehen, daß die übernommenen Altöle aus unterschiedlichen Produktionsstufen stammen und mit verschiedenen Fremdstoffanteilen versehen sind. Nach dem Altölgesetz der Bundesrepublik sind nur gebrauchs- und betriebsbedingte Fremdstoffe bis zu 10 % zulässig. Darüber hinausgehende und artfremde Fremdstoffe müssen der Anfallstelle in Rechnung gestellt werden.

Das Altölgesetz zwingt die Zweitraffinerien zu einer absolut wirtschaftlichen Optimierung der Ergebnisse. Letztlich haben innerhalb von 10 Jahren von 1969 bis 1978 10 Unternehmen ihre Arbeit eingestellt. Dies zwar aus unterschiedlichen Gründen, aber es zeigt deutlich, daß der vom Gesetzgeber geforderte Wettbewerb stattfindet.

Die Zweitraffinerien in der Bundesrepublik Deutschland haben sich auf diese speziellen Verhältnisse rechtzeitig eingestellt und kommen ihrer Aufgabe verantwortungsvoll und erfolgreich nach. Daß die Unternehmen mit gewissen Rahmenbedingungen unzufrieden sind, ist bereits an anderer Stelle vorgetragen worden.

3. Die Reststoffbeseitigung

Die Zweitraffinerien, die sich überwiegend mit der Herstellung von Grundölen und in der Weiterverarbeitung zu Motorenölen, Getriebeölen, Hydraulikölen und den unterschiedlichsten Industrieölen befassen, haben Reststoffprobleme zu lösen.

Die Säureharzentsorgung in der Bundesrepublik geschieht durch:

a) Eigene Verbrennung
b) Verbringung auf Hausmülldeponien
c) Verbrennung in Drehrohröfen der Zementindustrie
d) Verbringung in Aufarbeitungsfirmen

Alle 4 Wege sind kostenintensiv und bieten keine 100 %ige Entsorgungssicherheit.

Zu a) Eigene Verbrennung:

Die eigenen Verbrennungsanlagen sind gewöhnlich aufgrund der Gesetzgebung und der örtlichen Verhältnisse bezüglich der Verbrennung von Säureharzen eingeengt, d.h. Zweitraffinerien, die selbst über Spezialverbrennungsanlagen verfügen, müssen trotzdem Säureharze auch noch anderweitig verbringen.

Zu b) Verbringung auf Hausmülldeponien:

Die Entsorgung über Hausmülldeponien ist ebenfalls mengenmäßig limitiert und aufgrund der Strukturen der einzelnen Deponien nicht gesichert, da bei Anlegen von neuen Poldern und sonstigen Deponieveränderungen die Einlagerung oft über Monate aus Sicherheitsgründen nicht möglich ist.

Zu c) Verbrennung in Drehrohröfen der Zementindustrie:

Die Zementindustrie hat die Möglichkeit, wenn sie ihre Anlagen voll fährt, Säureharze als Energieträger einzusetzen. Aber viele Anlagen werden mehrmals im Jahr über Monate abgestellt.

Zu d) Verbringung in Aufarbeitungsfirmen:

Die Entsorgung über Aufbereiter - hier wird aus den Säureharzen Schwefelsäure zurückgewonnen - ist durch betriebstechnische Vorgänge begrenzt.

Eine weitere Möglichkeit sei hier noch angeschnitten. Das wäre die Neutralisation der Säureharze und deren Verbringung auf Sondermülldeponien. Auch hier ergeben sich, was die Neutralisation angeht, Kapazitätsschwierigkeiten und auch Einlagerungsprobleme. Darüber hinaus ist dieser Weg aber so kostenintensiv, daß er schon aus Wettbewerbsgründen ausscheidet.

Die Entsorgung auf der Bleicherdeseite ist weniger problematisch. Die in den Betrieben anfallende Bleicherde wird sowohl von der Zementindustrie als auch von Kraftwerken aufgenommen und wirtschaftlich genutzt. Sie kann auch von Ziegeleien dem Rohziegel beigegeben werden und führt zur Energieeinsparung und einem besseren Ausbrennen der Klinker. Steigende Energiepreise haben Bleicherde zu einem Rohstoff werden lassen.

4. Die Marktstrategie

Da die Aufbereiter von Altöl, die Zweitraffineure also, durch Gesetz verpflichtet sind, sämtliche Altöle aufzunehmen, müssen diese Betriebe auch diese Altöle verarbeiten. Natürlich verfügen alle Zweitraffineure über ausreichendes Lagervolumen, aber sicherlich kann niemand unendlich Altöl auf Lager legen. Das bedeutet wiederum, daß Schwierigkeiten in der Entsorgung bei den anfallenden Säureharzen oder Bleicherden direkte Auswirkungen auf die Herstellerbetriebe haben. Da man Altöl leider nicht wie Kohle auf Halde legen kann, haben verantwortungsvolle Unternehmen in dieser Branche in der Bundesrepublik andere Produkte, neue technische Verfahren und Absatzwege gesucht und auch gefunden.

Gestatten Sie mir einen Blick in die Vergangenheit, der einfach notwendig ist, um die Entscheidung einzelner Unternehmen zu verstehen, auch andere Produkte als ausschließlich Schmieröle zu fertigen. Es war nicht immer so, daß Altöl ein begehrter Rohstoff war, sondern wir hatten auch Situationen, wo es Altöl im Überfluß gab, weil der Absatz an Grundölen und legierten Produkten stagnierte. Auf die Problematik der Lagerhaltung hatte ich bereits hingewiesen. Seit dem Jahre 1974 stagniert der Schmierölmarkt. Die Absatzzahlen des Jahres 1973 wurden erst wieder im Jahre 1979 erreicht. Nach den Geschehnissen des Jahres 1979 ist auch nicht damit zu rechnen, daß ausgerechnet der Schmierölmarkt sich expansiv im Gegensatz zu den anderen Mineralölprodukten entwickeln wird.

Es ist eher damit zu rechnen, daß die wesentlichen Preiserhöhungen, die Verteuerungen also dazu führen, daß man mit diesen noch sparsamer als in der Vergangenheit umgehen wird. Wir haben also auch für die Zukunft begrenzte Marktverhältnisse zu erwarten und diesen gilt es Rechnung zu tragen.

Den Schmierölmarkt insgesamt bestimmen nicht die Zweitraffineure, sondern die Marktführer sind Großgesellschaften, d.h. die Zweitraffineure sind hier, was das Marktgeschehen angeht, in Entwicklungsprozesse eingebunden, die sie selbst nicht oder nur sehr wenig beeinflussen können. Daß ein stagnierender Markt, gleich wo auch immer, durch Preismanöver nicht zum Wachstum angeregt wird, hat wohl jeder von uns schon erfahren.

Dieser Umstand zwang schon damals verantwortungsvolle Unternehmer zum Handeln. Letztlich mußte Altöl gesammelt und umweltfreundlich beseitigt werden. Dieser Situation Rechnung tragend, haben ein Zweitraffineur und eine internationale Mineralölgroßgesellschaft gemeinsame Versuche unternommen, um aus Altöl Fluxöl herzustellen, welches dieses Unternehmen bei der Produktion von Oxydbitumen benötigt. Es ging hier darum, ein eigenes Produkt dieser Großgesellschaft durch das aus Altöl gewonnene Produkt zu ersetzen. Ich darf sagen, daß diese gemeinsamen Versuche, von großem beiderseitigen Vertrauen getragen, zum Ziele führten und heute mehrere Zweitraffineure Fluxöl herstellen und die Gesellschaften, die Oxydbitumen produzieren, mit diesem Produkt bedienen.

Das Fluxölgeschäft in der Bundesrepublik ist aus beiderlei Sicht gesehen, d.h. sowohl aus der Sicht der Abnehmer, der Großgesellschaften, als auch der Lieferer, der Zweitraffineure, nicht mehr wegzudenken.

Die Entscheidung der Zweitraffineure, neben Grundölen auch Fluxöl herzustellen, hat den Vorteil, daß sie nicht ausschließlich von der Schmierölfertigung, den Entsorgungsproblemen und vom Absatz her abhängig sind. Die Branche hat hierdurch nur an Vielfalt und Bedeutung gewonnen. Auch dies lassen Sie mich hier deutlich machen, die wirtschaftlichen Gründe, d.h. Preisoptimierung, haben bei diesem Entscheidungsprozeß keine Rolle gespielt. Es ist letztlich auch so, daß die Zweitraffineure, die Fluxöl herstellen, mit dem eigenen Produkt der Abnehmer, d.h. mit den hauseigenen Rohstoffen der Großgesellschaften, zu konkurrieren haben. Die Zweitraffineure stehen also mit ihrem Produkt Fluxöl im Wettbewerb zu dem eigenen Rohstoff ihrer Abnehmer. Dies trifft auch auf die Verfügbarkeit zu. Sie erwarten von ihren Lieferanten die jederzeitige Lieferfähigkeit und messen diese an ihren eigenen Versorgungsmaßstäben in ihren Raffinerien.

Unternehmen, die sich entschlossen haben, neben Schmieröl auch Fluxöl herzustellen, konnten nun mit einer gewissen Sicherheit sagen, wir stehen auf zwei gesunden Beinen, wir haben entsprechende Produktionssicherheiten, wir können die Anlagen unterschiedlich fahren, kommen somit zu besseren Durchsatzergebnissen und sind von Entsorgungsproblemen teilweise befreit.

Dies alles ist richtig, hat jedoch andere Zweitraffinerien nicht ruhen lassen, an einem dritten Absatzweg zu arbeiten, schon allein aufgrund der Qualität der aufzunehmenden Altöle.

Ob Altöl für die Fluxöl- oder Schmierölfertigung, es muß von gleicher Beschaffenheit sein. Die Qualitätsansprüche, die von den Abnehmern von Fluxöl gestellt werden, sind außerordentlich hoch und jeder Abnehmer hat andere Analysendaten, die es einzuhalten gilt.

Der Fluxölmarkt selber ist aber auch abhängig von der jeweiligen Baukonjunktur. Oxydbitumen wird an weiterverarbeitende Betriebe wie Dachpappen- und Isoliermaterialhersteller geliefert, die wiederum die Bauwirtschaft bedienen. Die Baukonjunktur bestimmt also hier den Absatz von Oxydbitumen und letztlich den mengenmäßigen Einsatz von Fluxöl.

Bei der Sammlung fallen aber auch Altöle an, aus denen sich nur unter hohen Schwierigkeiten und unter Mißachtung sämtlicher Kosten Schmieröl herstellen läßt. Für diese Altöle wurden Verarbeitungsverfahren und Absatzwege gesucht und nach vielen Versuchen auch gefunden. Ich spreche von dem Reduktionsöl, das einzelne Zweitraffineure der Stahlindustrie liefern. Auch von diesen Abnehmern werden qualitative Ansprüche ganz besonderer Art gestellt, d.h. auch hier gilt es Normen einzuhalten.

Auch bei der Suche nach der Verwendung von Altölen für diesen Zweck und dem Absatz waren preispolitische Überlegungen von untergeordneter Natur. Es ging darum, bestimmte Altöle, die aus dem Bereich der metallverarbeitenden Industrie und Tankreinigungen kommen, einem vernünftigen wirtschaftlichen Zweck zuzuführen, unter Beachtung des Gesetzesauftrages.

Diese Entwicklung, die wir in der Bundesrepublik Deutschland hatten, ist natürlich nicht typisch für andere Länder. Sie sollte nur deutlich machen, warum eine Produktionsvielfalt notwendig war und auch noch ist.

Die 3 Absatzsäulen

a) Grundöle/legierte Öle wie

    Motorenöle
    Getriebeöle
    Hydrauliköle und
    unterschiedlichste Industrieöle

b) Fluxöle

c) Reduktionsöle

haben aber in der Herstellung auch noch Nebenprodukte zur Folge, die es zu vermarkten gilt, und die mengenmäßig durchaus Bedeutung haben. Es handelt sich überwiegend um Arten von Gasöl und Schweröl. Diese Produkte stellen uns qualitativ vor keinerlei Probleme und lassen sich unterschiedlich vermarkten. In allen 4 Bereichen streben die Unternehmen in der Bundesrepublik nach höchster Qualität und Anerkennung.

So haben die am Markt tätigen Firmen für ihre Produktpalette ebenfalls die erforderlichen Freigaben der Automobilindustrie, wie z.B. von Daimler-Benz und MAN.

Wir wissen, daß auch Zweitraffinerien an den anspruchsvollen Feldversuchen von 30.000 km Ölwechselintervallen mitarbeiten und erste Ergebnisse und Auswertungen haben zu befriedigenden Resultaten geführt.

Auch Zulassungen nach den technischen Lieferbedingungen des Bundesamtes für Wehrtechnik und Beschaffung, die die Versorgung der Europäischen Streitkräfte und nachgeschalteter Behörden betreffen, werden von dem Unternehmen mit Nachdruck betrieben.

Vergleichsuntersuchungen beim Institut für Erdölforschung zwischen Erst- und Zweitraffinaten, ein ebenfalls anspruchsvolles Forschungsprogramm, haben gezeigt, daß Zweitraffinate gleichwertig sind. Es wird hier über dieses Forschungsprojekt gesondert berichtet.

Trotz all dieser Beweise, daß Zweitraffinate den Vergleich zu den Erstraffinaten nicht zu scheuen brauchen, werden einzelne Betriebe immer wieder aus den dargelegten Gründen sich für andere Fertigungslinien und Produkte zu entscheiden haben.

In der Bundesrepublik Deutschland sind ausschließlich unabhängig mittelständische Unternehmen in der Zweitraffination tätig. Diese Unternehmen haben durch Weitblick, Investitionsfreudigkeit, technischem Engagement und durch marktpolitische richtige Entscheidungen bislang den Gesetzesauftrag, alles Altöl zu sammeln und schadlos zu beseitigen, erfolgreich erfüllt.

Für die Zweitraffinerien ist Altöl ein bodenständiger Rohstoff, den es zu erhalten gilt. In dem Maße, in dem sich die Rohölreserven erschöpfen, wird der Rohstoff Altöl an Bedeutung zunehmen. Aus diesem Rohstoff stellen heute die Zweitraffineure hochwertige Produkte her, die jedem Vergleich standhalten und den internationalen und nationalen Normen entsprechen.

Daß Gebrauchtes nicht verbraucht ist, das beweisen diese Unternehmen seit Jahrzehnten jeden Tag.

DIE POSITION DER STAATLICHEN STELLEN IN DEN USA GEGENÜBER
ZWEITRAFFINATEN
OFFICIAL ATTITUDES IN THE UNITED STATES TOWARDS REGENERATED OILS
POSITIONS D'ORGANISMES OFFICIELS DES ETATS-UNIS A L'EGARD DES HUILES
REGENEREES
COMPORTAMENTO DEGLI ORGANISMI UFFICIALI DEGLI STATI UNITI NEI RIGUARDI
DEGLI OLI RIGENERATI

M.E. LEPERA, T.C. BOWEN
US Army Mobility Equipment Research & Development Command
J.F.   COLLINS
U.S. Department of Energy
D.A. BECKER
U.S. National Bureau of Standards

MILITARY ENGINE OIL SPECIFICATIONS MODIFIED FOR RECYCLED CONSTITUENTS

Maurice E. Lepera and Thomas C. Bowen
US Army Mobility Equipment Research & Development Command

SUMMARY

In the past, use of recycled constituents in military specification automotive engine oils had been restricted. As a result of recent governmental legislation and cooperative programs with Environmental Protection Agency and National Bureau of Standards, the technical feasibility for having oils formulated with recycled constituents meet military specification requirements was successfully demonstrated. The technology led to the development of MIL-L-46152A which was promulgated on 23 Jan 80. MIL-L-46152A, which describes the lubricating oil for administrative/commercial designed engine, now allows use of conventional, re-refined, or combinations thereof. The concern on base stock variability, which had previously been expressed, will be controlled due to a new requirement which mandates annual submission of base stock ingredients for monitoring purposes.

Department of Defense (DOD) has in the past placed emphasis on the potential use of recycled constituents in military specification products. Recently enacted legislation has resulted in further emphasis being directed on the recovery, reclamation and reuse of petroleum products. The two congressional bills having the greatest impact were: The Energy Policy and Conservation Act of 1975 (PL 94-163) and The Resource Conservation and Recovery Act of 1976 (PL 94-580). PL 94-580 directed federal agencies to initiate, where possible, procurement of products containing recycled materials starting in October 1978. DOD has recognized the need to recycle waste oil and has been active in encouraging programs to accomplish this goal. This position was reaffirmed in a Memorandum from the Assistant Secretary of Defense, issued June 1979 to Military Departments and Defense Agencies. This calls for action such as maximizing the recovery and collection of used oil, maximizing the sale of recovered used oil to re-refiners, encouraging Military and civilian employees who change oil in their personal vehicles, to use the Military recovery and collection system for disposal of the drained oil, and to also investigate other methods that would reduce generation of used lubricating oil.

Numerous governmental specifications exist for petroleum products which limit the material composition for the base product. However, the two specifications receiving the most attention have been the two describing automotive engine oils. These are: MIL-L-46152 - Lubricating Oil, Internal Combustion Engine, Administrative Service, and MIL-L-2104C - Lubricating Oil, Internal Combustion Engine, Tactical Service. MIL-L-46152 is designed soley for commercially designed vehicles and must meet or exceed the API SE/CC performance, whereas, MIL-L-2104C is intended for the tactical fleet operation and must meet or exceed API CD/SC performance level. The MIL-L-46152 and MIL-L-2104C requirements prohibited use of recycled ingredients in formulating products to meet the performance levels identified. Two actions were consequently required to accomplish the DOD objectives; namely -

- o Demonstrate the technical feasibility of lubricants formulated with re-refined material to meet specification requirements.

- o Develop a methodology and associated controls to allow the qualification of products containing re-refined material.

In conjunction with an EPA supported program, initiated in 1972-1978, the feasibility of using recycled materials in formulating MIL-L-46152 oils was completed. A similar program has been initiated to address the feasibility for using recycled materials in MIL-L-2104C formulations, and this is continuing. To implement changes to the two military specifications that would allow procurement of recycled products, our approach has been directed to modify MIL-L-46152 initially, and as the technology develops, subsequently modify MIL-L-2104C. However, we are more concerned with the MIL-L-2104C specification than the MIL-L-46152 because of its intended use and multiple applications. MIL-L-2104C products are designed for tactical engines which are considerably more expensive than smaller commercially designed administrative engines. MIL-L-2104C products must also staisfy a wide variety of power plant systems and types. These oils must satisfactorily lubricate small two- and four-cycle air and water cooled gasoline engines and large two- and four-cycle diesel engines which power armored combat vehicles and self-propelled guns. Further, MIL-L-2104C products are also used in tactical automatic transmissions, power transmission systems and hydraulic operations. Because of this use of MIL-L-2104C oils, any change to the specification requirements necessitates a thorough and comprehensive evaluation which we are currently conducting.

As was noted, our cooperative program with EPA demonstrated the technical feasibility for using recycled ingredients. Once this was demonstrated, we initiated a program to modify the MIL-L-46152 specification. Initially. ASTM activities were requested to assist our resolving the problems in establishing a base stock characterization methodology. Once established, it would apply to all formulations both recycled as well as those conventionally refined products. Following several meetings with ASTM, a draft to MIL-L-46152 was initially circulated for their input and concurrence. This led to the proposed "A" revision of MIL-L-46152 which was circulated on 25 April 1979. Following industry and governmental coordination, the final draft was developed and MIL-L-46152A was officially promulgated on 23 January 1980. The major changes in MIL-L-46152A are:
- o Restrictions prohibiting use of recycled oils removed.
- o Base stock requirement included with specific tests.
- o MS Engine Tests IIB & IIIC revised to IIC and IIID.

o Additional ASTM test procedures included.
o Complete engine performance test criteria now included.

As with previous military engine oil specifications, the MIL-L-46152A document is based on lubricant performance. As such, only minimal constraints are placed on lubricant composition and properties prior to qualification or approval. Materials to be used in formulating the lubricant are defined in general terms. For example, the base stock may consist of a virgin, re-refined or a combination virgin/re-refined product. In regard to property requirements, viscosity and viscosity index are used to define lubricant grade, and maximum pour and minimum flash point values are established to provide a general control of low-temperature handling and product volitality, respectively. The key to the specification is, however, the lubricant performance as defined by a series of engine dynamometer tests. These engine tests are L-38, MS IID, MS IIID, MS VC or VD, and CAT 1H2.

After qualification approval, the latitude allowed is removed as the manufacturer is restricted to the same base stock additive combination used for qualification approval. Tolerances are established for both the chemical and physical properties of the lubricant. In addition, the MIL-L-46152A specification now provides for annual monitoring and sampling of base stock materials. The requirements for base stock properties encompass viscosity, viscosity-index, gravity, pour point, carbon residue, sulfated ash, acid number, elemental content, color, and loading point distribution. It is noted that tolerances requirements are not currently assigned to these base stock properties. However, significant variation from the initial characteristics would require an explanation and could possible necessitate partial or complete retesting of a product.

As a result of SAE's recent approval of the new "SF" classification, a proposed MIL-L-46152B revision is now in process to align with new performance requirements. This draft will be circulated to industry in the very near future. To date, we have not obtained any approvals under MIL-L-46152A specification that utilize recycled ingredients. As soon as these types of products are approved, we plan to seek their procurement through the Defense Logistics Agency as further recognition of their utilization within the military.

USED OIL RECYCLING IN THE UNITED STATES

Jerome F. Collins, Ph.D.
Chief, Alternative Materials Utilization Branch
Office of Industrial Programs
Conservation and Solar Energy
U.S. Department of Energy

SUMMARY

Oil recycling firms in the U.S. are, with few exceptions, very small businesses in an industry which has declined sharply over the past two decades. Of the 150 firms refining 300 million gallons per year in 1960, less than ten remain with a capacity now of about 50 million gallons per year. The potential maximum amount of used automotive and industrial oils available for rerefining in the U.S. is about 1.6 billion gallons per year.

A coordinated national effort is underway to revitalize the rerefining industry. We feel that an annual rerefined oil production of 600 million gallons is attainable by the year 2000. Key elements of our effort include (1) the development by the National Bureau of Standards of practical tests to establish the substantial equivalence of rerefined oils with virgin stock oils, (2) action by the Environmental Protection Agency to tighten restrictions on the uncontrolled burning or other disposal of waste oils, (3) commercial demonstration projects supported by the Department of Energy to provide the industry with operating experience in the use of advanced, environmentally sound, rerefining technology, and (4) a program of applied research and development to address the problems of process waste disposal, used oil collection, and market acceptance.

Oil recycling has been practiced in the United States for almost fifty years, beginning in the airlines industry with the development in 1932, by American Airlines, of a closed-cycle oil rerefining system. Rerefined oil was produced in large quantities, however, only after the outbreak of World War II. In 1942 the Army Air Corps adopted a closed-cycle rerefining system similar to the American Airlines system. By 1946 the acid-clay rerefining process, in essentially its modern form, had been universally adopted as the basic technology of the new rerefining industry. Rerefined oil production reached a level of about 100 million gallons annually by the late 1940's. The bulk of this rerefined oil was returned to use as automotive crankcase lubricant.

Generally, the emergence of the rerefined oil industry paralleled the burgeoning growth of the automotive industry. The number of vehicle registrations grew from 33 million in 1940 to 74 million in 1960, when the rerefining industry reached its peak production of over 300 million gallons. At this time there were approximately 150 firms involved in the rerefining business. Perhaps the single most important boost to the industry during this period was the exemption in 1954 by the Federal government of rerefined oil from the six-cent per gallon oil tax. With this competitive advantage over virgin stock lubricants, rerefined oil production more than doubled between 1955 and 1960.

Several factors combined to reverse the growth of oil rerefining after 1960. These included the advent of discount store oil sales directly to "do-it-yourself" oil changers, reducing the quantity of crankcase drained oil which could be collected in bulk. In 1963, the Federal Trade Commission ruled that rerefined oil products must bear a prominent label identifying them as "previously used" oils, while the performance requirements of automotive lubricants continued to increase with greater engine horsepower and operating speeds. In 1964 the Department of Defense issued a procurement ban on rerefined oil products following an engine test failure. In 1968, rerefineries lost the six-cent per gallon tax advantage over virgin stock oil products in a tax ruling on the 1965 Federal Excise Tax Reduction Act. By 1970, annual rerefined oil production had fallen off to about 100 million gallons.

Through the 1970's the industry continued its decline, the number of active firms falling to fewer than ten with a capacity of about 50 million gallons per year. Recognizing that the industry's problem was in part technological, the Federal government initiated the Bartlesville Energy Technology Center oil recycling research and development program in 1971. The Bartlesville work has continued and expanded under the sponsorship of the Department of Energy and is indeed central to our goal of a revitalized rerefining industry.

Three major legislative changes during the 1970's have laid groundwork for a coordinated national effort to revitalize the rerefining industry. In 1975 the Energy Policy and Conservation Act (EPCA) directed the National Bureau of Standards to develop practical tests to establish the substantial equivalence of rerefined oils with virgin stock oils. The Federal Trade Commission must reexamine its mandatory labeling rule in light of these substantial equivalence tests. It is expected that such practical tests to establish substantial equivalence with virgin stock oils would result in large-scale purchases of rerefined oils by state and local, as well as federal, agencies. The Department of Defense has already moved to issue procurement regulations permitting the use of rerefined oil. The Resource Recovery and Reclamation Act (RCRA), passed in 1976, has been the vanguard of increasingly stringent environmental regulations controlling the handling and disposal of used lubricating oils. These controls will increase the cost of burning used oil as fuel, thereby improving the economics of rerefining. Finally, the National Energy Act (NEA) of 1978 provided for an additional ten percent investment tax credit on recycling equipment, and exempted rerefined oil product blends from the six-cent per gallon excise tax, thus restoring a formerly held tax advantage to rerefineries.

The long-term technology development program at the Bartlesville Energy Technology Center, the three major legislative initiatives, and a general resurgence of interest in oil recycling have all brightened the rerefining industry's prospects. Under the guidance of several vital interest groups, including the Association of Petroleum Rerefiners, the National Association of Oil Recovery Coordinators (NAORC), and the Interagency Committee on Oil Recovery and Recycling (ICORR), it is likely that rerefined oil production will increase rapidly over the coming decades. In our role as a coordinator and research, development, and demonstration sponsor, we at the Department of Energy have set a goal of 600 million gallons of rerefined automotive oil production by the year 2000. An industrial oil rerefining program is underway and is expected to have additional dramatic impact.

The current efforts being pursued include a study of rerefinery by-product uses -- for example, an environmentally safe building material made from acid sludge solids. Another effort, recently completed, involved the preparation of an engineering process design, specifications, and environmental information sufficient to serve as a bid package for construction of a solvent treatment, distillation process rerefinery. In addition, a study is underway to quantify the potential used oil recovery from do-it-yourself oil changers, and to identify the incentives necessary to encourage recycling. A project is now being planned to evaluate the viability of advanced lubricant filters which claim to eliminate the need for oil changes altogether.

Other projects address less widely known waste oil sources such as the many hundreds of sludge pits remaining from early petroleum refining operations. An industrial application study is planned to define the potential for oil-water emulsion recycling.

In this paper I have only touched upon some of the very exciting and interesting technical endeavors being carried out for the Department of Energy by the Bartlesville Energy Technology Center. These projects, and other important efforts orginating outside the Department of Energy, will be amplified elsewhere in these proceedings.

Re-refined Lube Oil Consistency and Quality: The Ultimate Question

Donald A. Becker
Recycled Oil Program
National Bureau of Standards
Washington, D.C. 20234

SUMMARY

It is well known that legislation and activities in the United States on the subject of oil recycling have increased dramatically in the past several years. Two bills are currently pending before the Congress (S.2412; H.R. 7833) and the Federal Trade Commission has just initiated actions leading to modification of the controversial labeling requirements for recycled motor oil (16 CFR Part 406; Federal Register, page 55223, August 19, 1980). The U.S. Army has recently issued their revised MIL-L-46152A specification, which permits qualification by re-refined motor oils. Many additional changes in the existing situation are occurring daily in Federal, State and local governments, in professional and trade associations, in both producer and consumer industries, and in the attitudes of the general public.

These changes and modifications are mostly positive in nature, encouraging the more efficient and effective utilization of a valuable natural resource, while minimizing the adverse environmental effects of improper reuse or disposal. However, I am aware that a substantial fraction of both industry and government in the U.S. have some concerns about the lack of scientific and technical data on certain aspects of the quality and consistency of recycled lubricating oils, particular re-refined engine oils. My attempts to locate such data in the U.S., Europe (particularly West Germany and England), India, South Africa and Australia have not met with total success, although Mr. Gairing, Mr. Wolf and the West German re-refiners have been particularly helpful, as has been the South African Bureau of Standards

Since 1976, the (U.S.) National Bureau of Standards (NBS) has had a legislatively mandated program to "...develop test procedures for the determination of substantial equivalency of re-refined or otherwise processed used oil ...wtih new oil for a particular end use" (42 U.S. Code 6363c). The first phase of the NBS program, which recommended a set of test procedures for the operational and environmental needs of recycled oil used as burner fuel, was completed and is published as NBS Technical Note 1130. Work on the second phase, test procedures for re-refined oil to be used as engine oil, was initiated in 1978.

The NBS technical effort is currently focussed on the development and evaluation of test procedures capable of monitoring the quality, consistency and additive response of a re-refined oil basestock in-between qualifications by means of the very expensive engine sequence tests. I feel these evaluated test procedures are crucial in order to allow full accommodation of re-refined oils within the existing API-SAE-ASTM classification system. The NBS research involves identification of the important characteristics which must be monitored, review and evaluation of existing tests for those required characteristics, and development of new or modified test procedures where necessary to adequately monitor an important characteristic or property. Procedures currently under evaluation include physical and chemical tests, hydrocarbon type characterization procedures, and bench scale performance tests for the evaluation of basestock additive response.

Of particular interest may be a recently initiated ASTM/NBS study of basestock consistency, on six re-refined lubricating basestocks and four virgin lubricating basestocks. Samples of these basestocks are submitted to NBS, where they are subdivided and coded, for analysis by over 40 different tests at 13 different laboratories. It is hoped that the results from this study will provide much of the data to indicate the appropriate characteristics necessary to adequately monitor the consistency of a re-refined engine oil.

The chemical property tests include: total acid number; saponification number; nitrogen, sulfur, chlorine and oxygen contents; metals content by emission spectroscopy; sunlight stability; additive compatibility; ethylene glycol content; differential infra-red analysis; nuclear magnetic resonance analysis; and compositional analysis by liquid chromatography and by low resolution mass spectrometry. Bench-scale tests being applied to these basestock consistency study oils include: oxidator oxidation response; extreme pressure wear test; rusting; LUBTOT oxidation test; Sequence IIIC simulator; rotating bomb oxidation test; turbine oil oxidation test; differential scanning calorimetry; and the Ford anti-oxidant titration test. It is hoped that the results from this study will provide much of the data to indicate the appropriate characteristics necessary to adequately monitor the consistency of a re-refined engine oil.

At NBS, in addition to work on the basestock consistency study described above, the Recycled Oil Program is currently evaluating the applicability and validity of the primary physical and chemical test procedures to re-refined oil. For example, we have spent a considerable effort on evaluating the sulfated ash procedure for use at very low ash levels. Additional effort is currently being placed on evaluating the saponification number and on test procedures for chlorine in re-refined basestocks.

We also have a number of research projects on the development of bench-scale performance tests for establishing basestock additive response. In particular, work on bench tests for evaluating the additive response of oils for wear and for oxidation stability appears to be promising, as does a test using differential scanning calorimetry (DSC). Additional bench-test R&D being performed outside NBS includes support for Professor Elmer Klaus' modification of his micro-oxidation test for use with petroleum lubricating oils, and a cooperative effort with the U.S. Army Fuel and Lubricants Research Laboratory on the development of background data and test methodology for engine deposits using engine parts obtained from both engine tests and a fleet test of new and re-refined oils.

NBS has also been active in evaluating hydrocarbon type characterization methods. Our initial long, activated clay columns for separating lube

oil fractions functioned reasonably well, but we are now using a preparative
scale high pressure liquid chromatograph to provide better separation in
much less time. We are particularly interested in polar components from
re-refined oils, including particularly chlorine compounds which are
found in some commercially re-refined U.S. oils at several hundred parts
per million. Some preliminary data indicate significant interactions
between selected separated lube oil fractions and some additives, which
may lead to identification of an additive response component in lubricating
oils. Our research in this area is continuing.

In conclusion, I believe there has been sufficient engine testing
and fleet testing of re-refined oils to establish that: (1) a high
quality re-refined lubricating oil which has been formulated with a high
quality additive package *can* be comparable to a virgin oil similarly
formulated, and (2) that both oils *can* provide adequate performance in
most types of automotive service. What is now required is a technically
accepted method for assuring the consistency of re-refined oil in-
between the engine sequence tests, i.e., a set of evaluated test procedures
for consistency. Development and acceptance of such consistency tests
is considerably more difficult because no such set of tests exists for
virgin lubricating oils. I believe that complete consumer acceptance of
re-refined oils, and thus parity in the marketplace, will not be obtained
until fully accepted test procedures for consistency are established.

Therefore, in my opinion, the Ultimate Question is: How consistent
are re-refined lubricating oils and which tests should be used to monitor
that consistency. While I agree that there are some very good re-
refined lube oils, I believe that, at least for U.S. re-refined oils,
the technical data does not exist at the present time to definitively
answer this question of consistency. However, evidence to date indicates
that this question can be answered, and NBS, along with the many cooperating
organizations and laboratories, is working diligently towards providing
these answers.

DIE TECHNOLOGIE DER ZWEITRAFFINATION
THE TECHNOLOGY OF REGENERATION
LA TECHNOLOGIE DE LA REGENERATION
LA TECNOLOGIA DELLA RIGENERAZIONE

Dr. C. LAFRENZ
Delegierter der Europäischen Kommission der Zweitraffineure beim
technischen Verband der europäischen Schmierstoffindustrie (ATIEL),
Bundesrepublik Deutschland
Delegate from the European Commission for Regeneration to the
Association of European Industry of Lubricants Technicians
(ATIEL), Germany
Délégué de la Commission Européenne de Régénération à l'Association
Technique de l'Industrie Européenne des Lubrifiants (ATIEL), Allemagne
Delegato della Commissione Europea della Rigenerazione presso
l'Associazione Tecnica dell'Industria Europea dei Lubrificanti
(ATIEL), Germania

ZUSAMMENFASSUNG

Die Aufarbeitung von Altölen zu Reraffinaten ist bereits in
vielen Ländern ein Wirtschaftszweig, dessen Bedeutung mit der
Rohstoffverknappung steigen wird. Das Ziel, qualitativ hochwertige Raffinate zurückzugewinnen, wird unter Anwendung sehr
unterschiedlicher Verfahrenstechnik erreicht. Ein vollständiger Vergleich bestehender resp. sich in der Entwicklung be-
findlicher Verfahren kann jedoch nicht auf die Bewertung der
Raffinatqualitäten beschränkt werden. Gleichermaßen müssen
heutzutage alle Hilfsmittel und resultierenden Nebenprodukte
inklusive Prozeßabwässer und die damit verbundenen Umwelt-

schutzaspekte berücksichtigt werden. Ferner ist als Folge der Weiterentwicklung der Schmierstoffe die Zusammensetzung und Verfügbarkeit der Altöle von Bedeutung. Aufgrund fehlender Daten ist jedoch ein detaillierter Vergleich der verschiedenen Verfahren zur Reraffination nicht immer möglich.

Die angesprochenen Verfahren lassen sich entsprechend der gewählten chemischen oder physikalischen Vorbehandlung der Altöle einteilen. Bei der chemischen Vorbehandlung finden stark saure bis stark alkalische Chemikalien Verwendung. Die auf der Solventextraktion oder Vacuumdestillation basierenden Verfahren nutzen physikalische Verfahrensschritte. Die Zukunft wird zeigen, welchem Prinzip langfristig der Vorrang zu geben ist.

SUMMARY

In many countries there is already an industry which specializes in the regeneration of used oils and which will grow in importance as raw materials become increasingly scarce. The objective, which is to recover high-quality raffinates, is attained through the use of widely differing techniques. A full comparison of the processes that already exist or are being developed cannot, however, be confined to an assessment of raffinate quality; consideration must nowadays be given also to the process inputs and the resulting by-products and effluents, as well as to the relevant environmental protection measures. Furthermore, as a result of the further development of lubricants, the composition and availability of used oils has become important. On account of the scarcity of information, however, it is not always possible to make a detailed comparison between different regeneration processes.

The processes concerned can be classified according to the chemical or physical method of used-oil pretreatment selected. In the case of chemical pretreatment, chemicals ranging from strong acids to strong alkalis are used, whereas processes based on solvent extraction or vacuum distillation employ physical methods. Time will tell which principle is to be given preference in the long term.

RESUME

La transformation des huiles usagées en produits régénérés est déjà dans de nombreux pays un secteur économique dont l'importance va s'accroître avec la raréfaction des matières premières. L'objectif qui consiste à récupérer des produits de raffinage de haute valeur peut être réalisé par l'application de techniques très diverses. Une comparaison complète des procédés existants ou en cours de développement ne peut toutefois se limiter à l'évaluation des qualités des produits raffinés. Il importe aujourd'hui de tenir compte également de toutes les matières et des sous-produits qui en résultent, notamment les eaux usées, et de réfléchir aux problèmes qu'ils posent pour l'environnement. Par ailleurs, du fait du perfectionnement des produits lubrifiants, la composition et la disponibilité des huiles usagées prennent de l'importance. Faute de données, il n'est toutefois pas toujours possible de procéder à une comparaison détaillée des différents procédés de régénération.

Ces procédés peuvent être classés en fonction du prétraitement chimique, ou physique appliqué aux huiles usagées. Pour le prétraitement chimique, on utilise des produits chimiques très acides ou très alcalins. Les méthodes fondées sur l'extraction des solvants ou la distillation sous vide appliquent des procédés physiques. C'est l'avenir qui nous dira quel procédé doit recevoir la priorité à long terme.

RIASSUNTO

La trasformazione degli oli usati in prodotti raffinati costituisce già in molti paesi un settore economico la cui importanza aumenterà con la rarefazione delle materie prime. L'obiettivo di ottenere dei prodotti raffinati di alta qualità viene raggiunto con l'utilizzazione di varie tecniche. Un confronto completo di tutti i processi attualmente in fase di sviluppo non può tuttavia limitarsi ad una valutazione della qualità dei prodotti ottenuti. Attualmente devono essere tenuti in considerazione tutti i mezzi ed i sottoprodotti (comprese le acque di processo ed i problemi ambientali che ne derivano). Inoltre come conseguenza del perfezionamento dei lubrificanti è estremamente importante la composizione e la disponibilità degli oli usati. In mancanza di dati non è tuttavia sempre possibile fare un raffronto particolareggiato dei vari processi di rigenerazione.

I processi cui si è fatto accenno possono essere suddivisi in base al trattamento preliminare che può essere chimico o fisico cui vengono sottoposti gli oli usati. In caso di trattamento chimico vengono utilizzate delle sostanze altamente acide o anche altamente alcaline. I processi basati sull'estrazione mediante solventi o sulla distillazione per depressione sono basati su soluzioni fisiche. Il futuro si indicherà quale principio si sarà rivelato più conveniente.

Die Aufarbeitung von Altölen zu Zweitraffinaten ist bereits
ein in vielen Ländern existierender Wirtschaftszweig. Wie
viele andere Recycling-Verfahren wird auch die Altölaufarbeitung an Bedeutung gewinnen, vor allem auch in jenen Ländern,
die kaum über Rohstoffe verfügen. Da heute eine Vielzahl
sehr unterschiedlicher Reraffinationsverfahren existiert bzw.
sich in der Entwicklung befindet, soll im folgenden außer der
Beschreibung ausgewählter Verfahren der Versuch gemacht
werden, gemeinsame Merkmale und Kriterien aufzuzeigen, die
beim Vergleich und bei der Auswahl passender Verfahrenstechnik beachtet werden sollten.

Allgemein betrachtet ist die Reraffination von gebrauchten
Schmierölen ein mineralölverarbeitender Prozeß mit dem Ziel
der maximal möglichen Rückgewinnung qualitativ hochwertiger
Basisöle (sog. Zweit- oder Reraffinate). Die Qualität der
zu erhaltenden Reraffinate muß vergleichbar sein derjenigen
von Erstraffinaten, für die Mindestanforderungen existieren
(in der BRD z.B. DIN 51517: Mindestanforderungen für unlegierte Schmieröle C). Es sind jedoch nicht alle Daten der
Basisöle festgelegt, so daß es der Praxis überlassen bleibt,
die Raffinatqualitäten für verschiedene Einsatzzwecke zu
prüfen und zu akzeptieren. Entscheidend ist, daß diverse
Fertigformulierungen wie Motorenöle etc. den nationalen und
internationalen Anforderungen genügen. Schon bei der Entwicklung neuer Verfahrenstechniken zur Reraffination von
Altöl ist es erforderlich, die Reraffinate - in Kombination
mit kommerziell erhältlichen Additiven - entsprechenden
Prüfstandstests zu unterziehen, sich also nicht auf die
analytische Beschreibung der Basisöle zu beschränken.
Eine vollständige Beurteilung eines Reraffinationsprozesses
ist mit der Bewertung der Raffinatdaten allein nicht möglich.
Zwangsläufig fallen weitere Produkte an, deren Zusammensetzung je nach verwendeter Technik sehr unterschiedlich sein
kann. Teilweise können sie lästiger Abfall, teilweise verkaufsfähige Nebenprodukte sein. Wenn auch die Nebenprodukte
vielfach nicht näher charakterisiert sind, so ist dennoch
das Augenmerk darauf zu lenken, da nur unter Betrachtung

des Bedarfs an Hilfsmitteln und der entstehenden Nebenprodukte eine Abschätzung der Wirtschaftlichkeit eines Verfahrens möglich ist. Schwierigkeiten bei der Verfahrensbewertung treten ferner dadurch auf, daß angegebene Ausbeuten sowie Hilfsmittelverbräuche sich auf unterschiedlich zusammengesetzte Altöle beziehen können.

Da heutzutage auch Umwelt- und Energieaspekte eine immer größere Rolle spielen, müßten Reraffinationsverfahren auch unter diesen Gesichtspunkten verglichen werden. Der unterschiedliche technische Standard der Reraffination sowie die in den einzelnen Ländern zu beobachtenden Weiterentwicklungen sind eng mit den jeweils bestehenden Umweltschutzgesetzen resp. -auflagen verbunden. Andererseits sollte bei der Verfahrensentwicklung und -auswahl sehr frühzeitig darauf geachtet werden, welche Emissions- und Abwasserstandards gelten oder zukünftig gelten werden und welche Entsorgungsmöglichkeiten für Nebenprodukte bestehen.

Zusammengefaßt beeinflussen folgende Faktoren die Verfahrenstechnik der Reraffination:

- Rohstoff Altöl (Zusammensetzung, Verfügbarkeit)
- Ausbeuten Reraffinate
- Qualität der Reraffinate
- Art und Ausbeuten von Neben- und Abfallprodukten
- Entsorgung und Verwendungsmöglichkeiten für Nebenprodukte
- Energiebedarf
- Umweltschutzauflagen (Emissionen, Abwasser).

Bei der Betrachtung des Rohstoffes Altöl ist auch darauf hinzuweisen, daß die diversen Syntheseöle langfristig nicht unbeachtet bleiben können. Ein Vergleich bestehender oder entwickelter Verfahren sollte unter Berücksichtigung obiger Gesichtspunkte erfolgen. Schon aufgrund fehlender Daten und der notwendigen Begrenzung des Themas ist dieser Vergleich im einzelnen nicht möglich, so daß im folgenden nur einige chemische und verfahrenstechnische Aspekte beschrieben werden.

Unter den gegenwärtig praktizierten Verfahren spielt nach wie vor die Reraffination mit konzentrierter Schwefelsäure und Bleicherde eine bedeutende Rolle. Das Verfahren, heute als sog. Kontaktdestillation betrieben, ist schematisch in Abb. 1 dargestellt.

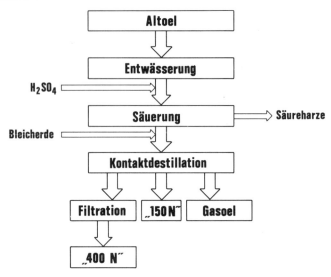

Abbildung 1: Kontaktdestillation

Danach werden ausgewählte Altöle einer Entwässerung unterzogen und anschließend mit konzentrierter Schwefelsäure behandelt. Nach Abtrennen der sich bildenden Säureharze wird das sog. Saueröl mit aktivierter Bleicherde versetzt und auf ca. 300°C unter Vacuum aufgeheizt, wobei Gasöl und Spindelöl abdestillieren. Das Sumpfprodukt des Verdampfers liefert nach Filtration ein Basisöl der Viskosität ca. 50 cSt/50°C (ca. 6,5°E). Bezüglich der Raffinatdaten ist auf die folgenden Präsentationen von Mr. C.J. Ratcliff resp. Herrn E. Wedepohl zu verweisen. Für das Abfallprodukt Säureharz, welches in erheblichem Maße die Wirtschaftlichkeit des Kontaktverfahrens beeinflussen kann, können im wesentlichen folgende Daten angegeben werden:

Typische Säureharz-Inspektionsdaten:

Schwefel:	7,5 - 11 %
Säurezahl:	450 - 550 mg KOH/g
Glührückstand:	ca. 11 %
Metallgehalte	
Zn:	ca. 4.600 mg/kg
Pb:	ca. 900 mg/kg
Kohlenwasserstoffe (Petroläther extrahierbar):	18 - 23 %
Heizwert:	ca. 16.700 - 23.000 KJ/kg
d/15:	1,28 - 1,32 g/cm^3
Viskositäten	
bei 20°C:	ca. 17.000 cP
50°C:	ca. 1.480 cP
90°C:	ca. 80 cP

Besonders in der Bundesrepublik Deutschland werden Säureharze mittlerweile in größerem Maße in der Zementindustrie eingesetzt bzw. durch Verbrennen auf Schwefelsäure und Oleum aufgearbeitet.

Die Effektivität des Säure-Bleicherde-Kontaktverfahrens hängt weitgehend von der Zusammensetzung des eingesetzten Altöles ab. Zu achten ist auf eine sorgfältige Altölauswahl, d.h. Vermeiden z.B. diverser Industrieöle als Feedstock. Desgleichen müssen Syntheseöle wie Phosphatester ausgeschlossen werden. Esteröle wie Adipate etc., für die nach einigen Prognosen [1] steigende Zuwachsraten erwartet werden und daher zukünftig vermehrt in Altölen auftreten dürften, werden nach eigenen systematischen Untersuchungen im Säure-Bleicherde-Verfahren zerstört. Es ist natürlich die Frage zu stellen, ob eine restlose Entfernung wertvoller Ester überhaupt zu fordern ist, wenn damit der Gebrauchswert der Raffinate nicht beeinträchtigt wird.

Aufgrund ihrer chemischen Struktur werden Syntheseöle wie Polyalphaolefine und Alkylaromaten nicht oder nur unwesentlich von Schwefelsäure entfernt und sollten in dem Maße in den

Reraffinaten zu finden sein, wie ihr Einsatz in Schmiermitteln zunimmt. Das gleiche gilt für die kohlenwasserstoffähnlichen OCP-VI-Verbesserer sowie Polybutene, die in kleineren Mengen in nach dem Kontakt-Verfahren hergestellten Zweitraffinaten anzutreffen sind.

Das Bestreben, den Schwefelsäure- und Bleicherdebedarf zu senken, hat in den letzten Jahren zu mehreren Verfahrensvarianten der sog. Wärmebehandlung, auch Thermoschock genannt, geführt. Die Wärmebehandlung des getrockneten Altöles ist ein der Schwefelsäurebehandlung vorgeschalteter Verfahrensschritt, der zu erheblichen Schwefelsäure- und Bleicherdeeinsparungen führen kann. Abhängig ist der Schwefelsäure- und Bleicherdebedarf vom Ausmaß der thermischen Zersetzung der Additive,

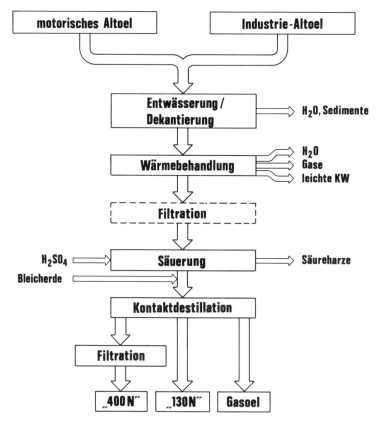

Abbildung 2 : Sopaluna-Verfahren

welcher durch geeignete Wahl der Prozeßvariablen Temperatur,
Verweilzeit und Druck beeinflußt wird.

Eine praktizierte Verfahrensausführung ist das in Frankreich
betriebene Sopaluna-Verfahren [2] , welches schematisch in
Abb. 2 dargestellt ist.

Die Wärmebehandlung wird bei 330 - 360°C und ca. 5 bar ausgeführt. Entstehende Gase und leichte Kohlenwasserstoffe werden
verbrannt und liefern die für die Wärmebehandlung nötige
Energie.

Durch Abänderung der Verfahrensbedingungen auf Temperaturen
von ca. 300°C und verlängerte Verweilzeiten sowie Betreiben
des Reaktors im Normaldruck- oder Unterdruckbereich sind z.B.
in Spanien und im Iran bestehende Kontakt-Destillationsanlagen
um Wärmebehandlungseinrichtungen erweitert worden mit dem
Effekt, daß der Schwefelsäurebedarf angeblich von 10 - 15 auf
4 - 6 % und der Bleicherdebedarf von 6 - 7 auf 2,5 - 3 % gesenkt werden konnte [3] . Vorausgesetzt, thermisches Cracken
der Mineralöl-Kohlenwasserstoffe wird vermieden, können
qualitativ gute, oxidationsstabile Reraffinate erhalten
werden.

Einen etwas anderen Weg der Anwendung des Thermoschock-Verfahrens ist in Frankreich Firma A. Mathys gegangen. Das
Schema des zur Zeit praktizierten Verfahrens ist in Abb. 3
dargestellt [4] .

Danach wird Altöl, welches im wesentlichen motorisches Altöl
mit einem konstanten Anteil ausgewählten Industrieöls darstellt und einen mittleren Wassergehalt von ca. 8 % hat, auf
90°C aufgewärmt und zentrifugiert. Etwa 95 % des vorhandenen
Wassers sowie ca. 80 % der Sedimente werden mit dieser Zentrifuge entfernt. Nach weiterem Erwärmen des verbleibenden Öls
auf 180°C werden das restliche Wasser sowie leichte Fraktionen
destillativ abgetrennt. Der Sumpf der Kolonne wird nach Erhitzen auf 360°C einer unter Vacuum betriebenen Destillations-Kolonne zugeführt, aus der ca. 3 % Gasöl sowie 2 Schmieröl-

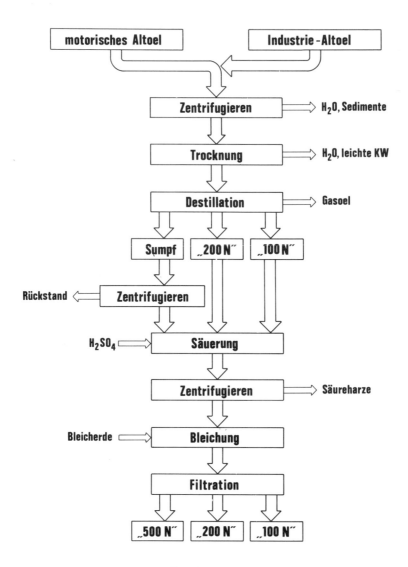

Abbildung 3 : Mathys-Verfahren

destillate (100 N und 200 N) abgezogen werden. Der Sumpf
dieser Kolonne wird kontinuierlich über eine bei 200°C arbeitende Zentrifuge gegeben, wobei ein im Öl unlöslicher Rückstand abgeschleudert wird. Das Zentrifugat sowie die Destillate werden gesäuert und mit ca. 5 % Bleicherde bei ca. 120°C
entfärbt.

Das Thermoschock-Verfahren á la Firma Mathys ähnelt also dem abgewandelten Sopaluna-Verfahren, indem im Vacuum unter Anwendung hoher Temperatur ein Teil des enthaltenen Schmieröls abdestilliert und gesondert konventionell mit Säure/Bleicherde nachraffiniert wird. Bedingt durch die hohe Temperatur werden weitere ölunlösliche Stoffe gebildet, die sich zentrifugieren lassen. Damit ist eine wirksame Säuerung des Zentrifugats bei geringerem Säureverbrauch möglich. Da das Zentrifugat praktisch metallfrei ist, plant Firma Mathys, ab 1982 dasselbe mittels Hydrierung zu raffinieren. Diese Maßnahme würde zu einer weiteren Säureharzverminderung und Raffinatausbeuteerhöhung führen.

Eine diskontinuierliche Form der Reraffination wird demnächst von der Atlas-Oil Recyclers in Irland betrieben, wo bislang im wesentlichen erst eine ausgebaute Erfassung der Altöle existiert [5] . Das Verfahrensschema der Atlas-Oil umfaßt die in Abb. 4 dargestellten Verfahrensschritte.

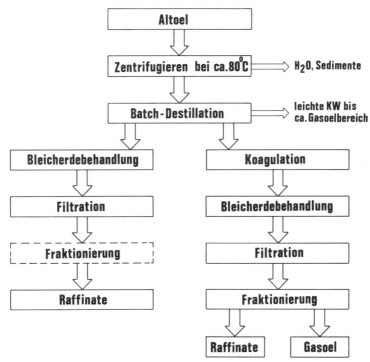

Abbildung 4 : ATLAS OIL RECYCLERS - Verfahren

Im Falle sehr gering verschmutzter Altöle wie gebr. Hydrauliköle ist nach Trocknung und Entfernen leichter Kohlenwasserstoffe eine Bleicherdebehandlung ausreichend. Stärker verschmutzte und höher additivierte Öle werden nach der Batch-Destillation einer Behandlung zur Koagulation der Begleitstoffe unterzogen mit anschließender Bleicherdebehandlung und evtl. Fraktionierung. Da die Batch-Destillation bezüglich Temperatur und Verweilzeit flexibel betrieben werden kann und damit auch Temperaturen im Bereich 300°C und darüber möglich sind, kann diese Destillation auch als Wärmebehandlung angesehen werden. Folglich sind auch in der Nachraffination Einsparungen an Hilfsstoffen zu erwarten.

Ein wesentlicher Teil des Konzepts ist die Altölseparierung, d.h. Übernahme von Altölen bekannter Zusammensetzung und Herkunft und die Abtrennung von Wasser und Sedimenten mittels Zentrifugen zur Senkung der Entwässerungs- und Hilfsmittelkosten.

In dem vorangegangenen Kapitel ist ein Überblick über verschiedene Variationen der in Europa praktizierten Säure-/Bleicherde-Reraffination von Altölen gegeben worden. Vom Prinzip her sind die angesprochenen Verfahren als chemische Raffinationsprozesse anzusehen, da die Reinigung der Altöle, d.h. das Entfernen von Additiven, Umwandlungsprodukten derselben, Oxydationsprodukten etc. meistens mit konzentrierter Schwefelsäure vorgenommen wird. In jedem Falle ist eine Nachraffination mit Bleicherde erforderlich.

Unter dem Gesichtspunkt der chemischen Raffination sind zwei weitere Verfahren zu betrachten, nämlich das von der Phillips Petroleum Co. entwickelte sog. PROP-Verfahren und das RECYCLON-Verfahren der Firma Aseol. Beim PROP-Verfahren [6] [7] [8] , welches bisher in Amerika und Kanada installiert wurde, handelt es sich um einen relativ neuen Prozeß, bei dem Altöle in einer Vorbehandlungsstufe mit Ammoniumsalzen soweit vorgereinigt werden, daß die verbleibende Ölphase anschließend hydriert werden kann. Sowohl die Vorbehandlungsstufe als auch die Hydrierstufe stellen chemische Raffinationsstufen dar.

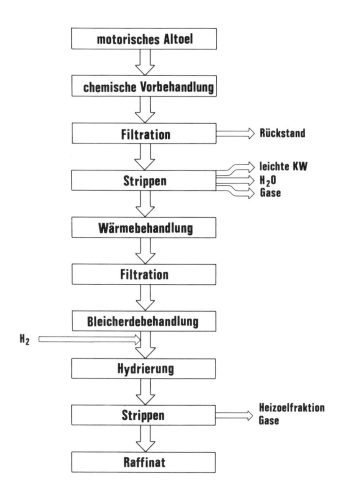

Abbildung 5: PROP-Verfahren

Schematisch umfaßt der PROP-Prozess folgende Stufen (Abb. 5):

Bezüglich der zu erhaltenden Raffinatdaten muß auf die angegebene Literatur verwiesen werden. Aufmerksam zu machen ist in diesem Zusammenhang auf die vergleichsweise geringen Schwefelgehalte der Raffinate [10] . Weiterführende Motorentests [9] zeigen, daß die Basisöle sich gut für die Formulierung hochwertiger Motorenöle eignen. Da polare Komponenten bei geeigneter Wahl der Bedingungen hydriert werden können, ist zu erwarten, daß auch Syntheseölkomponenten wie Esteröle eliminiert werden. Über Nebenprodukte der Vorbehandlung, Abwasserzusammensetzung sowie Off-gas-Zusammensetzung liegen keine Angaben vor.

Beim RECYCLON-Verfahren [11] der Firma Aseol wird zur Vorbehandlung der Altöle die chemische Reaktivität fein dispergierten metallischen Natriums genutzt, welches praktisch ausschließlich mit den polaren Additiven reagiert. Als durchschnittlicher Natriumverbrauch wird 0,8 % angegeben [12].
Nach beendeter Reaktion werden die Schmierölkomponenten im Hochvacuum mittels Dünnschichtverdampfer abdestilliert. Die weitere Fraktionierung kann ebenfalls mittels Dünnschichtverdampfer erfolgen.
Schematisch ist der Prozeß in Abb. 6 dargestellt.

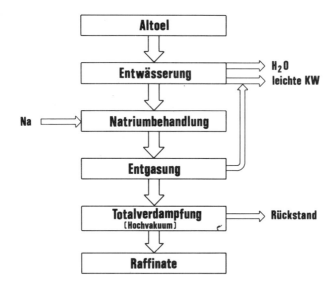

Abbildung 6: RECYCLON-Verfahren

Bezüglich der Raffinatdaten muß auf die angegebene Literatur verwiesen werden. Interessant ist bei diesem Verfahren der Rückstand (ca. 14 % des eingesetzten trockenen Altöls), der zumindest als Heizöl, z.B. im Gemisch mit den gleichzeitig anfallenden Leichtsiedern und Kühlfallenkondensaten, benutzt werden kann. Selbstverständlich enthält der Destillations-

rückstand alle aus den Additiven und Abrieb herrührenden
Metalle zuzüglich der für die Vorbehandlung eingesetzten
und im Rückstand angereicherten Menge Natrium. Demzufolge
liegt der Sulfataschegehalt in der Größenordnung 10 - 13 %.

Die bislang auf Pilotversuchen basierenden Ergebnisse sind
mit normalem motorischen Altöl erhalten worden. Es ist zu erwarten, daß ungünstiger zusammengesetzte Industrieöle zwar
ebenfalls verarbeitet werden können, sicherlich jedoch bei
erhöhtem Natriumverbrauch und geringerer Schmierölausbeute.
Unpolare Syntheseöle wie Polyalphaolefine und Alkylaromaten
durchlaufen die Natriumbehandlung unverändert, während
Esteröle etc. koaguliert werden.

Die bislang beschriebenen Verfahren lassen sich alle unter
dem Gesichtspunkt der chemischen Vorbehandlung der Altöle
betrachten, wobei stark saure bis stark alkalische Koagulier-
mittel Verwendung finden. Bezüglich der großen Zahl beschrie-
bener, hierfür geeigneter Chemikalien kann nur auf die Literatur verwiesen werden [13] [14].

Der Gruppe chemischer Vorbehandlungsverfahren steht eine Reihe
von Verfahren gegenüber, die unter dem Aspekt physikalischer
Trennmethoden gesehen werden können. Man kann im wesentlichen
die zwei Möglichkeiten a) Solventextraktion, b) Vacuumdestillation, wenn von dem IFP-Ultrafiltrationsverfahren [15] abgesehen wird, unterscheiden. Beiden Methoden gemeinsam ist die
physikalische Abscheidung eines Großteils der im Altöl vorhandenen Additive, Metalle, Oxidationsprodukte und Ruß. Ein
bekanntes Beispiel eines auf der Solventextraktion basierenden
Verfahrens ist das in Italien und Jugoslawien praktizierte
IFP-Verfahren, bei dem die Abscheidung der genannten Stoffe
durch Lösen des Altöls in flüssigem Propan bewirkt wird.
Das so vorgereinigte Altöl wird dann mit Säure-Bleicherde
nachraffiniert.

Schematisch ist das IFP-Verfahren in Abb. 7 dargestellt [16].

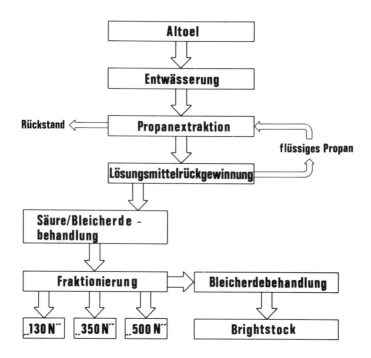

Abbildung 7 : IFP - Verfahren

Folgende Ausbeuten werden angegeben [17] :

Gasöl:	6,0 %
130 N:	ca. 14,0 %
350 N:)	ca. 42,0 %
550 N:)	
Brightstock:	ca. 14,0 %
Rückstand aus Propanfällung:	ca. 6,0 %

Bezüglich der Raffinatdaten darf auf die Literatur verwiesen werden.

Gegenwärtig findet der Rückstand der Propanfällung Verwendung als schweres Heizöl, z.B. im Gemisch mit Gasöl.

Ein besonderes Merkmal des IFP-Verfahrens ist die Rückgewinnung von Brightstock, wenn auch zu erwarten ist, daß sich Reste gewisser VI-Verbesserer in dieser Brightstockfraktion finden und dessen Zusammensetzung ungünstig beeinflussen können.

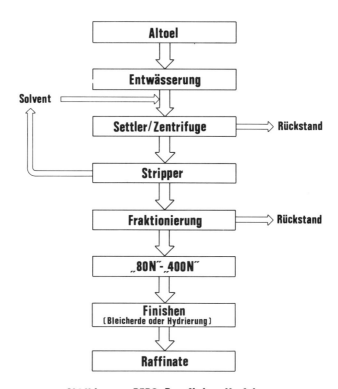

Abbildung 8 : BERC-Rerefining-Verfahren

Das IFP-Verfahren stellt insofern ein interessantes Verfahren dar, als damit frühzeitig der erfolgreiche Versuch unternommen wurde, durch eine der Säuerung vorgeschaltete Behandlungsstufe den Säureharzanfall zu vermindern, dafür ein zumindest als Heizöl verwertbares Nebenprodukt zu erzeugen und die Ausbeuten an Raffinat zu erhöhen. Durch die ebenfalls vom IFP geprüfte Nachraffination der extrahierten Öle mittels Hydrierung kann jeglicher Säureharzanfall ganz vermieden werden [18] [19] [20]

Ein dem IFP-Verfahren sehr ähnliches Verfahrensschema wird von Snamprogetti vorgeschlagen [21] . Es enthält ebenfalls eine Hydrofinishing-Stufe zur Nachraffination. Entsprechend der Literatur [22] bestehen Pläne, eine Anlage in Süditalien zu bauen.

Ebenfalls auf einer Vorbehandlung mit einem geeigneten Lö-

sungsmittelsystem beruht das in Abb. 8 dargestellte sog.
BERC-Verfahren [23] , in welchem in einer 1. Stufe getrocknetes Altöl mit einem Gemisch von Methyläthylketon, 2-Propanol und 1-Butanol vermischt und sich abscheidender Sludge abgezogen wird.

Die verbleibende Phase wird nach Rückgewinnung der eingesetzten Lösungsmittel einer Vacuumdestillation unterworfen, wobei ein zweites Rückstandsprodukt anfällt. Die Destillate werden konventionell mit Bleicherde oder durch Hydrierung nachraffiniert. Raffinatdaten und technische Details sind den vorhandenen Publikationen zu entnehmen.

Einen prinzipiell anderen Weg der physikalischen Trennung der zurückzugewinnenden Kohlenwasserstoffe von begleitenden Additiven, Abrieb, Ölharzen etc. stellt die Hochvacuumdestillation des getrockneten und zweckmäßig bereits von leichten Anteilen befreiten Altöls dar.

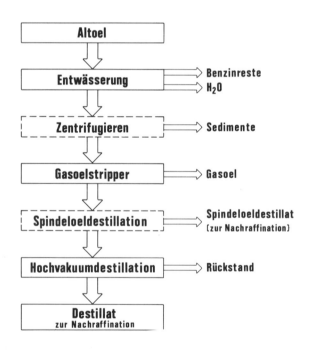

Abbildung 9 : Vakuum - Destillation von Altoel

Schematisch wird die Vacuum-Destillation von Altöl in Abb. 9 veranschaulicht:

Zwischen Gasölabtrieb und Hochvacuum-Destillation kann eine Spindelöldestillation geschaltet sein [24] . Destillat- und Rückstandsausbeuten sind dabei außer vom Altöl selbst von den Verfahrensvariablen Temperatur und Druck abhängig. Anders als beim IFP-Verfahren wird bei der Hochvacuum-Destillation auf eine vollständige Rückgewinnung einer Fraktion mit Brightstock-Viskosität verzichtet. Entsprechend einer Publikation der Motor Oils Refining Co. [25] kann anstelle des Gasölstrippers eine Vorbehandlungsstufe zur Verminderung von Fouling der nachfolgenden Destillationseinrichtungen geschaltet sein. Im Falle der MORCO findet die Vacuum-Destillation mittels Dünnschichtverdampfer Anwendung zur Vorreinigung von gebrauchten Motorenölen aus Diesellokomotiven. Desgleichen wird ein Dünnschichtverdampfer auch bei der Anglo-Pennsylvanian Oil Co. in Schottland verwendet. Über den Einsatz von Dünnschichtverdampfern zur Altöl-Destillation ist auf dem Altölkongreß in Houston berichtet worden [26] [27] . Auch das KTI-Verfahren [28] enthält als einen wesentlichen Verfahrensschritt eine effektive Hochvacuum-Destillation der Mineralölkomponenten selektierter Altöle. In Pilot-Anlage-Versuchen wurden Destillatausbeuten von ca. 90 %, bezogen auf entwässertes Altöl, realisiert. Der gleichzeitig anfallende Destillationsrückstand kann je nach Altöl folgendermaßen charakterisiert werden:

d/15:	0,976
Flammpunkt (COC) °C:	280 - 300
Viskositäten:	
V/70°C (cSt):	ca. 65.000 - 66.000
V/90°C (cSt):	12.300 - 29.300
V/120°C (cSt):	ca. 2.300 - 4.700
Schwefel:	ca. 0,5 %
Sulfatasche:	8,45 - 12,1 %
Blei:	0,66 - 1,15 %
Zink:	0,65 - 1,6 %
Calcium:	1,22 - 3,15 %

Barium: 0,27 - 0,5 %
Heizwert (KJ/kg): ca. 37.000

Derartige, allen Destillationsverfahren mehr oder weniger gemeinsame Rückstände eignen sich als schweres Heizöl bei entsprechender Rauchgasreinigung, als Bitumenkomponente, für die Blei-Zink-Rückgewinnung oder für untergeordnete Schmierzwecke.

Ein verfahrenstechnisch vom Verfahren der MORCO, der Anglo Pennsylvanian Oil Company oder auch vom KTI-Verfahren sich unterscheidendes Verdampferprinzip wurde in den USA von der VACSOL Corp. beschrieben [29] .

Außer beim Recyclon-Verfahren, welches sich ebenfalls bestimmter Dünnschichtverdampfertechnik bedient, bedürfen die Destillate aus Altöldestillationsverfahren wegen unvermeidlicher Additivreste der Nachraffination, welche mit Schwefelsäure/Bleicherde oder Bleicherde allein erfolgen kann.

Eine weitere interessante Möglichkeit der Destillatnachraffination ist die Solventextraktion mit selektiven Lösungsmitteln [30] [31] . Beim KTI-Verfahren schließlich wird zum Finishen der Destillate die Hydrierung vorgeschlagen, wobei die Qualität der Raffinate wiederum von der Wahl der Hydriervariablen beeinflußt werden kann. Vacuum-Destillate aus einer KTI-Pilotanlage liefern sowohl nach Hydrierung wie auch nach konventioneller Nachraffination mit wenig Schwefelsäure und Bleicherde hochwertige Raffinate. Daraus mit kommerziell erhältlichen Additiven formulierte Ein- und Mehrbereichsmotorenöle erbrachten in MWM-B- und Petter W 1 - Prüfläufen sehr zufriedenstellende Resultate.

Im vorangegangenen wurde versucht, in Europa und Amerika bestehende oder entwickelte Rerefining-Technologie unter gemeinsamen Aspekten zu betrachten. Ein vollständiger Vergleich ist nicht möglich, da Ausbeutedaten wegen unterschiedlicher Altöle nicht immer vergleichbar sind, ferner die verfahrenstechnisch bedingten Nebenprodukte, Art und Umfang potentieller Emittenten sowie diverser Abwässer in den wenigsten Fällen beschrieben und quantifiziert sind. Die Erfahrung zeigt, daß den Nebenprodukten und Umweltaspekten große Aufmerksamkeit zu widmen ist, da diese die Reraffination kostenmäßig stark beeinflussen können. Aufgezeigte Verfahrensentwicklungen folgen z.T. unmittelbar aus dem Bestreben, z.B. den Anfall solcher Abfallprodukte wie Säureharze zu vermindern oder ganz zu eliminieren.

Die zahlreichen bestehenden oder in der Literatur beschriebenen Verfahren lassen sich entsprechend der gewählten chemischen oder physikalischen Vorbehandlung der Altöle charakterisieren. Die Zukunft wird zeigen, welchem Prinzip auf die Dauer der Vorrang zu geben ist. Festzustellen bleibt, daß die Zweitraffination sich verfahrenstechnisch dem Rohstoff Altöl anpassen muß, wenn ein Optimum hochwertiger Reraffinate bei gleichzeitiger Minimierung des Energiebedarfs und der Umweltbelastung zurückgewonnen werden soll.

Literatur

[1] WILLIAMSON, E.I.: "Raw Materials for Synthetic Lubricants", Symposion "Chemicals for Lubricants & Functional Fluids", London, Nov. 8, 1979

[2] Dt. Patentschrift Nr. 2408240 vom 23.8.79; Franz. Patentschrift Nr. 811171 vom 15.3.74

[3] Private Mitteilungen von Firma Meinken, D 4358 Haltern, 1980

[4] Private Mitteilung von Firma Mathys, Paris; Mai 1980

[5] Private Mitteilung von Firma Atlas Oil Recyclers, Dublin; July 1980

[6] US Patent No. 3879282; April 1975
US Patent No. 3930988; January 1976
US Patent No. 4151072; April 1979

[7] LINNARD, R.E.; HENTON, C.M.: "Re-Refine Waste Oil with PROP"; Hydrocarbon Processing, Sept.79, p. 148

[8] WHISMAN, M.L.: "New Re-Refining Technologies in the Western World", Lubr.Eng. Vol. 35, No.5; May 79, p. 249

[9] LINNARD, R.E.: "PROP Re-Refined Oil-Engine Test Performance", Joint Conference on Measurements and Standards for Recycled Oils/Systems, NBS, October 23, 1979

[10] LINNARD, R.E.: "Phillips Re-Refined Program", Proceedings of the 3rd International Conference on Waste Oil Recovery and Reuse", Houston Oct. 16-18, 1978, p. 132

[11] ERDWEG, K.J.: "Recyclon - A New Process to Revert Spent Oils into Lubricants", Proceedings of the 3rd International Conference, Houston Oct. 16-18, 1978, p. 99

[12] Private Mitteilung Firma ASEOL, CH 3001 Bern, Juni 1980

[13] HESS, L.Y.: "Reprocessing and Disposal of Waste Petroleum Oils", Noyes Data Corporation 1979

[14] COTTON, F.O.: "Waste Lubricating Oil - An Annotated Review", United States Department of Energy, BETC/IC - 79/4

[15] AUDIBERT, F., et.al.: "Reclaiming of Spent Lubricating Oils by Ultrafiltration", Proc. of 3rd International Conference on Waste Oil Recovery, Houston, Oct. 16-18, 1978, p. 109

[16] VU QUANG, et.al.: "Rerefining Uses Propane Treat", Hydrocarbon Proc., April 1974, p. 129

[17] Private Mitteilung Firma Visculube, Italien, Juli 1980

[18] AUDIBERT, F.: "Used Lubes Rerefined - The Today IFP-Process", IFP-Report No. 26372

[19] VU QUANG, D., et.al.: "Spent Oil Reclaimed without Acid", Hydrocarbon Proc., Dec. 1976, p. 130

[20] BONNIFAY, P., DUTRIAY, R.: "A New Process for Reclaiming Spent Lubricating Oils", National Petroleum Refiners Association, Publication RN 1548 (T & L - 72 - 51), Sept. 1972

[21] ANTONELLI, S.: "Spent Oil Rerefining - A Proposed Technique", Petroleum Times, Sept. 17, 1976, p. 25, Ferner: Proceedings of the 3rd International Conference on Waste Oil Recovery and Reuse, Houston, Oct. 16-18, 1978, p. 121

[22] Chem.-Eng., July 3rd, 1978, p. 20C

[23] BRINKMAN, D.W. et. al., Department of Energy, Bartlesville Energy Technology Center, Bartlesville OK; BETC/RI-76/11; - 77/11; - 77/19; - 78/11; - 78/20; - 79/14; US Patents 4073719 und 4073720

[24] Private Mitteilung Anglo Pennsylvanian Oil Company Ltd., Glasgow, Scotland, July 1980

[25] BLATZ, F.J., PEDALL, R.F.: "Re-Refined Locomotive Engine Oils and Resource Conservation, Lubr.Eng., Nov. 1979, p. 168

[26] BISHOP, J., ARLIDGE, D.: "Recent Technology Development in Evaporative Re-Refining of Waste Oil", Proc. of 3rd Int.Conf. on Waste Oil Recovery and Reuse, Houston, Oct. 16-18, 1978, p. 137

[27] PAULEY, J.F.: "Thinfilm Distillation as a Tool in Re-Refining Used Oil", Proc. of 3rd Int.Conf. on Waste Oil Recovery and Reuse, Houston, Oct. 16-18, 1978, p. 151

[28] WESTERDUIN, R.F., HAZEWINKEL, J.H.O.: "Reraffinage van Gebruikte Smeerolien, Overwegingen en Rentabiliteit" Polytechnische Tijdschrift - Procestechniek 34, 1979, Nr. 5, S. 264-273

[29] Franz. Offenlegungsschrift Nr. 7716971, 3.6.1977 Dt. Offenlegungsschrift Nr. 2725132 (Anmeldetag: 3.6.77), US Patent 4140212, Febr. 20, 1979

[30] US Patent 4021333, May 3, 1977

[31] US Patent 4071438, Jan. 31, 1979

QUALITÄT VON ZWEITRAFFINATEN, HERGESTELLT NACH MODERNEN VERFAHREN
THE QUALITY OF PRODUCTS REGENERATED IN ACCORDANCE WITH MODERN
TECHNIQUES
QUALITES DES PRODUITS DE REGENERATION FABRIQUES SELON LES TECHNIQUES
MODERNES
QUALITA DEI PRODOTTI DELLA RIGENERAZIONE FABBRICATI SECONDO LE
TECNICHE MODERNE

C.J. RATCLIFFE
Operations Director, Anglo-Pennsylvanian Oil Comp. Ltd.,

United Kingdom

SUMMARY

Regenerated base oils from a variety of European sources were analysed to assess their suitability for use in lubricating oil. No attempt has been made to compare the results on a grade to grade basis. The samples taken from normal production materials are representative of regenerated oils currently available. Three samples were taken from pilot plants investigating modified processes, of these two used the acid/clay process for finishing and the other used an unspecified process which does not include the use of sulphuric acid.

The analytical results are tabulated as follows :-

Properties affected by physical separation processes.
Properties affected by finishing processes.
Properties determined by feedstock selection.
Oxidation characteristics.

The results of the analyses are discussed in the context of selection and suitability of the regenerated base oils for various uses in formulation of lubricating oils.

It is evident from the results that all the regenerated oils examined are paraffinic based and are suitable for a wide range of lubricating oils requiring this type of base.

ZUSAMMENFASSUNG

Zweitraffinat-Öle aus verschiedenen europäischen Quellen sind analysiert worden, um ihre Eignung zur Verwendung als Schmieröle festzustellen. Ein Versuch eines qualitativen Vergleichs der Ergebnisse wurde nicht unternommen. Die Proben von Stoffen aus normaler Produktion sind repräsentativ für die laufend verfügbaren Zweitraffinate. Drei Proben wurden in Pilotanlagen zur Untersuchung geänderter Verfahren entnommen, wovon bei zweien das Säure-Erde-Verfahren zur Endbearbeitung angewandt wurde; die dritte Probe wurde nach einem nichtgenannten Verfahren bearbeitet, bei dem kleine Schwefelsäure verwendet wurde.
Die analytischen Ergebnisse sind wie folgt eingeteilt:
- von physikalischen Trennverfahren beeinflusste Eigenschaften
- von Endbearbeitungsverfahren beeinflusste Eigenschaften
- durch die Auswahl des Brennstoffs beeinflusste Eigenschaften
- Oxidationseigenschaften

Die Ergebnisse der Analyse werden im Hinblick auf die Auswahl und Eignung wiederaufgearbeiteter Öle für verschiedene Zwecke zur Herstellung von Schmierölen untersucht.
Es geht klar aus den Ergebnissen hervor, dass alle untersuchten Zweitraffinate eine Alkanbasis haben und sich für eine grosse Zahl von entsprechende Schmieröltypen eignen.

RESUME

Les huiles de base régénérées à partir de diverses sources européennes ont été analysées afin d'évaluer si elles peuvent être utilisées comme huiles de graissage. Aucune tentative n'a été faite pour comparer les résultats obtenus pour les différents grades.
Les échantillons prélevés sur la production normale sont représentatifs des huiles régénérées actuellement disponibles. Trois échantillons proviennent d'installations pilotes qui expérimentent de nouveaux procédés; deux ont utilisé le procédé acide/argile pour le finissage et l'autre un procédé non spécifié qui n'implique pas l'utilisation d'acide sulfurique.

Les résultats des analyses sont classés comme suit:
propriétés affectées par les procédés de séparation physique;
propriétés affectées par les procédés de finissage;
propriétés déterminées par la sélection de la charge caractéristique
d'oxydation.
Les résultats des analyses sont examinés dans l'optique de la sélection
et de l'aptitude des huiles de base régénérées pour différentes utilisations dans la formulation d'huiles de graissage.
Les résultats prouvent que toutes les huiles régénérées étudiées sont à
base de paraffine et peuvent servir à produire une gamme très large
d'huiles de graissage exigeant ce type de base.

RIASSUNTO

Sono stati analizzati gli oli di base rigenerati di varie origine
europee per valutare la loro adeguatezza per l'impiego come oli lubrificanti. Non è stato tentato un raffronto tra i risultati sulla base
della loro qualità. I campioni prelevati dai normali materiali di
produzione sono rappresentativi degli oli rigenerati attualmente disponibili. Sono stati presi tre campioni dagli impianti pilota che sperimentavono processi modificati, tra cui due ricorrevano al processo basato su
acido/argilla per la finitura e l'altro usava un processo non meglio
specificato con esclusione dell'acido solforico.
I risultati analitici sono riportati in una tabella comprendente :
 - caratteristiche influenzate dai processi di separazione fisica,
 - caratteristiche influenzate dai processi di finitura,
 - caratteristiche determinate dalla scelta della carica,
 - caratteristiche di ossidazione.
I risultati delle analisi sono esaminati nel contesto della scelta e
della adeguatezza degli oli di base rigenerati per vari impieghi nella
composizione di oli lubrificanti.
Emerge dai risultati che tutti gli oli rigenerati esaminati sono a base
paraffinica e adatti per una vasta gamma di oli lubrificanti che
richiedono questo tipo di base.

When considering the quality of any material which may be used either in its basic form or used in combination with other materials to form an end product, one must be aware that any measurement of quality will be dependent on the intended use of the product. Also when examining a range of basically similar products from a variety of sources one must consider that the producer may have tailored his particular product for a specific use or section of the normal market. It is for this latter reason that no attempt has been made to compare the test samples on a grade to grade basis.

For the test programme, a total of 23 samples were received from 10 sources as follows :-

West Germany	9 samples	3 sources
France	4 samples	2 sources
United Kingdom	3 samples	1 source
Italy	6 samples	3 sources
Unknown	1 sample	1 source

In the majority of cases, two or more viscosity grades were supplied from each source, but in three cases only a single grade was supplied.

TABLE I

Shows the physical properties of the test samples and although the samples cover roughly the whole lubricating oil range of viscosity, samples from individual sources consisted mainly of one "light" and one "heavy" grade. This is the pattern that could be expected from European sources where most regeneration plants use the hot contact distillation process.

TABLE I

PROPERTIES AFFECTED BY PHYSICAL SEPARATION PROCESSES

Sample	Kinematic		Viscosity Index	Specific Gravity 15.5°C	P.M. Flash Point		P.M. Fire Point °C
	Viscosity 40°C	mm²/sec 100°C			Open °C	Closed °C	
1	19.4	4.18	120	0.868	170	174	194
2	19.85	4.05	101	0.862	188	196	216
3	22.34	4.57	120	0.866	200	218	230
4	22.96	4.35	93	0.870	190	204	218
5	23.97	4.69	114	0.868	204	210	226
6	31.40	5.38	105	0.874	176	216	224
7	32.72	5.45	101	0.874	176	216	224
8	36.35	6.24	121	0.882	196	204	240
9	40.62	6.5	111	0.876	208	212	235
10	54.06	7.51	100	0.881	218	230	258
11	57.21	7.74	98	0.883	200	230	250
12	62.39	8.34	103	0.878	240	258	268
13	71.52	9.40	108	0.882	212	240	254
14	76.93	9.79	106	0.883	218	226	252
15	80.36	10.34	111	0.883	210	246	260
16	84.37	10.55	108	0.882	218	240	262
17	86.32	10.77	109	0.879	218	230	268
18	88.89	10.76	105	0.886	216	244	260
19	91.89	11.14	107	0.874	212	238	254
20	92.16	11.07	106	0.884	220	246	254
21	96.28	11.35	105	0.886	218	240	262
22	104.67	11.68	99	0.885	240	262	280
23	433.79	29.77	97	0.900	266	302	330

Viscosity Index – the value showed more variation than expected. Regeneration can lead to a slight increase in VI, but the variation is of such a magnitude that one must suspect either that small amounts of VI improving polymers have been left in the regenerated oil or have been added to the oil after regeneration. It was not possible to detect the presence of reasonable amounts of polymers using infra red analysis, however as the concentration of polymers is likely to be very small they will not be easy to detect.

Flash and Fire Points – the values were as could be expected within the viscosity ranges and at a level suitable for virtually all uses.

TABLE II

Lists the properties that are largely dependent on finishing processes.

Colour itself is really no measure of quality, however, oils should be clear and bright and consistent in colour, as this would be a simple indication to a user that the product is consistent in quality.

Acidity is important in that it may well affect the additive requirements for the control of corrosion, oxidation, etc. The values recorded were uniformly low and at these levels would not cause any problems for blenders.

Copper corrosion – this test gives an excellent indication of the degree of removal of the various reactive sulphur compounds resulting from the breakdown of additives found in used oils. For lubricating oils the presence of reactive sulphur could cause rapid corrosion of copper or silver alloy bearings, etc., and hence serious equipment failure.

Of the 23 samples listed, 20 were completely free of reactive

TABLE II

PROPERTIES AFFECTED BY FINISHING PROCESSES

SAMPLE	COLOUR	ACIDITY Mgm/ KOH/gm	OXIDE ASH % wt	METALS ppm			COPPER CORROSION	DEMUL. NO.
				LEAD	CALCIUM	ZINC		
1	2.0	0.1	0.002	3	1	2	4a	60
2	2.5	0.05	0.003	7	3	ND	1a	300
3	1.5	0.05	0.02	2	9	6	1a	240
4	2.5	0.05	0.007	8	ND	ND	2c	60
5	2.0	0.05	0.002	3	8	10	1a	1200 +
6	2.0	0.05	0.03	2	3	ND	1a	510
7	2.0	0.05	0.03	10	ND	ND	1a	90
8	3.0	0.10	0.03	8	22	10	1a	1200 +
9	2.0	0.05	0.04	7	1	2	1a	1200 +
10	2.5	0.05	0.002	2	1	ND	1a	390
11	2.0	0.10	0.003	ND	1	ND	1a	60
12	2.0	0.05	0.004	5	4	5	1a	120
13	2.5	0.05	0.06	ND	13	15	1a	1200 +
14	3.0	0.05	0.04	8	12	6	1a	1200 +
15	2.5	0.10	0.005	8	1	ND	1a	1200 +
16	3.0	0.10	0.008	4	8	ND	1a	1200 +
17	2.5	0.05	0.005	9	ND	ND	1a	210
18	3.5	0.10	0.110	12	258	9	2c	1200 +
19	3.5	0.05	0.036	2	10	2	1a	1200 +
20	3.0	0.10	0.02	7	22	5	1b	1200 +
21	2.5	0.05	0.02	8	7	7	1a	1200 +
22	2.5	0.05	0.003	11	5	5	1a	150
23	4.5	0.05	0.08	9	189	5	1a	1200 +

ND = NOT DETECTED

sulphur, 2 gave slight corrosion but one sample gave severe corrosion. Based on this test, therefore, one sample would be quite unsuitable for use in lubricating oils. However, this sample was a light spindle oil and would be suitable for use in the formulation of metal working lubricants requiring the presence of reactive sulphur compounds. Very small amounts of reactive sulphur can be controlled by additives, but in general the refiner should aim to effect complete removal during refining.

Demulsification Number – this test demonstrates the ability of an oil to separate water, the result, expressed as a number, indicates the rate at which emulsified water will separate from a sample under specified conditions. The lower the number the faster the rate of separation.

The requirement for separation of water varies widely in each product use. For example, oils for use in compressors, turbines, and hydraulic systems should have good water separation characteristics, whilst oils for use in rock drill lubricants should have the ability to retain water in stable emulsion form. It would seem, therefore, that provided the end use is known the regenerator need not be concerned if the demulsification number of the regenerated oil is high. However, if the regenerated oil is to have the widest possible range application a low demulsification number is desirable.

Demulsification number can be adversely affected by the presence of even very small amounts of additives, acidity and oxidation products and one would expect therefore that this property would be determined by the efficiency of the finishing process in removing those contaminants. However, in general the results indicated that the lowest demulsification numbers were obtained from those products that had been distilled.

Ash Content – gives a simple and direct indication of the presence of metals and other ash forming materials in the

TABLE III

PROPERTIES DETERMINED BY CRUDE SELECTION

SAMPLE NO.	POUR POINT °C	SULPHUR % wt	ANILINE POINT °C	HYDROCARBON ANALYSIS %		
				C_A	C_P	C_N
1	− 14	0.56	95.0	7.95	56.32	35.73
2	− 16	0.44	97.4	8.34	59.45	32.21
3	− 18	0.50	98.8	6.93	60.47	32.60
4	− 24	0.46	95.6	7.16	55.96	36.87
5	− 20	0.36	96.4	7.27	56.43	36.3
6	− 12	0.62	97.8	6.87	57.59	35.54
7	− 10	0.60	98.2	5.81	56.93	37.27
8	− 32	0.41	99.2	7.91	57.25	34.85
9	− 14	0.70	99.4	6.55	58.23	35.23
10	− 30	0.37	103.1	5.96	56.29	37.76
11	− 12	0.76	103.6	5.82	57.39	36.79
12	− 12	0.61	107.6	5.57	59.77	34.67
13	− 14	0.52	105.6	6.77	56.3	36.94
14	− 22	0.68	105.2	6.84	56.95	36.21
15	− 20	0.52	105.5	6.44	57.29	36.27
16	− 24	0.67	107.8	5.95	63.9	30.15
17	− 12	0.54	104.2	6.37	59.89	33.74
18	− 16	0.71	107.8	5.25	57.80	36.94
19	− 22	0.15	107.5	6.83	54.07	39.10
20	− 16	0.67	110.1	6.09	60.93	32.98
21	− 14	0.61	107.0	6.07	56.82	37.11
22	− 16	0.66	111.7	6.63	60.65	32.72
23	− 12	1.06	120.0	6.13	59.79	34.07

regenerated oil. Comparison of the ash content of the waste oil feed with the ash content of the regenerated oil gives an indication of the efficiency of that part of a process designed to remove these components. Only three of the test samples showed a significant ash content, i.e. Numbers 13, 18 and 23.

Metals Content - metals are used in the preparation of many oil additives, Calcium, Barium, and Zinc, being the most widely used. Lead is also a component of some gear oil formulations but most of this metal found in used oil is derived from gasoline residues. Other metals may contaminate used oil as wear and corrosion debris. Iron, Chromium, Manganese, Tin, etc., may all be found.

Concentrations of lead, calcium and zinc are listed for each sample, traces of other metals were detected but were so low that they were not worth recording. Only samples 18 and 23 showed any significant metal concentration and in both samples this was confined to calcium. This high level of calcium would limit the use of these oils to a certain extent.

For example high quality engine oils have fairly low limits for ash content and therefore if high ash base oils were used in formulations against these grades, insufficient additives could be used to meet the performance requirements.

TABLE III

Lists the properties determined largely by selection of waste oil feed.

The Hydrocarbon Analysis - shows all the oils to be paraffinic based and this is perhaps not surprising as most of the waste oil arisings come from crankcase drainings or circulation systems, i.e. systems that require oils having good stability and oxidation resistance. The aromatic hydrocarbon contents

TABLE IV

OXIDATION CHARACTERISTICS

SAMPLE NO.	CONRADSON 'A' % wt	CONRADSON 'B' % wt	EVAPORATION LOSS % wt
1	0.01	0.93	16.4
2	0.035	0.71	16.9
3	0.03	1.12	17.2
4	0.016	1.02	18.1
5	0.14	1.05	16.7
6	0.07	2.08	18.3
7	0.02	2.36	17.0
8	0.08	3.5	20.9
9	0.074	1.08	17.1
10	0.055	1.24	9.5
11	0.08	1.95	7.9
12	0.17	1.14	5.2
13	0.16	2.26	13.2
14	0.17	1.87	10.6
15	0.14	2.82	8.8
16	0.17	3.65	12.1
17	0.22	1.88	6.9
18	0.32	3.16	10.6
19	0.195	3.15	11.7
20	0.18	2.18	7.6
21	0.16	1.64	9.4
22	0.16	1.40	4.2
23	0.87	2.55	1.4

of all the samples were in the range of 6% to 8%.

Considering the low aromatic hydrocarbon content in conjunction with the aniline points which were in the range 95°C to 120°C one would expect all the oils tested to show good compatability with the elastomatic sealing materials used in lubricating oil systems.

Sulphur Contents of the samples were all fairly low, and even if the reactive sulphur shown by the copper corrosion test was removed, the reduction in total sulphur content would be marginal, and would be unlikely to affect the range of applications.

Pour Point - the results of this test show a wide variation, this could be due to the following factors :-

> Pour point depressant additives are widely used in blended oils, the additives may be hydrocarbons and are extremely difficult to remove. The presence of even very small amounts of these additives remaining in the regenerated oil will lower the pour point.

> It is common practice to add pour point depressing additives to regenerated oils immediately after production.

TABLE IV

Shows the values for carbon residue formation before and after oxidation. The difference between the A and B figures is a measure of the resistance to oxidation. The lower the figure, the better the oil stability. The oxidation test was uncatalysed but was still severe. Of all the tests carried out these results show the greatest variation and vary from very good (sample 2) to very poor (samples 8 and 16).

CONCLUSIONS

The test results indicate that the general quality of the oils tested is good, with only four exceptions, i.e. samples 1, 4, 18 and 23, the oils could be used in a very wide range of lubricating oil applications, and with suitable additive treatment could be expected to meet even the most severe requirements.

Because of the presence of reactive sulphur compounds, samples 1, 4 and 18 would have a more limited application. However, it must be remembered that these oils are currently being sold and therefore must be considered suitable for their intended application.

Sample 23 would have restricted use in high quality hydraulic and engine oil formulations, but because of its high viscosity, it would probably be diluted with lighter oil and therefore the high calcium content would not be so critical.

The regenerated oils listed have been produced by a variety of processes, but it would not be possible from the results to say that any one process produces oil of markedly better quality than any of the other processes.

From this limited survey, involving single samples of each grade, it is not possible to comment on the consistency of quality of regenerated oils.

APPENDIX I
LIST OF TEST METHODS USED

TITLE	I.P.NO.	ASTM NO.	OTHER
Kinematic Viscosity	71/79		
Viscosity Index	224/78	D 2270-75	
Specific Gravity	160/68		
PM Flash Point Closed	34/75		
PM Open Flash and Fire Point	35/63		
Colour		D 1500	
Total Acidity	1/74		
Ash	4/75	D 482-74	
Metals Content	308/74 T		
Copper Corrosion	154/78	D 130-75	
Demulsification Number	19/76		
Pour Point	15/67	D 97-66	
Sulphur	63/55		
Aniline Point	2/78	D 611-78	
Hydrocarbon Analysis			Ford EU-AJ 51-1
Conradson Carbon	13/78		
Metals	308/74 T		
Oxidation	48/78		

ERGEBNISSE EINER VERGLEICHENDEN QUALITÄTSUNTERSUCHUNG VON ERST- UND
ZWEITRAFFINATEN
RESULTS OF COMPARATIVE RESEARCH INTO THE RESPECTIVE QUALITIES OF
REGENERATED OILS AND NEW OILS
RESULTATS DE RECHERCHES COMPARATIVES PORTANT SUR LES QUALITES
RESPECTIVES DES HUILES REGENEREES ET DES HUILES NEUVES
RISULTATI DELLE RICERCHE COMPARATE SULLE QUALITA RISPETTIVAMENTE
DEGLI OLI RIGENERATI E DEGLI OLI NUOVI

Dipl.-Ing. E. WEDEPOHL
Institut für Erdölforschung, Hannover
Prof. Dr.-Ing. W.J. BARTZ
Techn. Akademie Esslingen, Ostfildern (Nellingen)
Dr.-Ing. K. MÜLLER
Ciba Geigy AG, Basel

ZUSAMMENFASSUNG

Mit Erst- und Zweitraffinaten wurden Laboratoriumsversuche an Grundölen sowie motorische Prüfläufe mit Motorenölen, hergestellt aus diesen Grundölen durch Zugabe eines Additiv-Packages, durchgeführt.

Mit den Laboratoriumsuntersuchungen wurden die Verdampfungs- und Oxidationseigenschaften der Grundöle ermittelt. Für die Bestimmung des Verdampfungsverlustes wurden die Verfahren Wolf-Streifentest, Siede-Analyse nach Große-Oetringhaus und die Thermo-Analyse eingesetzt, wobei sich die Ergebnisse für die Zweitraffinate nicht durch Besonderheiten auszeichneten.

Die Versuche zur Ermittlung der Oxidationsstabilität mit den
Verfahren Wolf-Streifentest und Thermo-Analyse führten zu
keiner Korrelation der Ergebnisse. Für die motorischen Untersuchungen wurden dann Erst- und Zweitraffinate ausgewählt,
die - mit einem handelsüblichen Additiv-Package zu Motorenölen aufgemischt - im MWM-Prüfmotor getestet wurden.

Die bei reduzierter Prüfdauer im MWM-A durchgeführten motorischen Prüfungen lieferten den Nachweis der Ansprechbarkeit
auf Additive aller Erst- und Zweitraffinate. Ein unterschiedliches Verhalten der Zweitraffinate konnte aus diesen Ergebnissen nicht abgeleitet werden. Die Überprüfung der Leistungsfähigkeit der vollformulierten Motorenöle aus Erst- und Zweitraffinaten im MWM-B nach DIN 51 361 führte zu dem Ergebnis,
daß Zweitraffinate ebenso wie Erstraffinate zu leistungsfähigen Motorenölen aufgemischt werden können. Ein Einfluß speziell des Raffinattyps auf die erzielten Kolbennoten konnte
nicht nachgewiesen werden.

SUMMARY

Laboratory tests were carried out on both new and regenerated "regular"
oils and engine test runs were performed with additive-containing motor
oils derived from them.
The evaporation and oxidation properties of the regular oils were determined in the laboratory tests. The Wolf method, the Grosse-Oetringhaus
boiling analysis method and the thermal analysis method, were used to
determine the evaporation loss, for which the results obtained with regenerated raffinates did not differ substantially from those obtained with
new raffinates. Tests to determine the oxidation stability by means of
the Wolf method and by thermal analysis did not make it possible to establish a correlation between the results. New and regenerated raffinates
with an admixture of a standard commercial motor oil additive package were
then selected for testing in the MWM test engine.
Short tests conducted with the MWM-A engine demonstrated that both new
and regenerated raffinates responded to additives. No difference in behaviour between new and regenerated raffinates could be deduced from these
results. Tests conducted in the MWM-B engine in accordance with DIN 51 361
in order to determine the performance of full-specification motor oils produced from new and regenerated raffinates led to the conclusion that both
new and regenerated raffinates can be worked up into high-performance motor oils. There is no evidence that the type of raffinate has a specific
influence on the piston performance data obtained.

RESUME

Des essais de laboratoire sur des huiles de base ont été effectués sur des huiles régénérées et des huiles neuves ainsi que des essais de moteurs sur des huiles moteur produites à partir de ces huiles de base par l'ajout d'un certain nombre d'additifs.

Ces essais de laboratoire ont permis de déterminer les propriétés d'évaporation et d'oxydation des huiles de base. Pour déterminer la perte par évaporation, le test de Wolf-Streifen, l'analyse d'ébullition de Grosse - Oetringhaus et l'analyse thermique ont été utilisés; on n'observe pas de résultats particuliers pour les essais sur les huiles régénérées. Les essais visant à déterminer la stabilité à l'oxydation à l'aide du test Wolf-Streifen et de l'analyse thermique n'ont pas permis d'établir une corrélation des résultats. Des huiles neuves et des huiles régénérées ont en outre été choisies, mélangées avec un ensemble d'additifs courants pour en faire des huiles moteur, pour être essayées dans le moteur expérimental MWM.

Les essais effectués pendant une durée réduite sur le moteur MWM-A ont apporté la preuve de la sensibilité de toutes les huiles neuves et régénérées aux additifs. Les résultats ne permettent pas de déduire un comportement différent des huiles régénérées. La vérification de l'efficacité des huiles moteur entièrement formulées produites à partir d'huiles neuves et régénérées dans le MWM-B conformes à DIN 51 361 a permis de conclure que les huiles régénérées peuvent, comme les huiles neuves, donner des huiles moteur efficaces. Il n'a pas été possible de démontrer l'influence particulière du type de produit sur les résultats obtenus au fond du piston.

RIASSUNTO

Con gli oli ottenuti per raffinazione e per rigenerazione sono state svolte prove di laboratorio su oli di base e prove di funzionamento con oli per motori ottenuti da tali oli di base con l'aggiunta di additivi.

Per determinare le perdite per evaporazione sono stati utilizzati i processi Wolf-Streifentest, Siede-Analyse del Grosse-Oetringhaus e la termoanalisi ; i risultati relativi agli oli rigenerati non hanno messo in luce alcuna particolarità. Le prove per determinare la stabilità all'ossidazione con il processo Wolf-Streifentest e col sistema della termoanalisi non hanno portato ad alcune correlazione dei risultati. Per le prove sui motori sono stati quindi scelti degli oli nuovi e degli oli rigenerati che, trasformati in oli per motori comuni con l'aggiunta di additivi, sono stati provati all'interno di motori MWM.

Le prove di funzionamento di breve durata svolte negli MWM-A hanno dimostrato che tutti gli oli, nuovi o rigenerati, reagiscono agli additivi. Non è invece stata constatata alcuna differenza di comportamento da parte degli oli rigenerati. Il controllo delle prestazioni degli oli per motori ottenuti da oli nuovi o da oli rigenerati all'interno di motori MWM-B secondo le norme DIN 51 361, hanno dato come risultato che gli oli rigenerati, cosi come gli oli nuovi possono dar luogo ad ottimi oli per motori. Non è stato invece constatato alcun influsso specifico del tipo di raffinato sui risultati registrati sui pistoni.

In der Bundesrepublik Deutschland werden gebrauchte Schmieröle schon seit längerer Zeit gesammelt und zu sogenannten Zweitraffinaten wieder aufgearbeitet, die bei der Herstellung neuer Schmieröle wieder eingesetzt werden können.

Motorenöle auf der Basis von Zweitraffinaten sind bereits im Einsatz. Trotz zufriedenstellender Gebrauchseigenschaften ist jedoch noch nicht hinreichend geklärt, welcher Zusammenhang zwischen den Analysedaten des Grundöls und der Leistungsfähigkeit des Motorenöls besteht.

Ziel der Untersuchungen war die Überprüfung der Leistungsfähigkeit von Motorenölen auf der Basis von Zweitraffinaten durch Vergleich mit Motorenölen aus Erstraffinaten. Zu diesem Zweck wurde nach einem Arbeitsplan vorgegangen, wie in Bild 1 dargestellt.

In speziell ausgesuchten Laboratoriumsuntersuchungen wurden die Eigenschaften der Grundöle - Erst- und Zweitraffinate - ermittelt, wobei dem Verdampfungsverhalten sowie der Oxidationsneigung im Hinblick auf den Einsatz als Motorenöl besondere Bedeutung zukam. Diese Ergebnisse dienten dazu, Erst- und Zweitraffinate vergleichbarer Viskosität mit unterschiedlichem Oxidationsverhalten auszuwählen, aus denen durch Zusatz eines handelsüblichen Additiv-Packages Motorenöle hergestellt wurden, deren Gebrauchsfähigkeit in motorischen Prüfläufen ermittelt wurde.

Von den zur Verfügung stehenden Prüfverfahren für die Laboratoriumsuntersuchungen der Grundöle wurden die ausgewählt, die geeignet erschienen, das Verdampfungsverhalten und die Oxidationsneigung nachzuweisen (s. Bild 2):

      1. Vakuum-Destillation nach Große-Oetringhaus nach DIN 51 567,
      2. Wolf-Streifentest zur Bewertung des Oxidations- und Verdampfungsverhaltens,
      3. Thermo-Analyse in der Thermowaage unter Argon und technischer Luft,

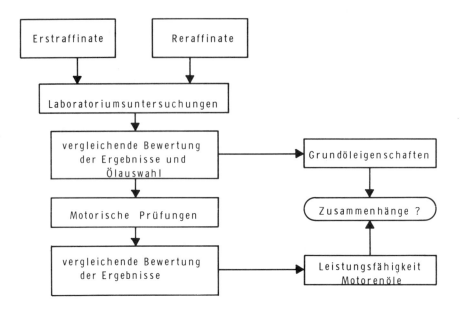

Bild 1: Arbeitsplan

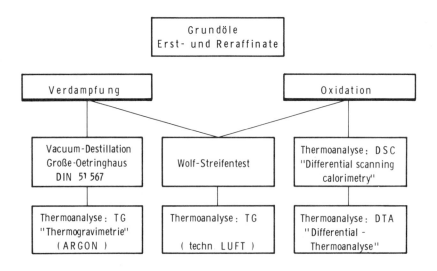

Bild 2: Laboratoriumsverfahren

4. Thermo-Analyse mit den Verfahren DSC (Differential Scanning Calorimetry) und DTA (Differential Thermo-Analyse) zur Beurteilung des Oxidationsverhaltens.

Als Prüfverfahren zur Ermittlung der Leistungsfähigkeit unter motorischen Bedingungen wurde der Motorenöltest im MWM-Ölprüfmotor nach DIN 51 361 gewählt, wobei unterschiedliche Prüfbelastungen, wie im Bild 3 dargestellt, eingesetzt wurden:

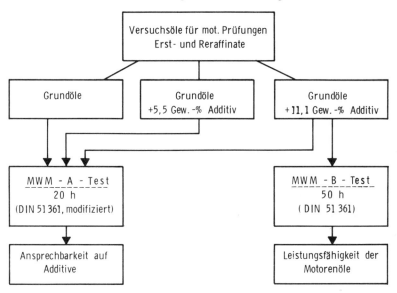

Bild 3: Motorische Prüfverfahren

1. MWM-A-Test nach DIN 51 361, modifiziert

Die Wahl des Verfahrens A sowie die Begrenzung der Prüfdauer auf 20 Stunden ermöglichte es, diese Versuche zum Nachweis der Wirksamkeit des Additiv-Packages bei stufenweiser Konzentrationssteigerung sowohl mit additivfreien Grundölen als auch teil- und vollformulierten Motorenölen durchführen zu können. Nach Rücksprache mit der Additiv-Industrie wurde für beide Raffinate ein Additiv-Package ausgewählt, mit dem in der empfohlenen Konzentration von 11,1 Gew.-% die Spezifikation MIL-L-2104-C erfüllt wird.

Es wurden folgende Additiv-Konzentrationen ausgewählt:

    I. Grundöl ohne Additiv-Package

    II. Grundöl mit 50 % der vom Additiv-Hersteller empfohlenen Konzentration entsprechend 5,5 Gew.-%

    III. Grundöl mit 100 % der vom Additiv-Hersteller empfohlenen Konzentration entsprechend 11,1 Gew.-%

2. MWM-B-Test nach DIN 51 361

Da die Leistungsfähigkeit eines Motorenöls nur unter den Prüfbedingungen, wie sie für den MWM-B-Test vorgeschrieben sind, überprüft werden kann, wurden die bereits im modifizierten MWM-A geprüften vollformulierten Motorenöle mit der Additiv-Konzentration von 11,1 Gew.-% noch einmal im MWM-B getestet.

Da die Ergebnisse der Laboratoriumsuntersuchungen an den Grundölen die Basis für die Auswahl von Grundölen zur weiteren Prüfung in motorischen Tests sein sollten, ist zunächst über diese Ergebnisse zu berichten.

Nach Ermittlung der Verdampfungsverluste mit den erwähnten unterschiedlichen Verfahren läßt sich zusammenfassend feststellen, daß der bei Mineralölen enge physikalische Zusammenhang zwischen Verdampfungsverhalten und Viskosität - wie zu erwarten - sowohl bei Erst- als auch Zweitraffinaten nachgewiesen wurde, wenn auch einzelne Öle offenbar gerätespezifisch unterschiedlich bewertet werden. Im Gegensatz hierzu zeigen die Ergebnisse bezüglich des Oxidationsverhaltens große Differenzen hinsichtlich der Bewertung in den einzelnen Verfahren.

Die Auswertung der Wolf-Teststreifen nach der Farbskala erwies sich wegen mangelnder Differenzierung der Ergebnisse als unbrauchbar. Eine Bewertung der Teststreifen nach dem Gewicht der Ablagerungen zeigte eine insgesamt bessere Differenzierung der Ergebnisse, wobei allerdings große Streuungen für hohe Ablagerungswerte nicht unerwähnt bleiben sollen.

Die unter technischer Luft in der Thermowaage ermittelten Ergebnisse lassen im Vergleich zu den unter Argon gefundenen Werten die Vermutung zu, daß auch unter Luft im wesentlichen die Verdampfungseigenschaften wiedergegeben wurden.
Zusammenfassend bleibt festzustellen, daß aus diesen Ergebnissen allein keine eindeutigen Aussagen bezüglich des Oxidationsverhaltens abgeleitet werden konnten. Trotz der zweifelhaften Aussagefähigkeit der Ergebnisse aus dem Wolf-Streifentest wurde die Menge der Ablagerung in diesem Test als Auswahlkriterium für Öle herangezogen, die in motorischen Testen weiter untersucht werden sollten.

Aus jeder Viskositätsklasse eines Raffinattyps wurden zwei Öle mit möglichst unterschiedlicher Bewertung ausgewählt (siehe Bild 4).

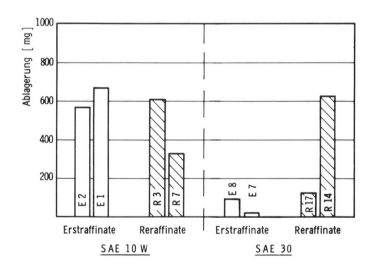

Bild 4: Wolf-Teststreifen-Versuch
- Ablagerungen -

Die an diesen Grundölen durchgeführten Thermo-Analysen mit den Verfahren DSC und DTA als zusätzliche Versuche zur Beschreibung des Oxidationsverhaltens führten zu folgenden Ergebnissen:

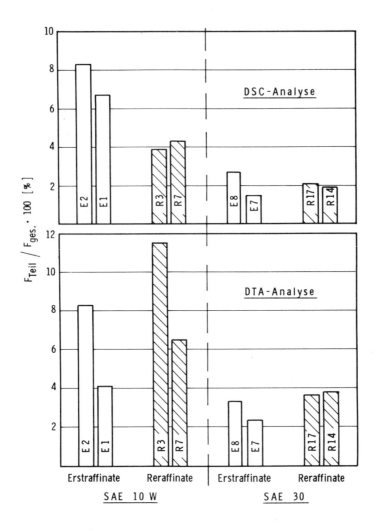

Bild 5: Thermoanalyse DSC und DTA

Im Bild 5 ist die Auswertung der registrierten Wärmeflußkurven aus der DSC-Analyse wiedergegeben. Anscheinend zeigen Öle der Viskositätsklasse 10W eine geringere Oxidationsstabilität als Öle der Viskositätsklasse SAE 30, wobei jedoch keine Korrelation zu den Ergebnissen aus dem Wolf-Streifentest erkennbar ist.

Bild 5 zeigt ebenfalls die Auswertung der Versuche in der DTA.

Die gemessenen großen Unterschiede im Verhalten der Zweitraffinate der SAE-Klasse 10W als auch die annähernd gleichen Werte für die Öle der Klasse 30 zeigen keine Korrelation zu den entsprechenden Ergebnissen aus der DSC-Analyse. Der Nachweis charakteristischer Unterschiede vergleichbarer Erst- und Zweitraffinate im Verhalten gegenüber Sauerstoff war nicht möglich.

Die Betrachtung aller Ergebnisse zum Oxidationsverhalten läßt erkennen, daß die Frage nach einem Laboratoriumsverfahren, welches die Oxidationseigenschaften eindeutig und wiederholbar beschreiben kann, mit der Wahl dieser Verfahren noch nicht beantwortet ist.

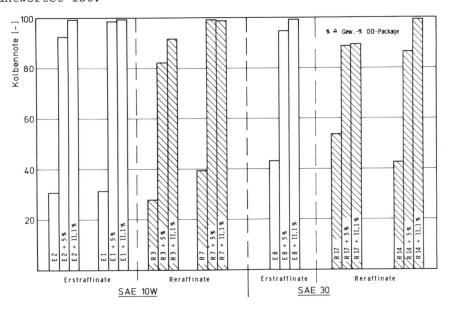

Bild 6: Prüflaufergebnisse MWM A (20h)
- Kolbenbewertung -

Die Kolbenbewertungen der MWM-A-Prüfläufe sind im Bild 6 wiedergegeben. Für alle untersuchten Öle, unabhängig von Viskositätslage und Raffinattyp, ist die Wirksamkeit des Additivpackages in Abhängigkeit von der Konzentration deutlich erkennbar. Daß in der SAE-Klasse 10W sowohl ein Erst- als auch ein Zweitraffinat schon in der Konzentrationsstufe 5-Gew.-% Additiv nahezu die höchstmögliche Bewertung erreicht, zeigt die Grenzen dieses Verfahrens. Ein Einfluß des Raffinattyps auf die erzielten Kolbennoten konnte nicht festgestellt werden.

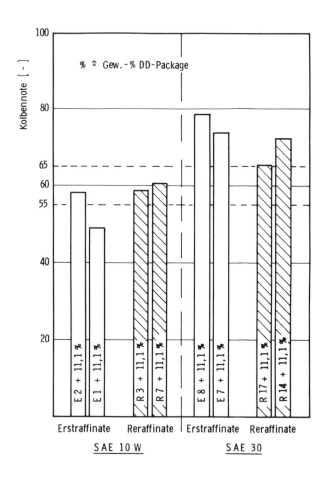

Bild 7: Prüflaufergebnisse MWM B
- Kolbenbewertung -

Die Durchführung der MWM-B-Prüfläufe führte zu den im Bild 7 dargestellten Ergebnissen. Selbst bei Berücksichtigung der geringen Anzahl der untersuchten Öle und der Verfahrens-Streubreite wird deutlich, daß alle Öle der Klasse 10W geringere Kolbennoten erreichen als alle Öle der Viskositätsklasse 30. Darüber hinaus scheinen die Zweitraffinate der Klasse 10W gleich gut oder geringfügig besser abzuschneiden als die Erstraffinate dieser Klasse, während die Zweitraffinate der Klasse 30 geringfügig unter der Bewertung der entsprechenden Erstraffinate liegen. Trotzdem wird mit den Zweitraffinaten der Klasse 30 im Mittel eine Kolbenbewertung von 69 Punkten erzielt. Das bedeutet, daß die gemäß CCMC-Spezifikation im MWM-B für Einbereichsöle für aufgeladene Dieselmotoren geforderten 65 Punkte mit allen Erst- und Zweitraffinaten der SAE-Klasse 30 erreicht wurden, während alle SAE 10W-Öle - mit einer Ausnahme bei den Erstraffinaten - die gemäß CCMC-Spezifikation für Öle für aufgeladene Dieselmotoren geforderten 55 Punkte erreichten.

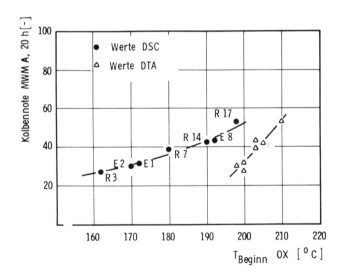

Bild 8: Korrelation Grundölergebnisse
Motorentest - Laboratoriumsversuch

Eine Gegenüberstellung der Ergebnisse aus motorischen Prüfläufen mit jenen aus Laboratoriumsuntersuchungen führte für die meisten Einzelergebnisse nicht zum Nachweis funktioneller Zusammenhänge in der Weise, daß für Grundöle gefundene Eigenschaften auch aus den Ergebnissen der motorischen Tests ablesbar waren, mit einer Ausnahme:

Die im Bild 8 wiedergegebene Darstellung der mit den <u>Grundölen</u> beider Raffinattypen erzielten Kolbennoten im MWM-A (20 h) in Abhängigkeit von der Temperatur des Oxidationsbeginns der <u>Grundöle</u>, ermittelt mit den Verfahren DSC und DTA unter Laboratoriumsbedingungen, läßt die Schlußfolgerung zu, daß hier zwischen Ergebnissen zur Ermittlung der Oxidationsstabilität im Motor (MWM-A) und im Laboratoriumsverfahren (DSC und DTA) eine Korrelation nachgewiesen wurde.

Auch in diesem Zusammenhang wurde kein spezielles Verhalten der Zweitraffinate (R) im Vergleich zu den Erstraffinaten (E) gefunden.

Zusammenfassend ergibt sich folgende Gesamtbewertung:

1. Die untersuchten Erst- und Zweitraffinatgrundöle lieferten innerhalb einer Viskositätsklasse sehr unterschiedliche Ergebnisse bei der Ermittlung der Oxidationsstabilität, wobei die Bewertung der Öle durch die gewählten Labormethoden als nicht eindeutig gesichert angesehen werden mußte.

2. Die ermittelten Differenzen im Oxidationsverhalten der Grundöle sind - abgesehen von der gefundenen Korrelation der Ergebnisse aus motorischen Tests und Laboratoriumsverfahren - nicht unmittelbar durch motorische Tests zu bestätigen, was ohne Einschränkung für Erst- als auch für Zweitraffinate gilt.

3. Die erzielten Kolbennoten im MWM-B machen deutlich, daß sowohl Erst- als auch Zweitraffinate (bei entsprechender Beachtung der Viskositätslage) für die Herstellung von Motorenölen mit Leistungsdaten - wie von den derzeitig gültigen Spezifikationen vorgeschrieben - eingesetzt werden können.

DIE VERWENDUNG VON ZWEITRAFFINATEN NACH DEM SOLVENTVERFAHREN BEI DER
FORMULIERUNG VON MOTORÖLEN
USE OF SOLVENT RE-REFINED OILS IN THE FORMULATION OF MOTOR OILS
UTILISATION D'HUILES REGENEREES A L'AIDE DE SOLVANTS POUR LA
FABRICATION D'HUILES DE MOTEUR
IMPIEGO DI OLI RIRAFFINATI CON SOLVENTE NELLA FORMULAZIONE DI OLI
MOTORE

A. MODENESI
Chef des technischen Dienstes der Clipper Oil Italiana, Italien
Chief of Technical Services of Clipper Oil Italiana, Italy
Chef des Services Techniques de la Clipper Oil Italiana, Italie
Responsabile dei Servizi Tecnici della Clipper Oil Italiana, Italia

RIASSUNTO

Viene esaminato il problema della riraffinazione degli olii lubrificanti esausti sia dal punto di vista ecologico che tecnico-economico.

In particolare sono illustrati i motivi che hanno portato alla scelta di un processo di riraffinazione a tecnologia avanzata basato sul seguente schema:

Distillazione atmosferica, estrazione al solvente, distillazione sottovuoto, trattamento al solvente del residuo vacuum, trattamento idrogeno.

Tale processo è destinato a sostituire quello all'acido solforico, in quanto consente rese-qualità del prodotto migliori ed evita i problemi connessi all'eliminazione dei fanghi acidi e delle terre.

Particolare attenzione è stata dedicata alle analisi fisico-chimiche e tecnologiche delle basi lubrificanti riraffinate.

Al fine di confermare la possibilità di impiego di dette basi nella formulazione di olii motore sono state eseguite prove a confronto con olii formulati con basi di prima raffinazione. Vengono esaminati i possibili utilizzi dell'asfalto ottenuto dalla estrazione al propano.

ZUSAMMENFASSUNG

Geprüft wird das Problem der Zweitraffination von Gebrauchtöl sowohl nach ökologischen als auch nach technisch-wirtschaftlichen Gesichtspunkten. Es wird insbesondere dargelegt, aus welchen Gründen ein fortgeschrittener Prozess der Zweitraffination gewählt wurde, der auf folgendem Schema basiert : Atmosphärische Destillation, Lösungsmittelextraktion, Vakuumdestillation, Lösungsmittelbehandlung des Rückstands im Vakuum, Wasserstoffbehandlung.
Dieser Prozess soll das mit Schwefelsäure arbeitende Verfahren ersetzen, weil er eine höhere Ausbeute und eine bessere Produktqualität ermöglicht und weil die mit der Beseitigung der sauren Schlämme und der Erden verbundene Probleme vermieden werden.
Besondere Aufmerksamkeit gilt den physikalish-chemischen und technologischen Analysen der Zweitraffinate.
Um die Möglichkeit der Verwendung dieser Zweitraffinate bei der Formulierung von Motorenölen zu bestätigen, werden Versuche zum Vergleich mit Ölen, die unter Verwendung von Erstraffinaten formuliert worden sind, durchgeführt. Geprüft wird die mögliche Verwendung von Asphalt aus der Propanextraktion.

SUMMARY

The problem of the re-refining of spent lubricating oils is examined both from the ecological and the technical and economic standpoint.
Special reference is made to the reasons for selecting an **advanced technology** re-refining process based on the following schedule:
atmospheric distillation, solvent extraction, vacuum distillation, solvent treatment of the vacuum residue, and hydrogen treatment.
This process is designed to replace the sulphuric acid process by offering improved yield and quality and avoiding the problems involved in the disposal of acidic sludge and earth.
Particular emphasis is given to the physico-chemical and technological analyses of the re-refined lubricating bases.
To confirm the possibility of using these bases in the formulation of motor oils, tests were carried out comparing them with oils formulated with **primary refining bases**. The possible uses of asphalt obtained from propane extraction will be examined.

RESUME

Le problème de la régénération des lubrifiants usagés est étudié du point de vue écologique et du point de vue technique et économique.

L'auteur commente en particulier les raisons qui ont incité à choisir un procédé de régénération de technologie avancée, basé sur les principales opérations suivantes : distillation atmosphérique, extraction au solvant, distillation sous vide, traitement au solvant du résidu sous vide, traitement à l'hydrogène.

Ce procédé est destiné à remplacer le procédé à l'acide sulfurique, dans la mesure où il permet d'obtenir un meilleur rapport rendement/qualité du produit, et évite les inconvénients liés à l'élimination des boues acides et des terres.

Une attention particulière a été accordée aux analyses physico-chimiques et technologiques des lubrifiants régénérés.

Afin de confirmer les possibilités d'utilisation de ces produits régénérés pour la fabrication d'huiles pour moteurs, des essais comparatifs ont été effectués avec des huiles obtenues à partir de bases d'un premier raffinage. Les possibilités d'utilisation de l'asphalte obtenu par l'extraction au propane sont étudiées.

In Italia a fronte di un consumo totale di circa 650.000 t/a di lubrificanti, sono potenzialmente disponibili oltre 220.000 tonn/a di olii esausti di cui solo 130.000 tonn risultano effettivamente raccolti.

Di questi, 70.000 vengono rigenerati e 60.000 destinati alla combustione diretta.

Sia gli oli lubrificanti esausti scaricati direttamente nel terreno o nelle fogne che quelli bruciati senza la depurazione dei fumi, recano notevoli danni all'ambiente per la presenza di prodotti chimici altamente inquinanti (presenza di Cl, P, Pb, S, Br). Inoltre, la combustione di oli esausti non deadditivati arreca danni per corrosione agli impianti.

Il problema ecologico risulta particolarmente e responsabilmente sentito a tutti i livelli specie nei Paesi industrializzati, in quanto è ormai accertato che gli inquinamenti condizionano la vita dell'uomo. E' quindi evidente che non si può permettere che migliaia di ton. di oli esausti, peraltro resi più nocivi per la presenza di additivi, vadano a procurare danni irreparabili.

A tale proposito si pone in evidenza l'interessante iniziativa della CEE, con la Direttiva del 16 giugno 1975 attinente alla raccolta e riutilizzazione di tutti gli oli esausti.

D'altra parte la particolare situazione energetica consiglia il riutilizzo dei prodotti di origine petrolifera per ridurre l'importazione di greggio ed il conseguente esborso di valuta pregiata.

La rigenerazione risponde pienamente sia alle esigenze energetiche che ecologiche ed è notevolmente più economica della raffinazione oli nuovi, sia per quanto riguarda il costo di investimento che il costo di esercizio, non essendo necessari per essa gli impianti di estrazione aromatici e di deparaffinazione.

Nei confronti poi della combustione di prodotto almeno parzialmente deadditivato, la rigenerazione risulta essere più conveniente consentendo un guadagno notevolmente apprezzabile dovuto alla migliore valorizzazione dell'olio lubrificante esausto.

Presso la CLIPPER OIL ITALIANA è stato utilizzato un processo di rigenerazione basato sul trattamento con acido solforico e terre.

Questo processo oltre a produrre olii rigenerati di qualità mediocre (quindi con limitazioni nell'impiego), con basse rese e alti costi, ha creato notevoli difficoltà nello smaltimento dei fanghi acidi, specialmente a seguito delle recenti leggi sull'inquinamento.

Si evidenzia che i tempi di reazione e di trattamento con acido sono aumentati, rendendo quindi anti-economico il processo, in quanto gli olii lubrificanti sono migliorati come prestazioni, ma peggiorati al fine della rigenerazione (qualità e quantità di additivi in essi contenuti). Ciò ha portato alla decisione di un sostanziale rinnovamento ed ampliamento dello stabilimento, allo scopo di rendere più economica e quindi competitiva la lavorazione dell'esausto, risolvendo anche i problemi relativi all'inquinamento.

Tra i vari processi di rigenerazione noti in campo internazionale è stato scelto quello messo a punto della SNAM-PROGETTI in collaborazione con l'ASSORENI (Gruppo ENI) per i seguenti motivi:

- Possibilità di trattare alte capacità con alte rese (in pratica viene tutto recuperato) e minori costi gestionali.
- Possibilità di ottenere prodotti nelle gradazioni e con caratteristiche del tutto equivalenti a quelle degli oli lubrificanti di prima raffinazione.
- Assenza di sottoprodotti inquinanti quali fanghi acidi e terre esauste; concentrazione di tutte le impurezze contenute nell'olio esausto nel residuo dell'estrazione con propano.

Il processo scelto, il cui schema è riportato in Fig. 1, è basato sui seguenti stadi:
- Preflash per eliminare acqua e idrocarburi leggeri con distillazione atmosferica
- Estrazione al propano per separare la maggior parte delle impurezze e degli additivi contenuti nell'olio di carica.
- Distillazione sotto-vuoto per frazionare la carica vacuum nei tagli laterali a viscosità richiesta.
- Trattamento termico ed estrazione con propano del residuo della colonna di frazionamento vacuum. Questo particolare trattamento permette di eliminare dal re-

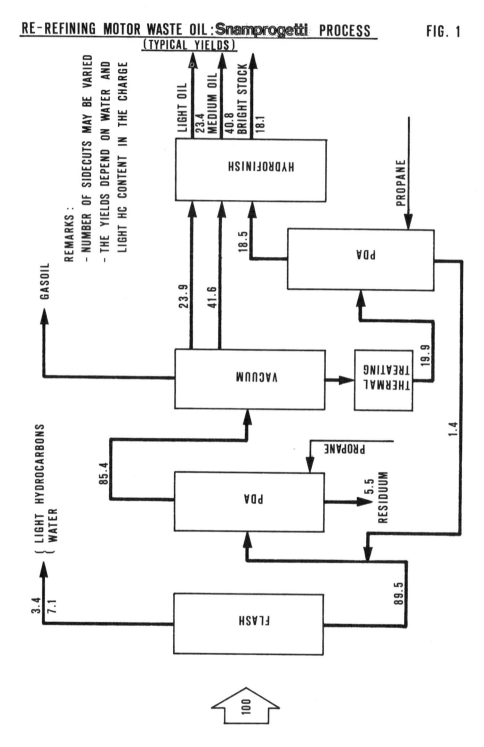

siduo le rimanenti impurezze ed ottenere un prodotto
che può essere trattato nell'impianto idrogeno senza
creare problemi per la vita del catalizzatore e di ottenere dopo idrogenazione un BS di eccellente qualità.
- Trattamento all'idrogeno delle basi ottenute da vacuum e da estrazione secondaria al propano.

Dal punto di vista economico risulta che un impianto da 50.000 t/a, il cui costo di investimento é di ca. 20.10^6 \$, ha un PAY OUT TIME intorno ai 3 anni.

I consumi per tonn/carica sono i seguenti:

Energia Elettrica	50 Kwh
Fuels	0.8 MM Kcal
Steam	0.9 ton
Cooling Water	65 m^3

Con oli lubrificanti (in particolare ex motori) disponibili presso la Clipper-Oil Italiana, sono state eseguite numerose prove di raffinazione su impianti pilota Assoreni che hanno consentito la messa a punto delle condizioni operative e la ottimizzazione delle caratteristiche dei prodotti (rese, caratteristiche cariche e prodotti, sono riportati nelle tabelle I e II).

Allo scopo di completare lo studio e di confermare le qualità dei prodotti finali ottenuti come componenti nella formulazione di oli motore, sono stati preparati su Impianto pilota ASSORENI secondo la sequenza SNAMPROGETTI, notevoli quantità di basi lubrificanti con cui sono state eseguite prove su motore in confronto con olio lubrificante di prima raffinazione, da grezzo Medio Oriente. Queste prove sono state effettuate nei due casi con oli lubrificanti di gradazione SAE-30 formulati entrambi con lo stesso pacchetto

TABELLA I

PRODOTTI	% p.	OLIO ESAUSTO	PREFLASH	I° ESTRAZ. PROPANO	VACUUM				II° ESTRAZ. PROPANO
					SN 150	SN 300	SN 500	BS	
RESA		100	89.5	85.4	38.4	6.18	20.92	19.9	18.5
CARATTERISTICHE:									
Peso specifico 15/4°C Kg/l		0.9110	0.9104	0.8886	0.8752	0.8806	0.8838	0.9101	0.8983
Viscosità a 40°C Cst.		112.34	104.04	75.61	24.34	58.72	88.94	532.6	375.8
" a 100°C Cst.		14.29	12.91	9.68	4.546	7.757	10.02	34.39	26.64
Indice di Viscosità		128.7	119.5	106.6	99.58	95.16	91.28	98.32	95.4
Numero di neut. mg KOH/g		3.39	3.16	0.81	1.14	1.04	0.89	2.1	0.22
H₂O Marcusson % V.		7.20							–
Ceneri solfatate % p.			1.34	0.144	0.002	0.002	0.005	0.0491	0.005
Colore ASTM			>8	>8	3	L3.5	7	>8	8
Idrocarburi leggeri % V.		2.10							
Metalli (Fluoresc. RX):									
Ba ppm.			400	15	<5	<5	<5	45	<5
Ca "			1560	80	<5	<5	<5	300	<5
Pb "			2600	160	<5	<5	<5	700	<5
Zn "			900	50	<5	25	<5	195	<5
P "			850	280	100	30	<20	1000	20
Cl "			650	460	980	40	20	25	20
Br "			580	20	40	<5	<5	<5	<5

TABELLA II

CARATTERISTICHE BASI RIGENERATE

PRODOTTI	BASE SN-150	BASE SN-300	BASE SN-500	BASE BS
RESA HYDROFINISH % p.	97.5	98.4	98.15	97.4
Peso specifico 15/4°C Kg/l	0.8698	0.8764	0.8828	0.8976
Viscosità a 40°C Cst.	24.63	57.84	89.70	340.1
Viscosità a 100°C Cst.	4.606	7.93	10.22	25.72
Indice di Viscosità	101.9	102.6	94.06	98.9
Colore ASTM	L 1	L 1	L 1.5	2.5
Colore AGED 3 ore a 100°C	L 1	1	1.5	3
Colore AGED 16 ore a 100°C	1	L 1.5	L 2	L 3.5
Solfo totale	0.27	0.28	0.51	0.68
Numero di neutr. mg KOH/g	< 0.02	< 0.02	0.029	0.04
Res. Carb. Conr. % p.	< 0.005	< 0.005	0.076	0.58
Res. Carb. Ramsb. "	0.023	0.037	0.074	0.38
Ceneri Solfate % p.	0.002	0.002	0.002	0.002
Infiamm. P.M. °C	205	248	240	266
Infiamm. V.A. °C	210	258	260	316
Pour Point °C	-9	-9	-9	-9
Prova di Ossid. IP-48:				
Aumento RC Ramsb. %p.	0.269	0.664	0.186	0.39
Rapporto viscosità a 100°F	1.290	1.254	1.097	1.14
Metalli (Fluoresc. RX)				
Ba	< 5	< 5	< 5	< 5
Ca	< 5	< 5	< 5	< 5
Pb	< 5	< 5	< 5	< 5
Zn	< 5	< 5	< 5	< 5
P	< 20	< 20	< 20	< 20
Cl	< 10	< 10	< 10	< 10
Br	< 5	< 5	< 5	< 5

e percentuale di additivi prodotti dalla Soc. OROGIL (Francia). Le prove motore utilizzate per la valutazione dello olio riraffinato in raffronto all'olio nuovo, sono state scelte in modo da avere dati rappresentativi per la qualificazione del lubrificante a livello internazionale (Nuova MIL-L-46152; API SE/CC o CCMC).

In particolare le prove effettuate presso qualificati Laboratori indipendenti sono le seguenti:

- <u>Caterpillar 1-H2</u>: Prove ad alta temperatura - motore Diesel. Indicative della tendenza a formazione di depositi.

- <u>Sequenza III D</u> : Prove ad alta temperatura - motore benzina. Indicative della tendenza alla ossidazione dell'olio ed alla formazione di depositi.

- <u>Ford Cortina</u> : Prove ad alta temperatura - motore benzina. Indicative della tendenza all'incollamento dei segmenti e formazione di depositi.

- <u>Petter W1</u> : Prove ad alta temperatura - motore benzina. Indicative della tendenza dell'olio ad ossidarsi ed a corrodere le bronzine.

I risultati ottenuti nelle singole prove sono riportati rispettivamente nelle tabelle III, IV, V, VI.

L'esame dei risultati consente di rilevare l'ottimo comportamento dell'olio riraffinato relativamente al superamento delle singole prove, in analogia alle stesse prove e risultati dell'olio di prima raffinazione.

Dalle dette prove come pure dalle caratteristiche fisico-chimiche rilevate sui vari campioni di partite degli impianti pilota, riteniamo molto affidabilmente che gli oli così raffinati sono equivalenti a quelli di prima raffinazione e quindi utilizzabili senza limitazioni.

CATERPILLAR 1 H2

TABELLA III

	LIMITI DI SPECIFICA	OLIO PRIMA RAFFINAZIONE	OLIO RIRAFFINATO
TGF %	45	1	1.5
WTD	140	110.5	131.7
RISULTATO		"PASS"	"PASS"

SEQUENZA III-D

TABELLA IV

	LIMITI DI SPECIFICA	OLIO PRIMA RAFFINAZIONE	OLIO RIRAFFINATO
CONSUMO D'OLIO		3.09 QTS	3.18 QTS
VISCOSITA' 40 h.	375% max	34.1 %	39.7 %
MORCHIE	9.2	9.5	9.34
LACCHE PISTONI	9.1	9.3	9.83
ZONE TRA I SEGMENTI (ORLF)	7.0	7.10	7.38
USURA MEDIA CAMME PIU' BILANCIERI pollici	0.0040	0.0011	0.0018
USURA MAX pollici	0.0080	0.0029	0.0034
RISULTATO		"PASS"	"PASS"

PETTER W 1

TABELLA V

	LIMITI DI SPECIFICA	OLIO PRIMA RAFFINAZIONE	OLIO RIRAFFINATO
MANTELLO ESTERNO		10/10	10/10
FONDO PISTONE		9.4/10	9.8/10
PERDITE PESO CUSCINETTO, mg	25	21	1
AUMENTO VISCOSITA' %	50	22	7.3
RISULTATO		"PASS"	"PASS"

FORD CORTINA

TABELLA VI

	LIMITI DI SPECIFICA	OLIO PRIMA RAFFINAZIONE	OLIO RIRAFFINATO
MANTELLO PISTONE		9.6/10	9.8/10
DEPOSITI PISTONE	8.7/10	9.9/10	9.9/10
INCOLLAMENTO SEGMENTI A FREDDO	9.8/10	10/10	9.9/10
RISULTATO		"PASS"	"PASS"

CARATTERISTICHE ESTRATTO

TABELLA VII

Peso specifico 15/4°C	Kg/l	1.225
Ceneri Solfatate	% p.	23.33
Cont. Olio (DIALISI)	"	4.3
Potere Calorifico Sup.	Cal/gr	6823
Punto di scorrimento	°C	indet.
Metalli (Fluoresc. Rx):		
Ba	ppm	6653
Ca	"	25597
Pb	"	42230
Zn	"	14710
P	"	10107
Cl	"	3736
Br	"	9675

Infine il residuo ottenuto come fondo nelle colonne di estrazione con propano è costituito essenzialmente da una miscela contenente tutti gli additivi impiegati nella formulazione degli olii motore, unitamente al Pb, derivante dalla combustione della benzina e da residui vari ed impurezze metalliche derivanti dall'usura del motore.

Nella tabella VII si riportano le caratteristiche principali del residuo.

Come noto, gli additivi, maggiormente presenti sono il polimero impiegato come "viscosity index improver" e gli additivi detergenti e disperdenti.

Per quanto concerne l'utilizzo del residuo si possono fare le seguenti considerazioni di carattere generale:

1 - L'impiego come componenti dei combustibili crea notevoli problemi non solo per la presenza di Cl, Br, P, Pb ecc. ma anche per i possibili depositi sulle tubazioni del forno e intasamento degli ugelli dei bruciatori.

    Minori inconvenienti ci sono se il combustibile contenente il residuo viene bruciato in forni per cemento, in quanto le "impurezze" vengono fissate nel clinker e quindi, si ottiene doppio beneficio: energetico ed ecologico.

2 - Utilizzando il residuo come componente per bitumi stradati si ha un miglioramento del comportamento dei bitumi stessi a freddo grazie alla presenza di polimero. Analogamente il residuo può essere utilizzato nella formulazione di bitumi speciali (impermeabilizzanti per edilizia, sigillante dei giunti di espansione in opere civili - tipo viadotti) ed anche in questo caso si ha un netto miglioramento del comportamento a freddo.

3 - Facendo riferimento all'alto contenuto in Pb del residuo si può inoltre prevedere di utilizzare il residuo stesso per il recupero di questo elemento. In tal caso è possibile concentrare il piombo fino al 10-12%p, trattando il residuo stesso con opportuni solventi.
La parte solubile, nella quale gli additivi sono più concentrati può essere utilizzata, in funzione dello alto contenuto in polimero ed additivi, in prodotti che non abbiano particolari esigenze per quanto riguarda il colore.
E' appena il caso di osservare che a differenza di altri processi quello selezionato dalla ns. Società consente il riutilizzo completo non solo delle basi lubrificanti ma anche di tutti i sottoprodotti ivi compreso questo residuo con ben evidenti benefici di carattere sia economico che ecologico.

DER EUROPÄISCHE SCHMIERSTOFFMARKT UND DIE ROLLE DER UNABHÄNGIGEN
FIRMEN
THE EUROPEAN LUBRICANT MARKET - THE ROLE OF INDEPENDENT COMPANIES
LE MARCHE EUROPEEN DES LUBRIFIANTS ET LE ROLE DES ENTREPRISES
INDEPENDANTES
IL MERCATO EUROPEO DEI LUBRIFICANTI E IL RUOLO DELLE IMPRESE
INDIPENDENTI

Dr. M. FUCHS
Geschäftsführender Gesellschafter der Rudolf Fuchs GmbH & Co.,
Mannheim

ZUSAMMENFASSUNG

Es wird auf der Grundlage umfangreicher Recherchen in 17 Ländern ein
zusammenfassender Bericht über den Schmierstoffmarkt Westeuropas gegeben. Der Bedarf und wichtige Produktgruppen werden dargestellt. Die
Schmierölraffinerien, das Recycling von Gebrauchtölen, die Additive
sowie der westeuropäische Binnen- und Außenhandel werden eingehend untersucht; des weiteren in einer Mengenbilanz der Versorgungsbeitrag
der einzelnen Quellen sowie Zahl, Betriebsgrößenverteilung, Kapazität
und Auslastung der Anlagen der Schmierölraffination und des Recycling.

Die Rolle der unabhängigen Firmen wird auf allen Verarbeitungs- und
Vertriebsstufen des westeuropäischen Schmierstoffmarkts behandelt
sowie das Ergebnis einer Strukturanalyse der Schmierölmischanlagen
und Fettfabriken vorgetragen. Es folgen eine Darstellung der Funktionen

sowie des Marktanteils der unabhängigen Schmierstoffirmen und schließlich ein Ausblick auf die in den 80er Jahren zu erwartende Entwicklung.

SUMMARY

A comprehensive report is given on the European lubricant market on the
basis of exhaustive research in seventeen countries. Demand and important
product groups are illustrated. Lubricating oil refineries, the recycling
of used oils, additives and Western European internal and external trade
are examined in detail, as are, by way of a quantitative comparison,
supplies from individual sources and the number, plant size distribution,
capacity and utilization of lubricating oil refineries and recycling
plants.
The role of independent companies is examined at all processing and sales
stages on the European lubricant market, and the result of a structural
analysis of lubricating oil mixing plants and lubricant factories is shown.
This is followed by an illustration of the functions and market share of
independent lubricant companies and finally prospective developments in the
80s.

RESUME

L'auteur présente une étude d'ensemble du marché des lubrifiants de
l'Europe de l'Ouest, réalisée sur la base de recherches importantes, effectuées dans 17 pays différents. Les besoins et les groupes de produits importants sont exposés. Les raffineries de lubrifiants, le recyclage des
huiles usagées, les additifs et le commerce intérieur et extérieur de
l'Europe occidentale sont étudiés en détail; l'auteur étudie ensuite, dans
le cadre d'un bilan quantitatif, la contribution à l'approvisionnement des
différentes sources, ainsi que le nombre, la répartition par dimensions,
la capacité et le taux d'utilisation des installations de raffinage de lubrifiants et de recyclage.
Le rôle des firmes indépendantes est étudié à tous les niveaux de traitement et d'écoulement du marché des lubrifiants de l'Europe occidentale, et
les résultats d'une analyse structurelle des installations de préparation
de lubrifiants et de fabrication de graisses sont exposés. L'auteur expose
ensuite les fonctions et les parts du marché des indépendants en lubrifiants, ainsi que les perspectives d'évolution au cours des années 80.

RIASSUNTO

In base ad esaurienti ricerche svolte in 17 paesi, è stata elaborata
una relazione riassuntiva sul mercato dei lubrificanti dell'Europa occidentale. Tale relazione descrive le esigenze e i principali gruppi di
prodotti. E' stata fatta un'indagine approfondita sulle raffinerie che
producono lubrificanti, sul riciclo degli oli usati, sugli additivi
nonché sul commercio interno ed estero dell'Europa occidentale ; inoltre,
in un bilancio quantitativo, è stato esaminato il contributo all'approvvigionamento delle singole fonti nonché il numero e la distribuzione secondo la dimensione, la capacità e il grado d'impiego degli impianti delle
raffinerie che producono lubrificanti o riciclano gli oli usati.

Il ruolo delle ditte indipendenti viene preso in considerazione a tutti i livelli di lavorazione e di commercializzazione sul mercato europeo dei lubrificanti e vengono indicati i risultati di un'analisi strutturale degli impianti per la mescolatura dei lubrificanti e delle fabbriche che producono grassi. Segue una descrizione delle funzioni e della quota di mercato delle imprese indipendenti di questo settore ed infine una panoramica sui suoi probabili sviluppi negli anni 1980.

Auf einem europäischen Gebrauchtöl-Recyclingkongreß erscheint es sinnvoll, daß wir uns den Schmierstoffmarkt als Ganzes ansehen, seine Struktur und Veränderungen in den vergangenen 2 Dekaden sichtbar machen sowie der Frage nachgehen, wohin er sich in den 80er Jahren entwickeln wird und welche Rolle dabei die unabhängigen Firmen spielen können. Wir beschränken uns mit dieser Betrachtung auf Westeuropa und haben versucht, hierfür aus den 9 EG- sowie 8 weiteren europäischen Ländern das erforderliche Material zu bekommen.

Der Schmierstoffbedarf ist in Westeuropa in den 60er Jahren noch stark gestiegen, ging jedoch in den 70er Jahren in die Phase einer Stagnation oder eines nur noch ganz geringen Wachstums über, wobei in vielen Ländern die Schrumpfung des Jahres 1975 erst gegen Ende der Dekade oder überhaupt nicht mehr ausgeglichen wurde. Die Ursachen dieser im Grunde weltweiten Entwicklung sind bekannt.

1979 dürfte der westeuropäische Schmierstoffverbrauch (ohne Bunker) knapp 5,7 Mio t gewesen sein und nimmt damit international eine bedeutende Stellung ein. Es gibt keine veröffentlichte Statistik des Weltschmierstoffkonsums, wir schätzen ihn jedoch für 1979 auf 29,8 bis 31,9 Mio t, so daß unter Hinzurechnung des Bunkerbedarfs von ca. 0,35 Mio t etwa 19 bis 20 % der Weltmenge auf Westeuropa entfallen. Dabei konzentriert sich der größte Teil der westeuropäischen Menge auf die EG (ca. 77 %) und deren 4 bedeutendste Märkte (ca. 67 %).

Bemerkenswert ist die Tatsache, daß der Anteil des Schmierstoffs am Mineralölgesamtverbrauch in den letzten 20 Jahren immer stärker zurückging. Für die wichtigsten Schmierstoffmärkte der freien Welt betrug er 1960 noch 1,4 %, 1979 jedoch nur noch 0,9 %. Vielfältige technologische Einflüsse und ein verändertes Konsumentenverhalten haben hier neben anderen Faktoren deutliche strukturelle Veränderungen bewirkt. So sank z. B. in Deutschland und Frankreich der Motorenölverbrauch pro KFZ von 27,5 bzw. 21,8 kg in 1965 auf nur noch 15,8 bzw. 18,1 kg in 1979 sowie von ca. 2,1 bzw. 2,5 % des Kraftstoffverbrauchs auf nur noch 1,2 bzw. 1,5 %, wobei er von Land zu Land ebenso auffallende Unterschiede zeigt wie der Schmierstoffverbrauch im Verhältnis zum Mineralölverbrauch und pro Kopf der Bevölkerung.

Eine Unterteilung des westeuropäischen Schmierstoffverbrauchs nach Produktgruppen zeigt, daß 1979 knapp 47 %, d.h. ca. 2.657.700 t auf Motorenöle und gut 53 %, d.h. ca. 3.008.300 t auf andere Schmierstoffe entfielen; dabei schwankt der Anteil der Motorenöle in den einzelnen Ländern je nach dem Motorisierungs- und Industrialisierungsgrad zwischen 37 und 60 %. Zwei Marktsegmente, in denen wegen großer Vielfalt der technischen Anforderungen und entsprechend hoher Spezialisierung die unabhängigen Schmierstoffirmen eine besonders wichtige Rolle spielen, sind Metallbearbeitungsflüssigkeiten und Korrosionsschutzöle sowie Schmierfette. Hier dürfte der westeuropäische Markt 1979 ein Bedarfsvolumen von ca. 396.000 t bzw. 172.000 t gehabt haben.

Die Versorgung des westeuropäischen Schmierstoffmarkts mit Grundölen kommt aus den in verschiedenen Ländern vorhandenen Schmierölraffinerien, aus dem Recycling von Gebrauchtölen und dem Import; hinzuzurechnen sind die Additive und sonstigen Chemiestoffe.

Sehen wir uns zuerst die westeuropäischen Schmierölraffinerien an. Es gibt davon 38 Stück, die 1979 eine Kapazität von ca. 7,9 bis 8,0 Mio t hatten. Das entspricht einer Durchschnittskapazität von ca. 208.000 bis 211.000 t pro Anlage, wobei sich für die 38 Schmierölraffinerien folgende Betriebsgrößenverteilung ergibt:

bis 100.000 t Jahreskapazität	12 Schmierölraffinerien
101 - 200.000 t Jahreskapazität	10 Schmierölraffinerien
201 - 300.000 t Jahreskapazität	7 Schmierölraffinerien
301 - 400.000 t Jahreskapazität	7 Schmierölraffinerien
> 400.000 t Jahreskapazität	2 Schmierölraffinerien

Es besteht also noch eine nennenswerte Zahl kleinerer Schmierölraffinerien, von denen einige nur naphthenbasische Rohöle bzw. feedstocks verarbeiten.

Die Grundölproduktion dürfte 6.417.000 t betragen haben, so daß die durchschnittliche Auslastung der Schmierölraffinerien 81 - 82 % war. In USA lag sie 1979 mit über 87 % deutlich höher. 6 westeuropäische Länder haben keine eigene Schmierölraffination, sind also ganz auf den

Import angewiesen; von den übrigen Ländern haben einige im Hinblick auf ihren Landesbedarf zu wenig, andere zuviel Kapazität.

Einen nicht unwesentlichen Beitrag zur Schmierstoffversorgung bringt auch das Recycling von Gebrauchtölen. 1979 wurden in Westeuropa etwa 467.000 t Regenerat gewonnen, wovon allerdings in Deutschland ein Teil als Destillat bzw. sog. "Fluxöl" in den Bitumensektor und als Reduktionsöl in die Stahlindustrie oder andere Verwendungen geht. Mit 395.000 t entfällt der größte Teil (85 %) der Regeneratproduktion auf Frankreich, Italien und Deutschland; in diesen Ländern bestehen auch entsprechende Subventionssysteme, die Sammlung und Recycling von Gebrauchtölen fördern. 8 der insgesamt 17 westeuropäischen Länder haben keine Altölaufbereitung. Der Versorgungsbeitrag der Regenerate betrug 1979 - bezogen auf die westeuropäische Schmierstoffgesamtproduktion - ca. 6,7 % (Frankreich 5,5 %, Deutschland inkl. Fluxöle ca. 14 % und Italien 8,0 %). Berücksichtigt man jedoch, daß vermutlich etwa 80 % der Regenerate in den Motorenölsektor gehen, so ergibt sich für 1979 - Exporte und Fluxöl vernachlässigt sowie auf Grundöl gerechnet - ein Anteil der Regenerate am westeuropäischen Motorenölmarkt von 10,7 % (Frankreich 18,4 %, Deutschland ohne Fluxöle 14,4 % und Italien 18,9 %).

Die westeuropäische Regeneratproduktion verteilte sich auf 40 Anlagen. Das entspricht einer jährlichen Durchschnittsproduktion von ca. 11.670 t pro Anlage, wobei hier erhebliche Unterschiede in den Betriebsgrößenklassen bestehen; es gibt Kleinanlagen mit nur ca. 2.000 t p.a. Regeneratproduktion ebenso wie solche mit 50.000 t und mehr. Die Auslastung der Altölaufbereitungsanlagen ist in einigen Ländern nur 50 bis 60 %, da nicht alle Altölmengen erfaßt, gesammelt und der Aufbereitung zugeführt werden, sondern in den Heizölsektor gehen oder ohne Aufarbeitung für andere Einsatzzwecke Verwendung finden. Im westeuropäischen Durchschnitt war die Kapazität der Zweitraffinerien von ca. 1 Mio t 1979 zu 67,5 % ausgelastet.

Eine immer größere Bedeutung am westeuropäischen Schmierstoffmarkt haben die Additive und sonstigen Zusätze erlangt. Ihr Anteil an Aufkommen und Verwendung des westeuropäischen Schmierstoffmarkts dürfte 1978 immerhin ca. 6,9 % betragen haben, d.h. er war allein schon mengenmäßig größer als der der Regenerate. Berücksichtigt man zusätzlich den Preis der Additive, so spielen diese für die Kalkulation eine entscheidende Rolle. Der Additivanteil steigt seit Jahren durch die fortschreitende Umstellung auf Mehrbereichsöle, Motor- und Getriebeöle mit hoher Leistung, auf höherwertige Hydraulik- und Korrosionsschutzöle etc. ständig an und hat sich z. B. in Frankreich von 1970 bis 1979 von etwa 6 % auf knapp 12 % fast verdoppelt; ähnliche Tendenzen sind für die USA veröffentlicht. Diese Entwicklung wird weitergehen. Für 1978 haben wir die in Westeuropa in den Schmierstoffabsatz (Inland und Bunker) gegangenen Mengen an Schmierstoffadditiven mit ca. 410.000 t und die sonstigen Zusätze (Fettstoffe, Emulgatoren, Verdicker, Chlorparaffin etc.) mit ca. 72.000 t ermittelt. Das hätte einer durchschnittlichen Legierungshöhe von 8,4 % entsprochen; für Motorenöle liegt sie in den einzelnen Ländern in der Regel zwischen etwa 10 % bis gut 13 %, in Frankreich darüber. Als durchschnittliche Legierungshöhe des westeuropäischen Motorenölmarkts haben wir für 1978 ca. 12,6 % berechnet. Etwa 335.000 t bzw. 82 % des gesamten Schmierstoffadditivverbrauchs von 410.000 t dürften auf Motoren- und Kompressorenöl entfallen sein.

Der westeuropäische Binnen- und Außenhandel mit Schmierstoffen hat ein erhebliches Ausmaß. 1978 wurden von den 17 Ländern 2.446.000 t im- und 3.511.000 t exportiert, insgesamt also 5.957.000 t bewegt. 1979 war es mehr (Deutschland, Frankreich und Italien ca. 30 % mehr Import und ca. 8 % mehr Export), es sind jedoch noch nicht alle Zahlen international veröffentlicht. Die Hauptim- und -exporteure waren die traditionellen "Transitländer" (Niederlande und Belgien mit 30 % des gesamten Handelsvolumens), die großen Schmierstoffländer (Großbritannien, Deutschland, Frankreich und Italien mit 50 %) und die reinen Importmärkte (Dänemark, Finnland, Norwegen, Schweiz, Irland und Luxemburg mit 8 %). Sieht man von den "Transithandelsländern" ab, so haben Großbritannien, Frankreich, Italien und Deutschland mit ihren relativ hohen Schmierstoffexporten einen bedeutenden Beitrag zur internationalen Versorgung erbracht, wobei Frankreich von 1970 bis 1979 um 78 % und Italien um 129 % gesteigert haben, während die deutschen und englischen Exporte im wesentlichen stagnierten.

Wichtig erscheint die Feststellung, daß der größte Teil des grenzüberschreitenden Handels westeuropäischer Binnenhandel ist (69 %), Westeuropa im Handel mit Drittländern aber dennoch ein bedeutender Schmierstoffnettoexporteur ist; es nahm 1978 mit einem Exportüberschuß von ca. 742.000 t in der Versorgung anderer Kontinente den 2. Platz hinter den USA als dem immer noch größten Schmierstoffnettoexporteur der Welt ein, wobei sich deren Exportüberschuß seit 1970 etwa halbiert hat und 1978 nur noch ca. 1.139.000 t betrug.

Der Mengenaustausch zwischen den USA und Westeuropa ist übrigens nicht besonders intensiv, d.h. der größte Teil ihrer jeweiligen Exporte geht in andere Erdteile; 1978 exportierten die USA nach Westeuropa ca. 249.000 t (etwa 19 % ihres Gesamtexports) und bezogen von dort ca. 122.000 t (das waren 9,4 % des westeuropäischen Exports in Drittländer).

Zusammenfassend ergibt sich für den westeuropäischen Schmierstoffmarkt in 1978 folgende Mengenbilanz:

Verwendung in 1.000 t	in %	
5.369	76,4	Gesamtablieferungen an das Inland
341	4,9	Bunker
1.315	18,7	Export an Drittländer
7.025	100,0	

Aufkommen		
573	8,2	Import aus Drittländern
467	6,6	Regenerate (inkl. Fluxöle in BRD)
482	6,9	Schmierstoffadditive und andere Zusätze
5.503	78,3	in die Schmierstoff-Gesamtablieferungen eingegangene Basisölmengen aus der Erstraffination
7.025	100,0	

Die Rolle der unabhängigen Firmen im europäischen Schmierstoffmarkt ist vielfältig und zumindest in einigen Ländern relativ bedeutend. Als unabhängige Schmierstoffirmen werden hier solche Gesellschaften bezeichnet, die über die Hälfte ihres Geschäfts mit Schmierstoffen machen und weder kapitalmäßig noch faktisch durch Markenverträge unter dem Einfluß der großen Mineralölkonzerne stehen; die Darstellung umfaßt also nicht den markengebundenen Schmierstoff- und Mineralölgroßhandel der Konzerne, die Tankstellen, Autowerkstätten etc.

Generell ist festzustellen, daß die unabhängigen Schmierstoffgesellschaften vor allem dort eine wichtige Position haben, wo es um höhere Stufen der oft stark spezialisierten Verarbeitung und Veredelung sowie um Dienstleistungen, Beratung und Vertrieb geht. So entfallen von der westeuropäischen Schmierölraffineriekapazität nur ca. 5,6 % auf unabhängige Firmen, während ganz oder beteiligungsanteilig 69 % von den "7 Schwestern" (Exxon/Esso, Shell, BP, Gulf, Texaco, Mobil und Socal/Chevron), 17,5 % von staatlichen oder halbstaatlichen und 7,8 % von privaten Mineralölkonzernen, die im breiten Publikumsbesitz sind, kontrolliert werden. Im Schmierstoffadditivegeschäft dürfte der Anteil der "7 Schwestern" (in diesem Bereich nur Exxon/Esso-Chemie und Oro-Gruppe/Chevron) demgegenüber weltweit und vermutlich auch in Westeuropa bei nur ca. 32 % liegen und an der Altölaufbereitung sind sie bisher in Westeuropa noch nicht beteiligt; lediglich andere private, halbstaatliche und staatliche Mineralölkonzerne sind an 4 der insgesamt 40 Zweitraffinerien, die 1979 allerdings etwa 20 % der westeuropäischen Regeneratproduktion erbrachten, beteiligt. Die Sammlung von Altöl und Herstellung von Regeneraten sind also in den meisten westeuropäischen Ländern noch eine reine Domäne der unabhängigen Schmierstoffirmen und werden es hoffentlich bleiben. Die meisten Unabhängigen sind jedoch in der Verarbeitung von Basisölen zu fertigen Schmierölen, Fetten und Spezialitäten tätig und vertreiben diese in den unterschiedlichsten Abpackungen an eine vielfältige Verbraucherschaft. Auf diese wollen wir uns im weiteren Verlauf konzentrieren.

Wir haben eine Analyse von Zahl, Betriebsgrößenklasse und Unabhängigkeit der Schmierölmischanlagen und Fettfabriken gemacht, die 15 Länder mit über 98 % des westeuropäischen Schmierstoffverbrauchs umfaßt. Das Ergebnis ist folgendermaßen:

In den genannten Ländern gibt es 390 Schmierölmischanlagen und 110 Fettfabriken.

282 Mischanlagen, d.h. über 72 % gehören unabhängigen Schmierstoffirmen; auf sie könnten gut 26 % des Gesamtvolumens der Betriebs-

größen entfallen sein. Bei den Fetten ist der Anteil der Unabhängigen größer; ihnen gehören 73 % der 110 Fettfabriken mit einem Volumenanteil von vermutlich gut 54 %.

Die Verteilung der Anlagen auf die verschiedenen Betriebsgrößenklassen sowie die Durchschnittsgröße differieren von Land zu Land. Die westeuropäische Durchschnittsanlage dürfte für Öl etwa 17.800 t und für Fett ca. 3.400 t "Kapazität" haben; nimmt man nur die Unabhängigen, so könnte ihre durchschnittliche Betriebsgröße bei etwa 6.500 t für Öl bzw. 2.540 t für Fett liegen. Die Verteilung auf Größenklassen ist:

Schmierölmischanlagen			Fettfabriken		
Größe in 1.000 t	Zahl in %	Volumen in %	Größe in 1.000 t	Zahl in %	Volumen in %
1	18,7	1,1	0,5	19,1	3,2
1 - 3	22,3	2,5	0,5 - 1	25,5	5,9
3 - 6	14,6	3,7	1 - 2	10,0	5,1
6 - 10	9,8	4,5	2 - 4	14,5	12,9
10 - 20	13,3	11,2	> 4	30,9	72,9
> 20	21,3	77,0			
	100,0	100,0		100,0	100,0

Viele der in der Regel in 1 Schicht arbeitenden Schmierölmischanlagen dürften schlecht ausgelastet sein. Eine einheitliche und zuverlässige Kapazitätsmessung ist zwar wegen der Unterschiedlichkeit der Anlagen, Verfahren, Produkte, Fertigungslosgrößen und Sortenwechselfolge unmöglich, wir schätzen jedoch die Auslastung auf nur ca. 2/3; die Hauptgründe hierfür liegen insbesondere in der Sortimentsbreite, der Kompliziertheit der Rezepturen, den oft kleinen Fertigungslosen und den teilweise zeitraubenden Qualitätskontrollarbeiten des Industrieölgeschäfts. Auch die Auslastung der Fettfabriken schätzen wir im westeuropäischen Durchschnitt auf nur etwa 2/3 der Kapazität (eine Repräsentativerhebung in Deutschland hatte allerdings für 1977 ca. 72 % ergeben).

Im Gegensatz zu den meisten der "7 Schwestern", die Schmierölmischanlagen und manchmal auch Fettfabriken in mehreren westeuropäischen

Ländern haben, betreiben die übrigen internationalen und nationalen Mineralölkonzerne derartige Anlagen teilweise nicht oder nur in einem Land, d.h. sie versorgen ihr europäisches Schmierstoffgeschäft zentral von einem Land aus und kaufen im übrigen zu, wobei nicht selten unabhängige Firmen die Lieferanten sind.

Das unabhängige Schmierstoffunternehmen ist kein uniformes Gebilde; es präsentiert sich in den unterschiedlichsten Formen und Dimensionen. Es wäre interessant, eine Typologie der unabhängigen Schmierstoffirmen in Westeuropa zu machen. Sie würde Firmen enthalten, die aus Rohöl oder feedstocks Basisöle bzw. aus Altöl Regenerate herstellen, andere wiederum haben sich auf die Herstellung von Motorenölen, Industrieölen oder Fetten spezialisiert; die einen produzieren, die anderen vertreiben nur - sei es regional, national oder international - und manche machen alles, d.h. sie arbeiten mit einem vollen Sortiment und Betriebsstätten in mehreren westeuropäischen Ländern auf den verschiedenen Stufen der Verarbeitung und des Vertriebs.

Die öffentliche Diskussion ist seit Jahren ganz von den energetischen Produkten beherrscht, aber fest steht: Wenn wir keine Schmierstoffe mehr hätten, bräuchten wir auch keine Kraftstoffe und Heizöle mehr. Und hier erfüllen die unabhängigen Schmierstoffirmen in ihrer Gesamtheit wichtige Funktionen, nämlich:

Angewandte Forschung und Entwicklung als Beitrag zum Gesundheits-, Arbeits- und Umweltschutz, zur Verminderung von Reibung und Verschleiß, zur Energieersparnis, zur rationellen Fertigung in der Metallbearbeitung und zum Korrosionsschutz etc.

Produktion und Weiterverarbeitung, für das Industrieschmierstoffgeschäft mit hoher Flexibilität in der Sortimentsbreite und der Deckung spezieller Bedarfsfälle.

Lagerhaltung für schnelle Belieferung und als Beitrag zur Versorgungssicherheit (die gesamten Schmierstofflagermengen der Unabhängigen dürften in einigen Ländern mindestens so groß sein wie die der Schmierölraffinerien).

Marktausgleich.

Technische und wirtschaftliche Beratung, Service.

Distribution, auch in kleinen Mengen.

Entsorgung durch Sammeln und Recycling von Gebrauchtölen als Beitrag zum Umweltschutz und zur Rohstoffersparnis.

Die Position der unabhängigen Schmierstoffirmen ist in den einzelnen westeuropäischen Ländern unterschiedlich stark. Ihr Gesamtmarktanteil - bezogen auf die Ablieferung fertiger Schmierstoffe in den Verbrauchersektor - beträgt in einigen Ländern bis zu ca. 45 %, in anderen nur etwa 10 bis 20 %. Den westeuropäischen Durchschnitt haben wir für 1979 mit ca. 27 bis 29 % berechnet, wobei für einige der kleineren Länder, aus denen nur begrenzte Informationen vorlagen, Schätzungen notwendig waren. Zum Vergleich seien hier die USA und Japan genannt, in denen der Anteil der unabhängigen Schmierstoffirmen bei ca. 20 % bzw. 40 - 50 % liegen dürfte. Bemerkenswert ist dabei die Tatsache, daß der Marktanteil der Unabhängigen in den wesentlichen Ländern seit Jahren gleich bleibt (z.B. Deutschland mit 43 - 46 %) oder steigt (z.B. Frankreich von 25,9 % in 1974 auf 28,6 % in 1979). Auch in den USA ist diese Entwicklung zu beobachten; so ist dort der Anteil der Mineralölkonzerne von 1965 - 75 zu Gunsten der Unabhängigen z.B. für Fette kontinuierlich von 76 % auf 64 %, für Industrieöle von 92 % auf ca. 85 % und für Motorenöle auf ca. 70 % zurückgegangen. Der Rückgang der Zahl unabhängiger Mitglieder in den Schmierstoffverbänden ist also kein Verlust der Marktposition, sondern Ausdruck einer vielfältigen Konzentration. Es besagt nichts über die Lage der unabhängigen Schmierstoffirmen, wenn z.B. die Mitgliederzahl des französischen SNICL von 350 Firmen in 1965 auf nur noch 220 Firmen in 1980 oder die der British Lubricants Federation von 339 Firmen in 1945 auf weniger als 100 zurückgegangen ist.

Wie wird nun die Entwicklung in den 80er Jahren weitergehen? Wir halten folgendes für relativ wahrscheinlich:

Der Weltschmierstoffverbrauch wird auch weiterhin nur noch in geringem Maß wachsen. Dieses Wachstum wird kaum in den großen Schmierstoffländern der freien Welt stattfinden, sondern im Ostblock sowie den Entwicklungs- und Schwellenländern. Vor allem bei Motorenölen wird man in Westeuropa und den USA mit Stagnation oder leichter Schrumpfung rechnen müssen.

In amerikanischen Veröffentlichungen wird beklagt, daß die Verfügbarkeit solcher Rohöle, die sich besonders leicht und gut auf Basisöle

verarbeiten lassen, zurückgeht. Wir haben den Eindruck, daß sich dies
vor allem auf die klassischen paraffinischen und naphthenischen US-
crudes bezieht, wobei dort erschwerend hinzukommt, daß der Anteil der
naphthenischen Öle 1978 immer noch bei ca. 25 % lag, während es in
Europa max. 15 % und in Japan 13 % waren. Westeuropa war für die Basis-
ölproduktion neben dem in Deutschland und Österreich verarbeiteten hei-
mischen Rohöl schon immer auf die traditionellen Schmierölimportprove-
nienzen aus Saudi-Arabien, Iran, Kuwait und Venezuela angewiesen und
nahm in jüngerer Zeit auch andere, insbesondere afrikanische Rohöle in
die Schmierölverarbeitung. Wir sehen also für Westeuropa das Qualitäts-
problem nicht so vorrangig, zumal es sich durch eine allerdings wesentlich aufwendigere, technische Auslegung der Schmierölraffinerien be-
wältigen läßt. Quantitativ wird Rohöl allerdings tendenziell knapp und
zunehmend teuer sein, wobei für die Schmierölerzeugung hinzukommen
kann, daß mit dem Ausbau der Konversionskapazitäten in Westeuropa atmosphärische Destillationsrückstände einen eigenen Marktwert bekommen.
Die "alternative Verwendung" wird dann für Basisöle eventuell zu einer
zusätzlichen Verteuerung führen, denn bisher liegt der Preis für
schweres Heizöl z.B. in Deutschland seit 1973 ständig unter dem Rohöl-
preis.
Die Kapazität der weltweiten Schmierölraffinerien liegt bei Ansatz einer relativ hohen Auslastung von 85 % nur noch um ca. 2 Mio t über dem
Bedarf, während es in früheren Jahren wesentlich höhere Überkapazitä-
ten gab. Neue Schmierölraffinerien werden in Westeuropa in Anbetracht
der extrem gestiegenen Investitionskosten, der hierdurch erforderlichen
Betriebsgrößen und des Preisdifferentials zwischen Basisölen und anderen Produkten kaum und in USA nur in wenigen Fällen gebaut werden.
Es wird neben dem in Westeuropa weitgehend abgeschlossenen debottle-
necking eher Umstellung und Ergänzung der Verfahren sowie generelle
Erweiterungen und in einigen Fällen eventuell auch Stillegung alter
Anlagen geben. Das alles spricht nicht nur für weiter steigende Betriebsgrößen in der Schmierölraffination, sondern auch für einen relativ festen Grundölmarkt. Ob die in einer Reihe von OPEC-Ländern, in
sonstigen afrikanischen Ländern und in Fernost geplanten Neubauten und
Erweiterungen von Schmierölexportraffinerien alle kommen und in der 2.
Hälfte der 80er Jahre international zu einem Überangebot von Basisölen
führen werden, erscheint zweifelhaft; immerhin müßten sie zunächst einmal den vermutlich weiter rückläufigen US-Schmierstoffexport sowie den

Anstieg des Weltschmierölverbrauchs ausgleichen.

Die Zweitraffination dürfte in der Lage sein, ihre Position in der Schmierölversorgung in Deutschland, Frankreich und Italien zu halten sowie in anderen westeuropäischen Ländern auszubauen, wenn angemessene Subventionssysteme bestehen bleiben bzw. entwickelt werden, um die Kosten der Altölsammlung zu decken, oder wenn die Verbrennung von Altöl in allen Ländern gesetzlich verboten sowie die unentgeltliche Abgabe an die Sammelorganisationen sichergestellt werden. Es gibt in verschiedenen westeuropäischen Ländern Neubau- und Erweiterungsplanungen sowie die Absicht, auf neue Verfahren umzustellen; andererseits sind weitere Schritte der Konzentration auch hier nicht auszuschließen.

Der mengen- und insbesondere wertmäßige Anteil der Additive und in begrenztem Umfang auch der Syntheseöle wird mit der Entwicklung zu immer höherwertigeren Schmierstoffen weitersteigen.

Die unabhängigen Schmierstoffirmen werden in der Lage sein, ihre Marktstellung zu behaupten sowie im Bereich der Industrieschmierstoffe und Spezialitäten eventuell sogar auszubauen. Im Bereich der Motorenöle wird dies schwieriger sein. Auf jeden Fall wird die Konzentration bei den Unabhängigen sowie ihren Mischanlagen und Fettfabriken weitergehen; die kleinen Firmen werden nur dort bestehen können, wo sie selektiv vorgehen und dann auf ihrem Arbeitsgebiet hochspezialisiert, flexibel und service-intensiv sind.

Schwierig zu beurteilen ist die künftige Entwicklung der erzielbaren Verkaufspreise für Schmierstoffe und damit der Wertschöpfung und Rentabilität der unabhängigen Firmen. Vergleicht man das Überangebot der 60er Jahre und die Zeiten der krisenhaften Mineralölversorgung in den 70er Jahren, so bestätigt sich die allgemeine Erfahrung, daß sich die Spielräume der Kostenüberwälzung bei relativ knapper Marktversorgung erweitern. Es ist aber zu berücksichtigen, daß dieses Geschäft ungewöhnlich kostenintensiv ist und der Aufwand für Personal, Verpackungsmittel, Frachten etc. schnell weitersteigen wird; im übrigen ist - insbesondere bei Motorenölen - nicht auszuschließen, daß trotz begrenzter Verfügbarkeit und relativ fester Verfassung des internationalen Grundölmarkts der Druck auf die Fertigproduktpreise anhält, da es an Misch- und Vertriebskapazitäten nicht fehlt. Hier wird es für die unabhängigen Schmierstoffirmen entscheidend auf Sortimentspolitik, Vertriebswege und Effizienz ihrer Organisationen ankommen.

ENERGIEEINSPARUNG DURCH WIEDERAUFARBEITEN VON GEBRAUCHTÖL
ENERGY SAVING BY RE-REFINING USED OILS
LES ECONOMIES D'ENERGIE PROCUREES PAR LA REGENERATION DES HUILES
USAGEES
LE ECONOMIE DI ENERGIA OTTENUTE ATTRAVERSO LA RIGENERAZIONE DEGLI OLI
USATI

C.J. THOMPSON, D.W. BRINKMAN
U.S. Department of Energy, Bartlesville Energy Technology Center
Bartlesville, Oklahoma, USA

SUMMARY

The case of burning of used automotive lubricating oil versus re-refining has been reevaluated based upon the 1980 American economic and energy conservation posture. Areas that have been reevaluated in the context of burning versus re-refining are the energy advantage to be realized in terms of resource conservation as well as an estimation of the economics and profit motives currently available in the disposition of used lube oil.

This study was undertaken because previous data and information related to the energy savings by re-refining used oils are no longer timely. Earlier reports were based upon acid/clay or other obsolete technology, and the disproportionate changes in costs of energy as compared to the general inflation rate have made simple extrapolation of earlier economic comparisons subject to major error.

The data show re-refining superior in comparison to burning in terms of resource conservation. This is particularly significant because of the continuing decline of the re-refining industry in the United States and the growing concern for American dependency upon petroleum imports.

ZUSAMMENFASSUNG

Eine neue vergleichende Untersuchung zwischen der Verbrennung von Gebrauchtöl aus Kraftfahrzeugen und der Zweitraffinierung ist unter Berücksichtigung der amerikanischen Wirtschaftslage des Jahres 1980 und der Anforderungen der Energieeinsparung durchgeführt worden. Bei dieser Untersuchung ist auch den Umweltbedingungen im Rahmen von vier Hypothesen Rechnung getragen worden, die von der restriktionslosen Verbrennung von Altschmieröl bis zum vollständigen Verbot der Verbrennung reichen, das die Zweitraffinierung aller Altschmieröle aus Kraftfahrzeugen zur Folge hat.

Gebiete, die im Rahmen der Gegenüberstellung von Verbrennung und Zweitraffinierung der Altschmieröle aus Kraftfahrzeugen neu beurteilt wurden, sind die Einsparung von Energieressourcen sowie die Wirtschaftlichkeit und die Gewinnlage bei der Beseitigung von Gebrauchtöl.

Diese Untersuchung wurde durchgeführt, weil die früheren Daten und Informationen über die Energieeinsparung durch Wiederraffinierung von Gebrauchtöl nicht mehr zeitgemäss sind. Den früheren Berichten lagen Säure/Erde- oder andere veraltete Technologien zugrunde, und eine einfache Extrapolation auf Grund früherer wirtschaftlicher Vergleiche war wegen der extrem hohen Schwankungen der Energiekosten im Vergleich zur allgemeinen Kostensteigerung unmöglich.

Nach den Ergebnissen der Untersuchung ist die Zweitraffinierung hinsichtlich der Einsparung von Ressourcen vorteilhafter als die Verbrennung. Dies ist angesichts der weiterhin rückläufigen Entwicklung der Raffinerie-Industrie in den Vereinigten Staaten und der zunehmenden Abhängigkeit Amerikas von Erdölimporten besonders wichtig.

RESUME

L'option entre le brûlage des huiles de graissage automobile et la régénération a été réévaluée compte tenu de la situation de l'Amérique en 1980 sur le plan économique et sur le plan de la conservation de l'énergie. Dans ces comparaisons, l'environnement est également pris en considération dans quatre scénarios allant du brûlage des huiles de graissage usagées, sans restrictions officielles, à l'interdiction totale du brûlage, impliquant que toutes les huiles de graissage pour automobiles sont destinées à être régénérées.

Les points qui ont été réévalués dans l'optique du choix entre brûlage et régénération des huiles de graissage pour automobile sont l'avantage que présente la conservation des ressources énergétiques ainsi que l'estimation des économies et des possibilités de profit qui existent actuellement dans l'élimination de ces huiles usagées.

Cette étude a été entreprise parce que les données précédentes et l'information sur les économies d'énergie produites par la régénération des huiles usagées ne sont plus à jour. Les rapports précédents étaient fondés sur l'acide/argile ou d'autres technologies désuètes, et les hausses disproportionnées des coûts de l'énergie par rapport au taux d'inflation général font que l'extrapolation pure et simple des comparaisons économiques antérieures peut donner lieu à des erreurs importantes.

Les données indiquent que la régénération est préférable au brûlage sous l'angle de la conservation.de ressources. Ce point est particulièrement intéressant au vu du déclin constant de l'industrie de la régénération aux Etats-Unis et de la préoccupation croissante devant la dépendance de ce pays à l'égard des importations de pétrole.

RIASSUNTO

Il problema della combustione di oli lubrificanti usati per motori rispetto alla loro rigenerazione è stato rivalutato sulla base di un orientamento americano adottato nel 1980 per la conservazione dell'energia e per fini economici. In questi raffronti, l'ambiente è considerato in quattro scenari che vanno dalla combustione illimitata di oli lubrificanti usati senza vincoli governativi al divieto assoluto di combustione con conseguente dirottamento di tutti gli oli combustibili per motori verso la rigenerazione.

I motivi di una rivalutazione della combustione rispetto alla rigenerazione di oli combustibili per motori sono costituiti dai vantaggi energetici realizzati in termini di conservazione delle risorse nonché da una stima degli aspetti economici e dei profitti attualmente esistenti nello smaltimento deli oli usati.

Questo studio è stato effettuato in quanto non sono più validi i dati e le informazioni precedenti sui risparmi di energia mediante rigenerazione degli oli usati. Le relazioni precedenti erano imperniate su una tecnologia obsoleta a base di acido/argilla o di altro tipo e i cambiamenti sproporzionati nei costi dell'energia rispetto all'inflazione generale hanno provocato errori basilari nel caso di una semplice estrapolazione dai precedenti raffronti economici.

I dati indicano che la rigenerazione è superiore rispetto alla combustione in termini di conservazione delle risorse. Questo aspetto è particolarmente importante a causa del continuo declino dell'industria di rigenerazione negli Stati Uniti e della crescente preoccupazione per la dipendenza americana dalle importazioni di petrolio.

Comparative energy balances related to the disposition of used lubricating oil through either burning to recover the heat value or re-refining for restoration to a useful role as a lubricant have been the subject of several recent studies. One of these was prepared for the U.S. Environmental Protection Agency in 1975. Unfortunately, the conclusions reported from that study are no longer timely, and simple extrapolations of previous economic comparisons are subject to major error.

The data that I will present today are related to the American economy and the American posture with regard to re-refining of used automotive lubricating oil.

I will discuss, in a comparative sense, the case for burning used lube oils versus the case for re-refining these same oils. In these comparisons, I will discuss resource conservation and economics.

Obviously prices, inflation rates, petroleum supplies, legislative mandates, and geographical considerations have a considerable impact. Therefore, it would be presumptious of me to attempt to extrapolate the conclusions of this study to the situation in most other countries of the world. However, many of the factors that are discussed may have worldwide application and from that view, I sincerely hope that a discussion of the American situation with regard to re-refining of used lube oils holds some interest for you.

In this updated study, which will be published early in 1981, the technology evaluated is that developed at the Bartlesville Energy Technology Center (BETC) of the U.S. Department of Energy and is the basis for economic comparisons between burning and re-refining. The BETC (as I shall refer to it) technology is essentially solvent treatment/vacuum distillation/hydrogenation, but we believe the general arguments could be used for any of the newer technologies.

With the depletion of naphthenic crude oils for the manufacture of lubricating oil basestocks, lube plants are using more and more paraffinic stocks which require additional costly processing steps and give generally lower yields of lube basestocks. Burning of used lubricating oils as a fuel oil supplement destroys high quality basestocks, and these must be replaced by the more costly production methods from crude oil at constantly escalating prices. The questions that I will address today involve the magnitude of these higher production costs; potential savings by re-refining in terms of energy; and the equivalence in crude oil imports to America.

Table I.--Lube Oil Production From Paraffinic Crude Oil

	Yield, Vol-Pct
Lube Oils	21.8
Fuels and Other Products	78.2
	100.0

Table II.--Used Oil Availability

	Vol-Pct
Lost in Service	40.0
Theoretically Collectable	60.0
	100.00

Table III.--Yield From BETC Re-Refining Technology
(Recoverable Oil Basis)

	Vol-Pct
Lube Oils	90.0
Fuel (Distillation Bottoms)	10.0
	100.0

Table IV.--Basic Assumptions for Material and Energy Balances

(1) Constant volume of reduced crude whether used oil is burned or re-refined.

(2) Reduced crude directly useable as fuel.

(3) By-products assumed useable as fuel without further treatment.

(4) Energy to remove water, fuel, additives and sludge from used oil included in total energy consumption for BETC Process.

(5) Distillation bottoms from vacuum distillation useable for fuel without further treatment.

The important assumptions related to production of lube oil and fuel that have formed the basis for subsequent energy calculations are shown in Tables I through IV.

Table I shows the yields of lube oil basestocks (21.8 percent) and by-products (78.2 percent) from a paraffinic crude oil that has been topped by atmospheric distillation.

Table II gives an estimate of the losses of lube oil basestock during use--some 40 percent. These losses are attributed to combustion, leakage, and handling.

Table III shows a yield of 90 percent of lube oil based upon recoverable oil using the BETC Technology. Water and fuel dilution of the used lube oil are neglected from consideration because they are common factors in our comparative cases of burning versus re-refining. To simplify material balances, we have chosen to eliminate additives from consideration. It is difficult to determine the energy requirement for production of a typical additive package, and at present we cannot define the embodied energy of additives which are removed primarily by the solvent treatment step or the heating value of these additives when they are burned. We have assumed, however, that the embodied energy of the additives either in the form of solvent-precipitated sludge or in used oil destined for fuel use is essentially equal to the energy required to manufacture the additives. Therefore, additives can be eliminated from material balance comparisons.

The more important assumptions that were used in our resource conservation estimates are included in Table IV.

The first of these premises is that a constant volume of reduced crude is processed regardless of the final role for used lube oil. We must make this assumption because gasoline production is the driving force that controls the crude oil processed in the United States and, further, we wish to maintain constant volumes of total lube production and total fuel production for purposes of energy comparisons.

Two--To maintain a fuel material balance we have assumed that reduced crude can be diverted from the manufacture of lube oil directly to use as a burner fuel supplement without further processing.

Three--Each of the by-products from a lube plant has embodied energy, and the heating values of these streams are treated as fuel that requires no further treatment prior to use.

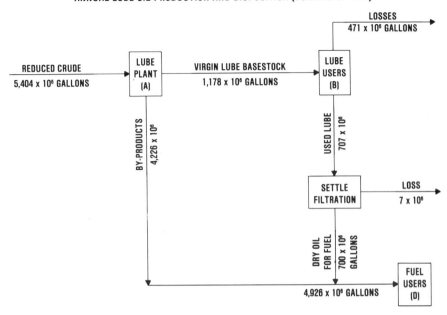

Figure 1

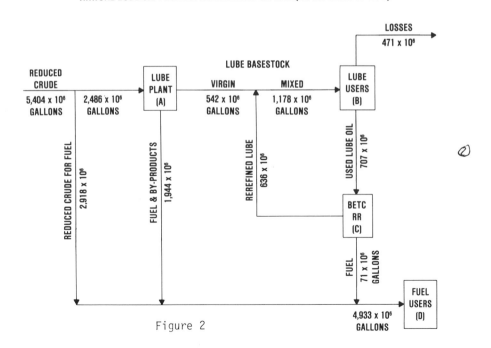

Figure 2

Table V.--Energy Consumed in Process Units Per Gallon Feed

Process	Energy Consumption BTU/Gal
Virgin Lube Plant	
Total Energy Per Gallon.................	23,000
BETC Re-Refining Technology	
Total Energy Per Gallon.................	20,000
Settling and Filtration	
Water Removal From Used Oil.............	0.3

Table VI.--Heating Values For Various Refining Streams

Material	Heating Value, BTU/Gal
By-Products, Weighted Average............	152,828
Reduced Crude............................	150,750
Re-Refining Fuel (Dist. Bottoms).........	160,000
Virgin and Re-Refined Lube Oil...........	142,800
Solvent Make-Up..........................	100,000

Table VII.--Annual Energy Consumption and Potential Savings

	Burning Scenario	Re-Refining Scenario
Virgin Lube Production		
$(5.404 \times 10^9)(23{,}000 \text{ BTU/Gal}) =$	124.3×10^{12}	
$(2.486 \times 10^9)(23{,}000 \text{ BTU/Gal}) =$		57.2×10^{12}
Lube by BETC Re-Refining		
None	--	
$(707 \times 10^6 \text{ Gal})(20{,}000 \text{ BTU/Gal}) =$		14.0×10^{12}
Fuels by Settling and Filtering		
$(700 \times 10^6)(.32 \text{ BTU/Gal}) =$	224×10^6	
Annual Total Energy Consumption	124.3×10^{12}	71.2×10^{12}
Annual Total Energy Advantage, Re-Refining Over Burning		53.1×10^{12}
Crude Oil Equivalence		9.1×10^6 BBL
Cost Advantage		$\$299 \times 10^6$

Four--Water contamination, fuel dilution, additive, and sludge from used lube oil are eliminated from our material balances because they represent identical quantities whether the used oil is burned or re-refined. However, the energy required to remove these materials is included in the total energy consumption for the BETC Re-Refining Process which is discussed later.

Five--Finally, we assume that distillation bottoms from the vacuum distillation step of the BETC Re-Refining Process are useable for fuel without further treatment other than blending with appropriate fuel oils.

Figure 1 shows a schematic drawing in which all used automotive lube oil in the United States is burned. The volume of reduced crude needed to provide the required 1.2 billion gallons of basestock is obtained by processing 5.4 billion gallons of reduced crude. Based upon our assumed lube oil yield of 21.8 percent from lube plant (A) and 78.2 percent by-products, the equivalent fuel production of the lube plant would exceed 4.2 billion gallons annually. Simultaneously, directing the total used automotive lube oil generated by lube users (B) provides another 700 million gallons of fuel oil. The total contribution of fuel to users (D) is more than 4.9 billion gallons annually.

Figure 2 is related to the BETC Re-Refining Scenario. For comparative purposes, the national demand of 1.2 billion gallons of lube oil is maintained. By re-refining all of the used oil generated by lube users (B) and returning it to lube users through re-refining, the required production of lube oil from virgin materials is reduced substantially.

The material balance advantage in this scenario can be viewed in the context that since gasoline production is currently the prime driving force in the American refining industry, total crude processing would not be changed by the conservation and recycling of 636 million gallons of re-refined lube oil. We can reason that the contribution of 636 million gallons of lube oil basestock from re-refining would decrease the output requirement of lube plant (A) from 1.2 billion gallons of lube oil to 542 million. Based upon 21.8 percent yield of lube oil basestock from lube plant (A), the feedstock requirement of reduced crude directed to the manufacture of lube oil would be changed from 5.4 billion gallons to 2.5 billion. This allows almost 3 billion gallons of reduced crude to be diverted to fuel use. Our material balance is maintained by combining the diverted reduced crude, the lube plant by-products, and the fuel from the re-refining process.

From published data, the process energy consumed per gallon, in a virgin lube plant, in a plant with the BETC Technology and in a settling and filtration plant is shown in Table V. It is assumed that the reduced crude has two possible uses--that is, fuel or lube oil, with the implication that reduced crude goes to one use or the other. Accepting this interpretation, then the difference in energy requirements for re-refining versus burning of recovered used automotive lube oils becomes an important factor in our comparative analysis.

The energy requirement to process reduced crude in a lube plant is estimated to be about 23,000 BTU per gallon of plant throughput. The value for re-refining is lower at about 20,000 BTU per gallon. Settling and filtering of used lube oil require less than one BTU per gallon. These then are numbers of significance as they are used in future energy calculations.

Table VI shows the heating values that were used for energy comparisons. Each by-product has embodied energy which can be equated to fuel without further processing. Based upon the yields of each from the lube plant a weighted average heating value was calculated of 152,828 BTU per gallon. The other numbers in this Table relate to the various other feedstocks and products of the systems which are being compared.

Table VII contains the material balances and the energy consumption derived from previously presented data. The importance of this summation is reflected in the energy consumption for similar volumes of automotive lube oil when used oil is re-refined and when used oil is burned for fuel. The difference in annual energy consumption is about 53 trillion BTU's. Based upon an average heating value of 5.86 million BTU per barrel of crude oil, the advantage for converting used lubricating oil to reuseable lube oil rather than a burner fuel is equivalent to nearly 9 million barrels of crude oil annually with a monetary savings of more than a quarter billion dollars.

In the consideration of economics of re-refining used lubricating oils versus burning, a different set of assumptions was required from those set down for resource conservation material balances.

We have based our cost comparisons on a basic volume of 10 million gallons per year of used automotive lube oil and since most areas in the United States currently burn used lube oil in combination with fuel oils with no treatment other than settling and filtration we have considered only the case of re-refining versus settling and filtering.

Table VIII is based upon the premise that raw used lube oil can be collected and delivered to a processing site for about 37 cents per gallon.

Used lube oil that has had the free water and nonsuspended solids removed by settling and filtration is assumed to be very competitive as a burner fuel supplement at 45 cents per gallon with low sulfur No. 6 fuel oil selling at about 70 cents per gallon. The distillation bottoms from the BETC Re-Refining Process are assumed to be marketable at near 50 cents per gallon as a burner fuel supplement.

The investment for process facilities to handle 10 million gallons per year of used lube oil input is shown to range from a nominal $265,000 for a settling and filtration facility to $5.2 million dollars for a re-refinery designed around the BETC Technology.

Product volumes from each facility are also shown ranging from 99 percent fuel production by settling and filtration to the generation of 6.6 million gallons of high quality lube oil basestock by re-refining with additional amounts of fuel and by-products.

The sludge produced from the BETC solvent treatment is currently being evaluated for its utility. At the present, it appears promising that a useful role for this sludge exists that will lend embodied energy to this stream. Our current position is a middle-of-the-road stance assuming neither embodied energy nor cost for disposal.

Table IX shows the process costs in thousands of dollars per year to convert 10 MM Gal of used automotive lube oil to the designated products.

These process costs are compared with revenues shown in Table X to provide an estimate of profits from the two operations--almost $650,000 in the fuel mode and over 2 million by re-refining.

Re-refining yields a lower return upon investment than does a settling and filtration plant, but as shown in Table XI, re-refining produces a greater gross and net income based upon handling 10 million gallons per year of used lube oil than settling and filtering. Return on investment is about 20 percent after taxes for re-refining. Although, this is a very respectable return, it does not compare favorably with that realized by diverting used automotive lube oil to fuel with only a cursory treatment of settling and filtering.

Conclusions that we have drawn from this study are then set forth in Table XII.

(1) Re-refining finds its superiority in comparison with burning in terms of resource conservation. The potential savings of energy is sub-

Table VIII.--Cost Basis

	Used Automotive Lube Oil	
	Settled & Filtered	Re-Refined
Selling Price for:		
Raw Used Lube Oil for Fuel, ¢/Gal	45	-
Fuel From Re-Refining, ¢/Gal	-	50
Re-Refined Lube Basestock, $/Gal	-	1.12
Facility Investment, MM $	0.265	5.2
Assumed Product from 10 MM Gal Per Yr Raw Used Oil		
Fuel, MM Gal	9.9	1.6
Lube, MM Gal	-	6.6
Sludge, MM Gal	-	1.0
Other, MM Gal	0.1	.8

Table IX.--Economic Comparisons

	Used Automotive Lube Oil, 10 MM Gal	
	Settled & Filtered For Fuel	Re-Refined To Lube Oil
Process Costs, $1000/year		
Raw Used Lube Oil @ 37¢ Gal	3700.	3700.
Power @ 6¢/KWH	19.	246.
Steam @ $5.05/1000 Lbs	-	98.
Solvent @ $2/Gal	-	108.
Fuel @ 40¢/Gal	-	308.
Catalyst @ $125/CF	-	17.
Hydrogen @ $6/1000 SCF	-	162.
Operating Labor @ $20,000	20.	240.
Overhead @ 100% of Op. Labor	20.	240.
Maintenance @ 5% of Investment	13.	260.
Ins. & Taxes @ 3% of Inv	8.	156.
Depreciation @ 10% of Inv	27.	520.
Total Process Costs	3807.	6055.

Table X.--Economic Comparisons

	Used Automotive Lube Oil, 10 MM Gal	
	Settled & Filtered For Fuel	Re-Refined To Lube Oil
Revenues, $1000/Year		
Lube Oil @ $1.12/Gal.....	-	7392.
Fuel @ 50¢/Gal...........	-	800.
Fuel @ 45¢/Gal...........	4455.	-
Total Revenues...........	4455.	8192.
Total Process Costs......	3807.	6055.
Profits	648	2137

Table XI.--Economic Comparisons

	Used Automotive Lube Oil, 10 MM Gal	
	Settled & Filtered For Fuel	Re-Refined To Lube Oil
Profits/Loss Statement, $1000/Year		
Profits, Before Taxes........	648.	2,137.
Income Tax @ 50%.............	324.	1,069.
After Tax Profit.............	324.	1,068.
After Tax Return on Investment, %.	122.	19.8

Table XII.--Conclusions

(1) In terms of resource conservation, re-refining is superior to burning and conserves an estimated 53 trillion BTU's per year. This is the energy equivalent of nearly 9 million barrels of crude oil or a savings of about $300 million annually.

(2) Used lube oil that has been settled and filtered as a burner fuel supplement shows the greatest percent return on investment.

(3) Re-refining provides an attractive return based upon 1980 markets, costs, and dollars.

stantial, and it represents about the amount that the United States imports in 1 1/2 days in 1980 from OPEC Nations.

(2) Using essentially untreated used automotive lube oils as burner fuel supplement is wasteful in terms of resource conservation, but in today's economic and social climate in the United States burning represents the largest return upon investment in the disposition of used lube oil.

(3) Re-refining using the BETC Technology will show a very attractive return on investment and would represent resource conservation advantages to benefit the Nation's energy shortages. It appears that the re-refiner can compete in the current United States market for used oil feedstock, but the investment required to re-refine makes re-refining considerably less attractive than settling and filtering for fuel.

It appears doubtful that the current trends in America will change without incentives to the re-refiner in the form of government grants, low-interest loans or other financial aids, and without comprehensive legislation to discourage the burning option of used lube oils while providing a less competitive climate for used oil feedstock from lube users.

EMULSIONSBEHANDLUNG
TREATMENT OF EMULSIONS
TRAITEMENT DES EMULSIONS
TRATTAMENTO DELLE EMULSIONI

J. DUMORTIER
Vizepräsident der S.N.F.R.H.G., Paris
Vice President of S.N.F.R.H.G., Paris
Vice-Président du S.N.F.R.H.G., Paris
Vice Presidente del S.N.F.R.H.G., Parigi

RESUME

L'exposé de la méthode MATTHYS/G.A.R.A.P. concernant le traitement des émulsions et éventuellement des solutions met en évidence le fait que ladite méthode se recommande à un triple point de vue :

- Sur le plan écologique : rejets d'excellente qualité ou même suppression totale des rejets.

- Sur le plan des économies d'énergie : dépense beaucoup plus faible que dans le cas des autres méthodes.

- Sur le plan du recyclage : récupération pratiquement intégrale de la totalité des hydrocarbures que l'on ne détruit jamais.

ZUSAMMENFASSUNG

Die Erläuterung der Methode MATTHYS/GARAP zur Behandlung der Emulsionen und gegebenenfalls der Lösungen lässt erkennen, dass sich diese Methode unter drei Gesichtspunkten empfiehlt:
- vom ökologischen Standpunkt: Rückstände ausgezeichneter Qualität oder sogar völliges Fehlen von Rückständen;
- in Bezug auf Energieeinsparungen: sehr viel geringerer Aufwand als bei den anderen Methoden;
- unter dem Aspekt des Recycling: praktisch völlige Rückgewinnung der Gesamtheit der Kohlenwasserstoffe, die niemals zerstört werden.

SUMMARY

The report on the **MATTHYS/GARAP** method for the treatment of emulsions and possibly of solutions illustrates the fact that this method is suitable from three points of view:
- ecology: high quality waste or even total elimination of waste;
- energy saving: far less expense than with other methods;
- recycling: practically complete recovery of all hydrocarbons, which are therefore never destroyed.

RIASSUNTO

L'esposizione del metodo MATTHYS/GARAP sul trattamento delle emulsioni ed eventualmente, delle soluzioni mette in evidenza il fatto che tale metodo si riferisce ad un triplice punto di vista :
- dal punto di vista ecologico : rifiuti di eccellente qualità oppure eliminazione totale degli residui ;
- dal punto di vista del risparmio energetico : minor spesa che con altri metodi ;
- dal punto di vista del riciclo : ricupero praticamente integrale di tutti gli idrocarburi che non vengono mai distrutti.

L'industrie mécanique française, au même titre que celle des autres pays industrialisés, se heurte au problème posé par l'élimination des huiles solubles.

Ce problème est d'autant plus important que cette catégorie d'huiles, compte tenu des avantages qu'elle présente pour l'utilisateur, tend à se développer beaucoup plus rapidement encore que la production industrielle puisqu'elle se substitue de plus en plus à l'utilisation des huiles entières dans un certain nombre de domaines.

En général, la destruction de ces émulsions entraîne la consommation de quantités considérables d'énergie d'origine thermique, qu'elles soient détruites par évaporation ou par incinération.

D'autre part, la fraction huileuse se trouve perdue alors que dans le même temps l'industrie de la régénération se heurte actuellement en FRANCE à l'insuffisance des ressources en huiles usées.

Compte tenu de ces circonstances, les Sociétés MATTHYS et G.A.R.A.P. ont donc étudié en commun s'il ne serait pas possible de détruire ces émulsions tout en récupérant la fraction huileuse en vue de sa régénération.

C'est l'étude de ce procédé, actuellement exploité industriellement depuis plus de 3 ans, qui va être entreprise ci-après (on notera que ce procédé est d'ores et déjà exploité dans trois centres différents -un quatrième centre étant actuellement en projet et plusieurs consultations étant en cours d'examen-).

I - <u>TRAITEMENT DES HUILES SOLUBLES</u>

Il faut distinguer deux grandes catégories de produits, qui sont respectivement :

- les émulsions,
- les solutions.

A/ **Les Emulsions**

Les émulsions d'huiles minérales ou d'huiles minérales mélangées à des produits de synthèse, dites "huiles semi-synthétiques", sont traitées de la façon suivante :

1°) Chauffage entre 60 et 80° C, au moyen d'huile chaude circulant dans un serpentin et en provenance de la distillation sous vide (ou au moyen de vapeur pour les centres indépendants de l'usine de régénération).

2°) Extraction des hydrocarbures par centrifugation dans une centrifugeuse G.A.R.A.P., donnant une accélération de 7 500 g et permettant de séparer en continu deux phases, à savoir :

- une phase hydrocarbures contenant 90 à 95 % des huiles,
- une phase eau contenant les sédiments, ayant une DCO comprise entre 8 000 et 15 000 et pouvant être facilement floculée.

3°) Floculation. Cette opération comporte :

a) un ajustement du pH,
b) la floculation proprement dite, au départ d'un produit préparé en solution,
c) un réajustement du pH à 6,5, la floculation pouvant éventuellement faire descendre le pH jusqu'à 3.
d) une finition de la formation du floc en utilisant un poly-électrolyte.

Toutes ces opérations sont exécutées au moyen d'un mélangeur statique en ligne et entièrement en continu.

L'eau ainsi floculée est envoyée dans un décanteur où l'on obtient deux phases :

- de l'eau à moins de 5 000 de DCO, représentant 85 % du volume traité, qui est envoyée dans un bassin d'observation,
- une boue représentant environ 15 % du volume traité et dont la teneur en matières sèches est comprise entre 2 et 5 %.

Concentrée ultérieurement dans un séparateur centrifuge, cette boue qui contient alors 15 % de matières sèches peut être facilement incinérée dans les incinérateurs de l'usine ou d'un centre agréé.

En ce qui concerne l'eau, celle-ci, à la sortie du bassin d'observation, est envoyée vers un traitement bactérien qui fait descendre la DCO en-dessous de 500 et permet son rejet sans problème à l'égout public.

On notera qu'est actuellement à l'étude, en vue de son installation dans les usines de régénération ou dans les centres incinérateurs ayant des excès de calories (ceci est le cas de l'usine MATTHYS de LILLEBONNE), un procédé qui substitue au traitement bactériologique qui intervient à la sortie du bassin d'observation un évaporateur à triple effet.

Ce dernier permettra non seulement de supprimer tout rejet, mais même de récupérer les huiles de base dans le cas des solutions vraies.

B/ Les Solutions

En-dehors de la variante que l'on vient d'examiner et qui permettra la récupération des bases de solutions vraies, les solutions qui sont actuellement envoyées à la destruction sont des produits qui ont gardé la plupart de leurs propriétés et qu'il est facile de recycler.

Vidangées parce qu'elles sont polluées par des hydrocarbures étrangers (fuites de circuits hydrauliques, par exemple), des sédiments ou même des bactéries qui donnent des odeurs, on peut, en fait, les recycler grâce aux traitements suivants :

a) Centrifugation à 80° C pour obtenir la séparation des hydrocarbures étrangers des sédiments et de la majeure partie des bactéries (qui partent dans la phase des hydrocarbures étrangers).

b) Chauffage en ligne à 140° C sous 3 bars, suivi d'un refroidissement et d'une détente.

Le produit ainsi traité est complètement débarrassé :
- des hydrocarbures étrangers,
- des bactéries,
- des sédiments,

et peut être à nouveau employé, soit directement, soit après réajustement de son dopage (par l'utilisateur lui-même ou par le fabricant) si le cas le requiert.

Compte tenu de la valeur des produits et des coûts de destruction actuellement requis, il est inutile de souligner l'intérêt de cette technique de recyclage.

II - UTILISATIONS AUTRES DE CES CENTRES

Les centres de traitement des huiles solubles et des solutions faisant appel à la technique MATTHYS/G.A.R.A.P. présentent d'autre part de nombreux avantages.

Ils permettent en effet :

1°) De traiter, outre les huiles solubles, de nombreuses boues huileuses en provenance de l'industrie ou même d'autres stations de traitement d'huiles solubles dont les procédés ne présentent pas la même efficacité que le procédé MATTHYS/G.A.R.A.P.

C'est ainsi que sont actuellement traités les inter-faces de deux autres centres qui, autrement, ne pourraient se débarraser de ceux-ci que par incinération.

2°) Dans le cas de centres adjoints à des usines de régénération, de mettre les huiles usées collectées à l'état standard.

Cette mise à l'état standard (et notamment la phase de déshydratation) présente en effet l'intérêt de s'effectuer pratiquement en économisant la chaleur latente de vaporisation de l'eau, puisque cette dernière est séparée par voie mécanique et non par évaporation (l'élimination ultérieure de cette eau par incinération dans un four brûlant des déchets présente le double avantage d'être beaucoup moins coûteuse tout en augmentant considérablement la productivité de la distillation atmosphérique).

Enfin, l'extraction de l'eau par voie mécanique entraîne également l'évacuation de la totalité des sels solubles qui, en d'autres circonstances, seraient restés dans l'huile après évaporation de la phase aqueuse.

Comme ces sels solubles sont à l'origine de multiples difficultés de traitement (filtration, couleur), il est inutile d'insister sur l'intérêt présenté par cette méthode de déshydratation.

CONCLUSION

On voit que la méthode de traitement des émulsions et éventuellement des solutions par le procédé MATTHYS/G.A.R.A.P. permet :

- La récupération des hydrocarbures au lieu de leur destruction.

- L'évacuation des eaux à l'égout sans problème pour l'environnement.

- Le recyclage des solutions ou même la récupération des produits de base (ce qui est très supérieur à leur destruction par incinération).

Cette méthode se trouve donc être parfaitement satisfaisante sur les triples plans :

- Ecologie.
- Economies d'énergie.
- Recyclage des produits.

UMWELTBELASTUNG BEI VERBRENNUNG VON GEBRAUCHTÖLEN
POLLUTION OF THE ENVIRONMENT BY THE BURNING OF WASTE OILS
LA POLLUTION DE L'ENVIRONNEMENT PAR LE BRULAGE DES HUILES USAGEES
L'INQUINAMENTO DELL'AMBIENTE ATTRAVERSO LA COMBUSTIONE DEGLI OLI USATI

W.B. WALKER
Director, Braybrooke Chemical Services Ltd.,
United Kingdom

SUMMARY

Ever increasing quantities of waste oil are being burnt in small furnaces for the space heating of garages and warehouses.

There is strong evidence to suggest that pollution of the atmosphere occurs when waste oil is used as a fuel. In the research three different types of fuel oil have been burnt in a space heater of a type sold throughout the United Kingdom. The three fuels used were :-
    Untreated waste oil.
    Waste oil after the removal of water and volatile
      products.
    Gas oil.

The physical characteristics of the fuels used have been tabulated and the volume of gases passing through the stack, the gross particulate emissions and the analysis of the particulate samples have been recorded.

The stack gases were similar for all three fuels but the solid contaminants were much higher from the waste oil fuels than those from the gas oil. The solids were largely below 5/10 micron in size and therefore in the respirable range. Lead was one of the major solids passing from the stack into the atmosphere and calculations show a significant amount of atmospheric pollution at ground level.

ZUSAMMENFASSUNG

In kleinen Öfen für die Raumheizung von Garagen und Lagerhäusern werden immer grössere Mengen Altöl verbrannt.
Es gibt triftige Gründe zur Annahme, dass die Luft durch die Verbrennung von Gebrauchtöl verschmutzt wird. Bei unseren Forschungen wurden drei verschiedene Brennöltypen in einer Raumheizanlage von einem im ganzen Vereinigten Königreich verkauften Typ verbrannt:
- unbehandeltes Altöl
- Gebrauchtöl nach Entfernung von Wasser und flüchtigen Stoffen
- Gasöl
Die physikalischen Eigenschaften der verwendeten Brennstoffe sind in Tabellen wiedergegeben und das Volumen der den Schornstein durchlaufenden Gase und die grobkörnigen Feststoffemissionen sowie die Analyse der Emissionsproben aufgezeichnet worden.
Die Schornsteingase waren bei allen drei Brennstoffen ähnlich, doch war der Anteil an festen Schadstoffen bei Gebrauchtöl viel höher als bei Gasöl. Die Abmessungen der Feststoffe waren grösstenteils unter 5/10 um und lagen somit im einatembaren Bereich. Blei war einer der häufigsten Feststoffe, die aus dem Kamin in die Atmosphäre gelangten, und die Berechnungen ergeben eine beträchtliche Luftverschmutzung auf Bodenniveau.

RESUME

Des quantités croissantes d'huiles usagées sont brûlées dans de petites chaudières pour le chauffage de parcs de voitures ou de grands magasins.
Il semble certain que l'utilisation des huiles usagées comme combustible pollue l'atmosphère. Pour ces recherches, trois types différents d'huile ont été brûlés dans une chaudière de chauffage du type que l'on trouve couramment dans le commerce au Royaume-Uni.

Il s'agit des trois combustibles suivants:
- huiles usagées non retraitées;
- huiles usagées après élimination de l'eau et des produits volatils;
- gasoil.

Les caractéristiques physiques des combustibles utilisés ont été classées et le volume des gaz sortant de la cheminée, les émissions particulaires brutes et l'analyse des échantillons particulaires ont été enregistrés.

Les gaz de cheminée sont similaires pour les trois combustibles, mais les produits de contamination solides sont plus importants pour les ombusti- bles provenant des huiles usagées que pour ceux à base de gasoil. Les solides sont largement inférieurs à cinq dixièmes de micron et donc dans la gamme respirable. Le plomb est l'un des principaux produits soli- des passant par la cheminée dans l'atmosphère, et les calculs indiquent un degré élevé de pollution atmosphérique au niveau du sol.

RIASSUNTO

Quantità sempre crescenti di oli usati vengono bruciate in piccoli forni per il riscaldamento di garages e magazzini.

Esistono motivi validi per credere che l'olio usato impiegato come combustibile contribuisca all'inquinamento dell'atmosfera. Ai fini della ricerca sono stati bruciati tre diversi tipi di olio combustibile in una stufa di un tipo venduto in tutto il Regno Unito.

I tre combustibili impiegati erano :
- olio usato non trattato ;
- olio usato previa rimozione dell'acqua e dei prodotti volatili ;
- gasolio.

Sono state indicate in una tabella le caratteristiche fisiche dei combustibili impiegati registrando anche il volume dei gas attraverso il camino, le emissioni di grosso pulviscolo e l'analisi dei campioni di pulviscolo.

I gas al camino sono risultati simili per tutti e tre i combustibili mentre i contaminanti solidi erano molto più elevati per l'olio usato rispetto al gasolio. I solidi erano notevolmente al di sotto di 5-10 micron e pertanto nella gamma respirabile. Il piombo era uno dei solidi principali passati dal camino all'atmosfera e i calcoli indicano una notevole percentuale di inquinamento atmosferico a livello del terreno.

There is strong evidence to suggest that pollution of the atmosphere occurs when waste oil is used as fuel in conventional equipment.

Waste oil fuel is normally burnt in large or medium sized industrial boiler plants or the smaller space heaters. It is the pollution from these space heaters using waste oil as fuel that has been examined in this first series of tests.

The space heater used in the research is sold throughout the United Kingdom and is rated at 49 k.w.h. and is represented in Diagram I.

The flame of the air atomized burner heats the combustion chamber, and the products of combustion pass directly to the stack.

Hot air for space heating is obtained by a fan blowing air past the hot combustion chamber and out through the ducts at the top of the heater. The combustion chamber is completely separated from the air circulation system and products of combustion cannot contaminate the air for space heating. The unit is fully automatic and is equipped with a stack, the top of which is 6.5 metres from the ground and has a diameter of 178 mm.

The manufacturers recommend as fuel, either waste sump oil, recycled waste oil or gas oil.

The fuels used were :-

      a)   Untreated waste oil.
      b)   Recycled waste oil.
      c)   Gas oil.

DIAGRAM I

BURNER INSTALLATION

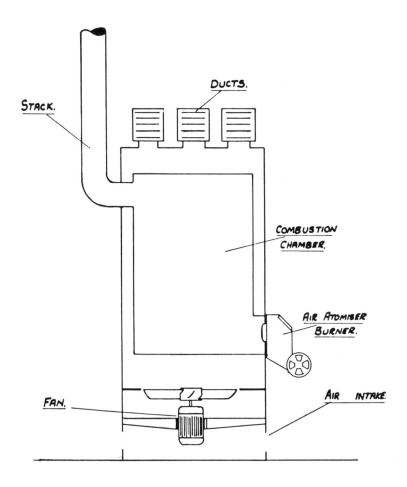

TABLE I

PHYSICAL CHARACTERISTICS OF THE FUELS USED

	UNTREATED WASTE OIL	RECYCLED WASTE OIL	GAS OIL
Specific Gravity at 15.5°C	0.902	0.896	0.849
Viscosity at 40°C c.s.	75	74	3.3
Water vol. %	4.2	0.2	Nil
Distillate to 100°C vol. %	4.5	Nil	Nil
Sulphur wt %	0.98	1.02	0.37
Chlorine wt %	0.37	0.05	Nil
Ash wt %	1.2	1.2	0.008
Gross Calorific Value cals/g	10,120	10,480	10,980
Metals Reported – mg/kg			
Lead	2,018	2,128	N.D.
Calcium	1,312	1,287	30
Zinc	937	919	5

Also present in the waste oils, in trace amounts, were :-

Copper,   Nickel,   Cadmium,   Chromium
Manganese, Antimony,   Tin,   Barium

TABLE II

FUEL CONSUMPTION AND STACK GAS ANALYSES AND QUANTITIES

	UNTREATED WASTE OIL	RECYCLED WASTE OIL	GAS OIL
Fuel Consumption l/hr	5.10	4.72	6.50
Orsat gas analysis Vol. %			
Carbon dioxide	7.2	8.4	8.2
Carbon monoxide	0.2	0.1	0.3
Unsaturated Hydrocarbons	Nil	Nil	Nil
Volume of gas passing through stack m^3/min. (n.t.p. $(H_2O)_S$ $(CO_2)_S$.)	4.4	4.2	4.0
Flue gas temperature °C	240/250	240/250	265/280

The untreated waste oil was collected from various sources in the normal manner, allowed to settle, and the free water removed. The recycled waste oil was similarly collected from various sources, dried by heating with circulation to 115°C and centrifuged.

The physical characteristics of these fuels are listed in Table I.

Clearly the water content of the untreated waste oil could cause uneven burning and the presence of very volatile materials shown in Table I could be an explosive risk on relighting. For comparison purposes gas oil was used as this product was also recommended by the equipment manufacturer.

The samples were withdrawn from the stack using an Airflow Developments Ltd., flue gas sampling apparatus, manufactured to the B.C.U.R.A. specification incorporated in B.S.3405.(1). This consists of a nozzle pointed into the direction of flow in the stack which is connected by a stainless steel tube to a cyclone and filter. This is followed by condensers, absorption tubes, control valve and thence to the metering suction unit.

In taking samples from the stack preliminary work indicated that the results of sampling at the centre of the stack gave almost identical results to those obtained by sampling at various points across the stack diameter, hence centre sampling was used throughout the tests.

Table II shows the hourly consumption of the test fuels together with the stack gas analyses and the volume of gases passing through the stack.

The high consumption of gas oil was due to using the same burner settings for all three fuels. It is of interest to note that the stack gases for all three fuels were similar.

TABLE III

GROSS PARTICULATE EMISSIONS

	UNTREATED WASTE OIL	RECYCLED WASTE OIL	GAS OIL
Mean particulate concentrations (including probe deposits) mg/m³			
n.t.p. $(H_2O)_S$ $(CO_2)_S$.	201	207	15
n.t.p. dry $(CO_2)_S$.	220	226	17
n.t.p. dry 10% CO_2.	305	269	20
Soot % of total particulates	62	57	84
Mean particulate soot concentration mg/m³ at n.t.p. dry 10% CO_2	189	153	17

The gross results for particulate emissions are given in Table III and are presented for three standard conditions.

a) Concentrations per cubic metre of stack gases at 0°C at 1 atmosphere pressure (101.3 KNm^{-2}) and with the water and carbon dioxide at the same proportions as they were in the stack - referred to as n.t.p. $(H_2O)_S$ $(CO_2)_S$.

b) Concentrations per cubic metre of stack gases at 0°C and 1 atmosphere, dry and with carbon dioxide at the same level as in the stack - n.t.p. $(CO_2)_S$.

c) The normal method of presenting stack concentrations is by standardization to 10% carbon dioxide - Hawksley, Badzioch and Blackett (2). This is the data given under the heading - n.t.p. dry 10% CO_2.

The greater part of the gross particulate emission was soot (i.e. particles of less than 5/10 microns) in each case, which would be expected in view of the low stack velocities and the low viscosities of the fuels.

The particulate emissions were then analyzed and the main materials detected are given in Table IV, the results have been calculated to conform to n.t.p. dry 10% CO_2.

Table V shows the input and output masses for the untreated and recycled waste oils of total particulates, lead, calcium and zinc calculated from results shown in previous tables. The figures for gas oil have not been included as they are extremely low.

TABLE IV

ANALYSES OF PARTICULATE SAMPLES MG/M³ AT N.T.P. DRY 10% CO_2

	UNTREATED WASTE OIL		RECYCLED WASTE OIL		GAS OIL	
	TOTAL	% SOOT	TOTAL	% SOOT	TOTAL	% SOOT
Phosphorus Pentoxide	53	40	49	39	2	66
Sulphate	11	>60	8	>70	4	>60
Chloride	<1.0	–	<1.0	–	<0.4	–
Lead	26	78	38	89	1	81
Calcium	29	10	24	5	0.75	35
Zinc	14	79	14	83	0.6	89

Also present in the gases from the waste oils in trace amounts were nickel, cadmium, chromium, manganese, antimony, tin, iron, barium and copper.

Phosphorus pentoxide and the chloride were also determined in the condensate from the stack but these quantities were very small and did not affect the results.

TABLE V

COMPARATIVE INPUT/OUTPUT MASSES OF PARTICULATES AND METALS IN G/HR

	UNTREATED WASTE OIL		RECYCLED WASTE OIL	
	INPUT EX FUEL OIL ASH	OUTPUT EX STACK	INPUT EX FUEL OIL ASH	OUTPUT EX STACK
Total Particulates	55.5	53.1	50.8	51.9
Lead	9.3	4.6	9.0	7.4
Calcium	6.1	5.1	5.4	4.6
Zinc	4.3	2.7	3.9	2.7

CONCLUSIONS

There was no significant difference between the emission results of the sample of untreated waste and the sample of the recycled waste oil.

It can be seen from the results for the total emissions from the burning of waste oils and gas oil that the dust and soot emissions from the former are ten times that of the latter. The amount of lead emitted from the waste oil fuels, probably as oxide, is appreciable at 26 and 38 mg/m^3 and as this is an emotive subject no Official is prepared to declare whether it is safe or unsafe.

However, according to the formula by A.W.C.Keddie (3), ground level concentration of lead arising from the burning of the waste oils tested, assuming a wind speed of 5m/sec., would be 6 and 9.7 micrograms per cu. metre respectively.

These lead emissions are much higher than the maximum mean concentration of 2 micrograms per cu. metre as recommended in Lead & Health, 1980, (4) a United Kingdom Government publication.

Ever increasing quantities of waste oils are being burnt in small furnaces for the space heating of garages, warehouses and small factories.

Therefore, it would appear that unless a standard of treatment for waste oil is specified and strict control of the suitability of burner installations implemented, there will be a high probability of serious atmospheric pollution in localized areas. Every effort must be made to minimize this risk. Where the design of the system burning the waste oil ensures that the majority of particulates are retained as ash within the system, there is probably an even greater risk that this ash is likely to be dumped in a concentrated form and will present a real pollution hazard to ground waters. It is therefore of utmost importance that the disposal of such ash deposits are identified and controlled.

In the test programme waste oils from many different sources were combined and blended. The waste oils were average samples and not from single sources. The burner used was new and was the latest model of a standard type in common use, but similar results would not be expected if different burners and conditions were used.

It is to be hoped that environmental authorities will enlarge this area of research into the burning of waste oils and in particular the atmospheric pollution caused by medium and large sized installations.

Results from such programmes are likely to vary considerably since waste oil arisings are such a variable commodity.

Where the purchase of waste oil is left to normal market forces the companies that recycle waste oil to fuel and those who regenerate waste oil to base lubricants pay the same price for their raw material - waste oil. The installation, overheads and treatment costs of the former are only a fraction of the corresponding costs of the latter.

Hence, in countries like the United Kingdom where purchase of waste oil is left to normal market forces, the profitable method of recycling waste oil is to burn it with minimal treatment thus causing additional atmospheric pollution. The re-refiners who do not cause this atmospheric pollution are thus being economically strangled.

The author wishes to acknowledge with thanks the valuable co-operation and analytical work carried out by the Department of the Regional Chemist, Strathclyde Regional Council.

REFERENCES

(1)   British Standard 3405

(2)   Hawksley, Badzioch and Blackett, <u>Measurement of Solids in Flue Gases</u>, 1977 Edition.

(3)   <u>Odour Control - A Concise Guide</u>, chapter XI by A.W.C.Keddie

(4)   Department of Health & Social Security.
        <u>Lead & Health - The Report of the Department of Health and Social Security Working Party on Lead in the Environment - 1980.</u>

ASPEKTE DES IMMISSIONSSCHUTZES BEIM BETRIEB VON ZWEITRAFFINERIEN
ASPECTS OF THE PROTECTION OF THE ENVIRONMENT IN THE OPERATION OF
USED OIL REFINERIES
ASPECTS DE LA PROTECTION DE L'ENVIRONNEMENT DANS L'EXPLOITATION
D'USINES DE REGENERATION
ASPETTI DELLA PROTEZIONE DELL'AMBIENTE NELL'ESERCIZIO DEGLI
STABILIMENTI DI RIGENERAZIONE

Dipl.-Ing. G. STEINMETZGER
Haberland Engineering GmbH, Dollbergen

Arbeiten von
Research by
Recherches de
Ricerche di

F.K. KRENTEL
Leitender Gewerbedirektor, Bezirksregierung Hannover

ZUSAMMENFASSUNG

Die Bedingungen, unter denen Zweitraffinerien arbeiten, unterscheiden sich in einigen wesentlichen Merkmalen von denen, die für Schmierstoffraffinerien oder ganz allgemein Erdölraffinerien gelten.

Dies trifft vor allem auf die unterschiedlichen Rohstoffvoraussetzungen - Zweitraffinerien verarbeiten Altöle unterschiedlicher Herkunft, die dabei sehr unterschiedliche Fremdstoffe enthalten können, Schmierstoffraffinerien verarbeiten dagegen einen im allgemeinen sehr gleichmäßig zusammengesetzten Rohstoff - zu.

Dies hat Auswirkungen auf die konstruktive Auslegung der Anlagen, die anzuwendenden Prozeßschritte sowie einzusetzenden Prozeßhilfsmittel beim Verarbeiten von Altöl und daher auch auf die Art und Konzentration der zumeist gas- oder dampfförmig dabei auftretenden Schadstoffe.

Aufgabe eines funktionierenden Immissionsschutzes ist es, den Austritt solcher Schadstoffe in die Umwelt auf ein Minimum zu begrenzen. Dem Ingenieur sind hierzu vielfältige Möglichkeiten je nach spezieller Aufgabenstellung an die Hand gegeben, einige dieser Möglichkeiten und Grenzen werden aufgezeigt. Ein ganz wesentlicher Beitrag zum Immissionsschutz besteht darin, durch technologische Verbesserungen schadstoffbildende Komponenten so früh wie möglich aus dem Prozeß zu entfernen und die Zugabe schadstoffbildender Prozeßhilfsmittel einzuschränken oder diese ganz zu eliminieren.

Nach den gesetzlichen Regeln und Vorschriften, die im Bereich der Bundesrepublik Deutschland zur Begrenzung der maximal zulässigen Immissionswerte, d.h. in erster Linie der zulässigen Schadstoffkonzentrationen in der Luft gelten, überwachen die Aufsichtsbehörden Planung, Bau und Betrieb von Zweitraffinerien. Dabei werden heute schon automatisch arbeitende, komputerunterstützte Systeme zur Luftüberwachung gefährdeter Siedlungsräume eingesetzt.

Am Beispiel einer großen Zweitraffinerie wird deutlich, daß die von den Überwachungsbehörden durch systematische Meßerfassung festgesetzten Schadstoffkonzentrationen in der Luft im Einklang mit den maximal zulässigen und als umweltverträglich erkannten Kurz- und Langzeit-Immissionswerten stehen.

Dies beweist, daß bei gezieltem Einsatz der richtigen Umweltschutztechnologie von Zweitraffinerien ebensowenig Gefahren für Mensch, Tier und Vegetation ausgehen wie von anderen vergleichbaren Industrieanlagen. Dies gilt sinngemäß auch für den Lärmschutz sowie den vorbeugenden Gewässerschutz und die hierzu anzuwendenden Maßnahmen.

SUMMARY

The conditions under which re-refining plants operate differ in several important respects from those governing lubricant refineries and oil refineries in general.
This applies above all to the differing raw-material requirements: used-oil refineries process used oils of mixed origin, which can therefore contain a wide variety of foreign bodies, whereas lubricant refineries process a raw material that, generally speaking, has an extremely uniform composition.
This fact has repercussions on plant design and on the process steps applied and the catalysts used in the processing of used oils, and hence also on the type and concentration of the pollutants produced, which occur mainly as gases or aerosols.
The function of effective immission-control is to minimize the escape of such pollutants into the environment. A variety of means depending on specific requirements, is available to the engineer, and some of these means and their limitations are described. An extremely important form of immission control is, by means of technological improvements, to remove pollutant-forming constituants from the process as soon as possible and to cut down or eliminate altogether the addition of pollutant-forming catalysts.
The supervisory authorities exercise surveillance over the planning, construction and operation of used-oil refineries in pursuance of the laws and regulations in force in the Federal Republic of Germany relating to the observance of maximum permissible immission levels, i.e., primarily to the permissible atmospheric pollutant concentrations. Automatic, computer-aided systems are already in use for monitoring the atmosphere in populated areas at risk.
In the case of a large used-oil refinery it is shown that the atmospheric pollutant concentrations ascertained by the supervisory authorities from systematic measurements are compatible with the maximum permissible short- and long-term immission levels that are known to be tolerated by the environment.
This proves that when appropriate use is made of the correct environmental protection technologies, the hazards to man, fauna and flora arising from used-oil refineries are no greater than those from other, comparable, industrial plants. By analogy the same holds true for noise prevention, the prevention of water pollution and the measures that are to be taken in connection with the latter.

RESUME

Les conditions de fonctionnement des usines de régénération diffèrent sur quelques points essentiels de celles qui caractérisent les raffineries de lubrifiants ou les raffineries de pétrole en général.
Ces différences sont surtout liées aux matières premières : les usines de régénération transforment des huiles usagées d'origine diverse, qui peuvent donc contenir des substances étrangères très différentes, alors que les usines de production de lubrifiants n'utilisent généralement qu'une matière première très homogène.
Cette particularité se traduit dans la conception des installations, dans les procédés appliqués ainsi que dans les matières auxiliaires utilisées pour la transformation des huiles usagées et donc également dans la nature et la concentration des éléments polluants qui en résultent, généralement sous forme de gaz et de vapeur.

Une politique efficace de protection de l'environnement doit limiter au maximum les rejets de ces éléments polluants dans l'environnement. L'ingénieur dispose à cet effet, selon le cas, de possibilités diverses; nous exposerons certaines de ces possibilités et leurs limites. L'une de ces mesures consiste par des perfectionnements technologiques à éliminer au stade le plus précoce du processus les composants susceptibles de donner naissance à des éléments polluants et de limiter ou même de supprimer totalement l'ajout de matières auxiliaires produisant des éléments polluants.

Conformément aux règles et aux prescriptions légales qui fixent sur le territoire de la République Fédérale d'Allemagne les taux de pollution maximaux admissibles, et tout d'abord les concentrations admissibles de polluants dans l'air, les autorités de surveillance contrôlent la conception, la construction et le fonctionnement des usines de régénération. Des systèmes automatiques informatisés de contrôle de l'air dans les zones à risque sont déjà en service à l'heure actuelle.

Le cas d'une importante usine de régénération révèle que les concentrations de pollution constatées par les autorités de contrôle par des mesures régulières sont conformes aux valeurs maximales admissibles d'immission à court et à long terme considérées comme tolérables pour l'environnement.

Cet exemple démontre que si l'on utilise à bon escient une technologie adéquate de protection de l'environnement dans les usines de régénération, celles-ci ne représentent pas plus de risques pour l'homme, l'animal et la végétation que d'autres industries comparables. Il en va de même pour la protection contre la pollution acoustique et la pollution des eaux et les mesures à prendre à cet égard.

RIASSUNTO

Le condizioni in cui lavorano le imprese di rigenerazione si differenziano per alcuni aspetti essenziali da quelle delle raffinerie che producono oli lubrificanti e da quelle delle raffinerie in generale.

Tale differenza è dovuta in primo luogo alle materie prime in quanto per la rigenerazione vengono utilizzati oli usati di varia provenienza che possono contenere delle sostanze estranee molto diverse mentre le raffinerie che producono lubrificanti lavorano in generale con un prodotto che ha una composizione estremamente uniforme.

Per tale motivo gli impianti utilizzati per la rigenerazione degli oli usati si differenziano dal punto di vista costruttivo nonché per i processi e le sostanze utilizzate che possono dar luogo all'emissione di sostanze nocive di tipo e di concentrazione diversa, nella maggior parte dei casi sotto forma di gas o di vapore.

Il compito di un dispositivo antinquinamento è di limitare al minimo la diffusione nell'ambiente di tali sostanze nocive. L'ingegnere dispone a tal fine di numerose possibilità che sono diverse a seconda dello scopo che si vuole raggiungere. Un contributo particolarmente importante per la protezione ambientale consiste nella tempestiva eliminazione dal processo dei componenti dannosi attraverso un perfezionamento tecnologico e limitando o eliminando completamente l'utilizzazione di sostanze nocive.

In conformità delle disposizioni legislative e delle regolamentazioni in vigore nella Repubblica federale di Germania per quanto riguarda la limitazione delle immissioni nocive ossia soprattutto la concentrazione ammessa di sostanze inquinanti nell'aria, gli organi responsabili controllano i progetti, la costruzione e l'esercizio delle industrie di rigenerazione. In proposito sono già stati messi in funzione nelle zone più colpite, dei sistemi automatici computerizzati per il controllo del grado di inquinamento dell'aria.

Dall'esempio di una grossa industria di rigenerazione si può dedurre che la concentrazione di sostanze nocive constatata dagli organi di controllo attraverso misurazioni sistematiche risulta conforme ai valori massimi ammissibili e ritunti sopportabili a breve ed a lunga scadenza.

Ciò dimostra che utilizzando una tecnologia adeguata per la rigenerazione non si crea alcun pericolo per l'uomo, per gli animali o per la vegetazione, e si rimane su valori di industrie comparabili. Lo stesso vale per l'inquinamento fonico e per la protezione preventiva delle acque.

Um auf die mit dem Betrieb von Zweitraffinerien zusammenhängenden Immissionsfragen einzugehen, drängt es sich auf, zunächst die im Vergleich zu Erdölraffinerien hierfür wesentlichen unterschiedlichen Rohstoffvoraussetzungen zu betrachten.

Erdölraffinerien verarbeiten Rohöle, die aus komplexen Gemischen von Kohlenwasserstoffen mit niedrigen bis hohen Siedepunkten und Molekulargewichten bestehen, zumeist aus Paraffinen, Naphthenen, Aromaten und Olefinen bzw. Derivaten derselben, wie z.B. Iso-Paraffinen. Daneben treten im Rohöl selbst bzw. als Folge der Verarbeitung in der Raffinerie Begleitstoffe auf, die - nicht nur im Zusammenhang mit Emissions- bzw. Immissionsfragen - in Erdölraffinerien zu beachten sind. Hierzu zählen z.B. ungesättigte Kohlenwasserstoffverbindungen, Schwefelverbindungen sowohl in Form von Schwefelwasserstoff und Merkaptanen als auch in Form anderer Gase. Einige dieser Stoffe, wie z.B. Merkaptane oder Schwefelwasserstoff, fallen, noch in geringen Konzentrationen, durch ihren belästigenden Geruch auf, worauf später noch eingegangen wird.

Die Rohstoffbasis von Zweitraffinerien bilden hingegen gebrauchte Mineralöle, die unterschiedlicher Herkunft entstammen.
Die nebenstehende (Tabelle I) zeigt eine Prognose der Altölherkunft am Beispiel der Bundesrepublik Deutschland, bezogen auf den Gesamtaltölanfall. Zweitraffinerien streben im allgemeinen an, in erster Linie solche Siedefraktionen aus dem Mineralölbereich einzusetzen, die für ihre Produktpalette die besten Voraussetzungen bieten. Befaßt man sich mit der Herstellung hochwertiger Schmierstoffe, so bedeutet das, daß man nach Altölen mit hohem Anteil an paraffinbasischen Mineralölen Ausschau hält. In der Praxis ist diese Idealvorstellung selten zu verwirklichen, da die Altöle eine Reihe typischer wie auch atypischer Fremdstoffe enthalten können, erstere sind gebrauchs- und betriebsbedingt, letztere zumeist beigemischte Abfallstoffe.
(Tabelle II) Zu den typischen Fremdstoffen zählen z.B. Metall-

Bezeichnung	Anteil am Gesamtaltöl [in Gew.%]
Motorenaltöle	53,6
Maschinenöle	14,6
Metallbearbeitungsöle	14,2
Getriebeöle	10,9
Spindelöle	3,5
Achsen- und Dunkelöle	1,6
Turbinenöle	1,4
übrige Mineralöle	0,2

Quelle: Umweltbundesamt (UBA) Berlin 1979
Basis: 1977, Bundesrepublik Deutschland

Haberland Engineering 1980 — Altölherkunft bezogen auf den Gesamtaltölanfall — Tabelle I

Altöle enthalten:

a) typische Fremdstoffe

Metallabrieb	Fe	Si
Benzinreste	Cr	Ca
Ruß	Zn	Ph
Staub	Cu	S
Wasser	Al	Cl
	Mg	Pb
	Mn	Ti
	Ni	

b) atypische Fremdstoffe

chlorierte Kohlenwasserstoffe (Kaltreiniger)

Ketone ⎫
Ester ⎭ organische Lösungsmittel

Nitroverdünner

Glykole (Frostschutzmittel)

Plastik, Lackreste

Sedimente

Haberland Engineering 1980 — Typische und atypische Fremdstoffe im Altöl — Tabelle II

abrieb, Kraftstoffreste, Kohlenstoff in Form von Ruß sowie
Wasser, zu den atypischen zählen z.B. organische Lösungsmittel, chlorierte Kohlenwasserstoffe und Frostschutzmittel. Die
in Tabelle II gezeigten Metalle stammen ihrer Herkunft nach
teilweise aus den Additiven, teilweise aus dem Abrieb gleitender Teile in Motoren, Getrieben und Maschinen bzw. aus
Kraftstoffen.

Die nachfolgende (Tabelle III) zeigt typische Metallgehalte
in der Sulfatasche von Motorenablaßölen, wie sie im Durchschnitt in den Jahren 1975 - 1979 bei durchgeführten Untersuchungen festgestellt wurden. Diese Stoffe können bei ungenügender Beachtung Quellen erheblicher Schadstoffentwicklung
bei der Verarbeitung dieser Altöle werden.

Daneben haben sich in den Altölen, bedingt durch ihren vorangehenden Einsatz in Maschinen, Getrieben und Motoren Oxidations- und Zerfallsprodukte aus den Additiven gebildet.

(Tabelle IV) Bekannt ist auch, daß Motorenöle, je nach Kilometerlaufzeit und Betriebsbedingung, wie das nebenstehende
Bild zeigt, polyzyklische aromatische Kohlenwasserstoffe enthalten, deren Konzentration mit steigender Kilometerlaufleistung zunimmt.

Berücksichtigt man, daß alle diese Fremdstoffe in ständig
wechselnder Konzentration im Altöl vorkommen können bzw. sich
während der Verarbeitungsschritte daraus Schadstoffe bilden
können, so wird klar, daß wirksame Maßnahmen zur Emissionsbegrenzung, d.h. solche, die den Austritt von Schadstoffen in
die Umwelt auf ein Minimum zu begrenzen haben, mit hohen Anforderungen an das Ingenieurwissen verbunden sind. Das oft
anzutreffende "Zerrbild vom Ingenieur zweiter Klasse in Zweitraffinerien" muß auch in diesem Zusammenhang nachhaltig korrigiert werden. Das Gegenteil ist notwendig und richtig.

Die eingangs geschilderten Voraussetzungen bedingen spezielle

Element	Gew. %	Element	Gew. %
Calcium	6 – 12	Mangan	0 – 2
Barium	0 – 3	Eisen	5 – 8
Magnesium	1 – 5	Nickel	1 – 4
Zink	12 – 18	Chrom	3 – 7
Aluminium	0 – 2	Silicium	0 – 3
Kupfer	2 – 6	Titan	0 – 3
Blei	35 – 50		

Quelle: Umweltbundesamt (UBA) Berlin 1979
(Untersuchungen TÜV-Rheinland 1975 – 1979)

Haberland Engineering 1980 — Metallgehalte in der Sulfatasche von Motorenablaßölen — Tabelle III

	Frischöl	nach 18 Europa-Testen	nach 5000 km	nach 10000 km	nach 10000 km plus 18 Europa-Testen
Fluoranthen	0,11	5,4	173,0	270,0	302,1
Pyren	0,29	10,7	450,0	700,0	743,0
Chrysen/Cyclopentenopyren	0,56	6,3	190,4	236,6	255,0
Benzo(b/j/k) fluoranthen	0,08	3,3	82,2	141,0	166,4
Benzo(e)pyren	0,14	5,6	182,0	278,4	282,0
Benzo(a)pyren	0,045	5,5	162,0	242,4	271,0
Perylen	0,03	1,5	35,6	57,4	65,6
Indeno (1,2,3-cd) pyren	0,03	2,2	59,4	83,2	89,0
Benzo (ghi) perylen	0,12	6,7	207,6	289,4	300,2
Coronen	0	1,3	36,7	63,0	72,1

Angaben in mg/kg   Quelle: Umweltbundesamt (UBA) Berlin 1979

Haberland Engineering 1980 — Betriebsabhängigkeit der Konzentration von Polycyclischen aromatischen Kohlenwasserstoffen (PAH) in Motorenölen — Tabelle IV

Verarbeitungsschritte und hierauf zugeschnittene Anlagen, abhängig natürlich vom jeweiligen Endziel, dem gewünschten Fertigprodukt.

Die hierbei am häufigsten auftretenden dampf- oder gasförmigen Schadstoffverbindungen zeigt (Tabelle V), wobei wir zwischen organischen und anorganischen Verbindungen zu unterscheiden haben.
Die Art der emissionsvermindernden Maßnahmen hängt davon ab, unter welchen Bedingungen, in welcher Kombination und Konzentration diese Schadstoffverbindungen auftreten.

Grundsätzlich gilt, schadstoffbildende Komponenten so früh wie möglich aus dem Verarbeitungsprozeß bzw. schon aus dem Altöl zu eliminieren bzw. die Zugabe schadstoffbildender Hilfsmittel einzuschränken. Dies wirft selbstverständlich eine Reihe verfahrenstechnischer Grundsatzfragen auf, wie sie auch in den Vorträgen meiner verehrten Herren Vorredner heute zu erkennen waren. Darüber ist ernsthaft nachzudenken.

Die bekanntesten Verfahren zur praktischen Reduzierung oder Eliminierung von Schadstoffkonzentrationen zeigt (Bild 1). Es sind die Verfahrensprinzipien

Kondensation
Absorption
Adsorption
Filtration
Sedimentation
Verbrennung

die in der Praxis oft in Kombination angewandt werden müssen.

(Bild 2) zeigt dazu Möglichkeiten, Schadstoffe zurückzuhalten, z.B. nach dem Prinzip der

Pendelung
Anwendung von Schwimmdachkonstruktionen bei Tankbehältern

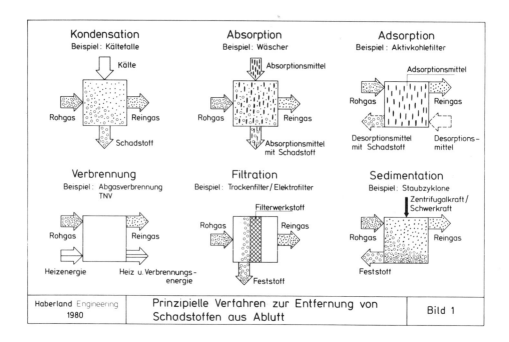

Bild 1: Prinzipielle Verfahren zur Entfernung von Schadstoffen aus Abluft
Haberland Engineering 1980

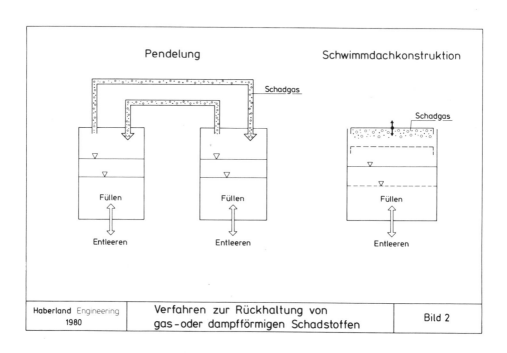

Bild 2: Verfahren zur Rückhaltung von gas- oder dampfförmigen Schadstoffen
Haberland Engineering 1980

Als Ergänzung wäre noch zu erwähnen die Methode der unverminderten Abgabe von Schadstoffen in großen Quellhöhen über Schornsteine mit dem Ziel, jene luftverdünnt auf größere Flächen zu verteilen. Die ökologischen Grenzen dieser Methode, vornehmlich für vorbelastete Gebiete, sind nach neueren Untersuchungen bereits sichtbar.

Die wesentlichen Anwendungskriterien der vorgenannten Verfahren ersehen Sie aus nebenstehender (Tabelle VI). Welches Verfahren dabei im einzelnen Erfolg verspricht, muß durch kritische Analyse der auftretenden Stoffe einschließlich ihrer Randbedingungen ermittelt werden. Entscheidend für die endgültige Anwendung des einen oder anderen Verfahrens oder einer Kombination ist das geforderte Ziel auf Einhaltung der Emissionsgrenzwerte sowie die Wirtschaftlichkeit der Maßnahme. Falsche Entscheidungen können dazu führen, daß das Problem der Schadstoffbeseitigung nur verlagert wird, beispielsweise aus der Luft in das Wasser bzw. andere Schadstoffe (Absorptionsträger) neu gebildet werden.

Mitunter ist es sinnvoll, schadstoffhaltige Nebenprodukte einer weiteren Behandlung außerhalb des Betriebes zuzuführen, als Beispiel seien genannt die Verbrennung von Kohlenwasserstofffraktionen mit hohem Chloranteil in speziellen Anlagen, wie Verbrennungsanlagen auf hoher See oder die Aufarbeitung von Säureharzen zu Schwefelsäure. Es besteht kein Zweifel, daß die Summe solcher emissionsvermindernder Maßnahmen die Rentabilität der Altölaufarbeitung wegen der damit verbundenen, zumeist hohen Kosten, erheblich reduzieren kann. Auch aus diesem Grunde sollte der Vermeidung bzw. Reduzierung unnötiger Schadstoffbildner, wie sie z.B. in Form bestimmter Prozeßhilfsmittel existieren, Beachtung geschenkt werden. Hierzu sind entsprechende Verfahrenstechniken nötig.

Einige typische praktische Anwendungsbeispiele emissionsvermindernder Maßnahmen in Zweitraffinerien zeigt das nebenstehende (Bild 3). Am Beispiel der Absorption von SO_2 und Chloriden

anorganische Verbindungen	Schwefeldioxid	SO_2
	Stickoxide	NO_x
	Kohlenmonoxid	CO
organische Verbindungen	Kohlenwasserstoff	$C_m H_n$
	aromatische Kohlenwasserstoffe	$C_m H_n$
	Chlorkohlenwasserstoffe	$C_m H_n Cl_x$
	Merkaptane	$C_m H_n SH$
	Schwefelwasserstoff	H_2S
	Phenole	$C_m H_n OH$
	Ester	$C_m H_n COO\ C_{m'} H_{n'}$
	Nitroverbindungen	$C_m H_n (NO_2)$
Stäube	schwermetallhaltige Stäube	

Haberland Engineering 1980 — Typische Schadstoffe beim Aufarbeiten von Altöl — Tabelle V

Maßnahmen Verfahren \ Wirkungen	Anwendung möglich bei		Kohlenwasserstoff-restgehalt			TA-Luft wird eingehalten		Restkonzentration unterhalb der Geruchsschwelle		Sicherheitstechnische Aspekte		Investitionskosten		Betriebskosten	
	Tanks	Prozeßanlagen	wie vor der Maßnahme	einige g/Nm³	einige mg/Nm³	ja	nein	ja	nein	günstig	problematisch	niedrig	hoch	niedrig	hoch
Pendelung															
Schwimmdachkonstruktion															
Absorption															
Adsorption															
Kondensation															
Verbrennung mit Wärmerückgewinnung															

Haberland Engineering 1980 — Verfahren zur Schadstoffentfernung in Zweitraffinerien — Tabelle VI

sowie anderer gasförmiger Verbindungen an Säuerungseinrichtungen, der thermischen Nachverbrennung von Ablüften und Abgasen schwankender Zusammensetzung und wechselnder Konzentrationen aus Destillationen und Behältern sowie der Abgabe von Verbrennungsabgasen in ausreichenden Quellhöhen über Kamine; besondere Sorgfalt ist auch überall dort zu beachten, wo Altöle, Neben- oder Fertigprodukte außerhalb geschlossener Anlagen auftreten, z.B. bei Verladevorgängen.

Bewertung von Emissionen:

Die Wirkungen luftfremder Stoffe sind sehr unterschiedlicher Natur und bisher nur teilweise erforscht.
Sie beziehen sich im wesentlichen auf:

klimatologische Veränderungen
Einwirkungen auf Sachgüter, wie z.B. Korrosionen
Einwirkung auf den Vegetationsmechanismus sowie den tierischen und menschlichen Organismus in Form von Belästigungen oder Gesundheitsgefährdung.

Zum Schutz vor schädlichen Umwelteinwirkungen wurde in der Bundesrepublik Deutschland 1974 das sogenannte Bundesimmissionsschutzgesetz (BImSchG) erlassen, es folgten verschiedene Verordnungen und Richtlinien, die (Tabelle VII) zeigt und die ganz allgemein auch für Zweitraffinerien Anwendung finden, so z.B.:

1. Die Bundesimmissionsschutzverwaltungsvorschrift (BImSchVwV) in Form der technischen Anleitung zur Reinhaltung der Luft - TA-Luft -. Diese dient als Rechts- und Fachgrundlage für die Überwachungstätigkeit der einschlägigen Behörden und enthält unter anderem Grenzwerte für Emissionen und Immissionen bestimmter Schadstoffgruppen, die, abhängig vom Grad der Umweltschädlichkeit des jeweiligen Schadstoffes in Stoffklassen eingeteilt sind. Das Gesetz wird außerdem ergänzt durch weitere diverse Fachrichtlinien.
Man bezeichnet als tolerierbare Grenzwerte, unterhalb derer nach dem heutigen Wissensstand Menschen, Tiere und Vegeta-

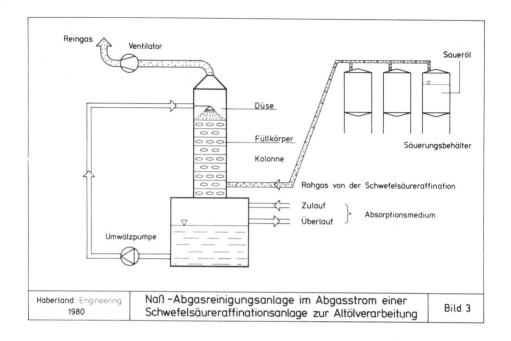

Haberland Engineering 1980 — Naß-Abgasreinigungsanlage im Abgasstrom einer Schwefelsäureraffinationsanlage zur Altölverarbeitung — Bild 3

Gesetze	Verordnungen	Richtlinien	
		Raffinerie-Richtlinie Nordrhein-Westfalen	
Bundes-Immissionsschutzgesetz (BImSchG)	TA-Luft (1. BImSchV)	VDI-Richtlinien:	
		VDI 2090	Quellen von Schadstoffen
		VDI 2104	Begriffe
BGB	TA-Lärm	VDI 2108	MIK, SO_2
StGB		VDI 2289	Stoff-Ausbreitung in die Atmosphäre
	Genehmigungspflichtige Anlagen (4. BImSchV)	VDI 2306	MIK organische Verbindungen
GewO		VDI 2310	max. Immissionswerte
		VDI 2880	Auswurfbegrenzungs- werte
		Meßverfahren zur Konzentrationsbestimmung von Schadstoffen in der Luft	

Haberland Engineering 1980 — Wesentliche Gesetze, Verordnungen und Richtlinien für den Immissionsschutz in der Bundesrepublik Deutschland — Tab VII

tion nicht gefährdet erscheinen neben den Grenzwerten, welche die TA-Luft enthält, die von der VDI-Kommission - Reinhaltung der Luft - erarbeiteten MIK-Werte (maximale Immissionskonzentrationen). Daneben gelten Werte für maximale Arbeitsplatzkonzentrationen (MAK). Ihre Bedeutung zeigt (<u>Bild 4</u>). Von besonderer Bedeutung für Zweitraffinerien ist die VDI-Richtlinie 2306 - MIK-Werte für organische Verbindungen. Alle Werte unterscheiden nach Kurzzeit- und Dauereinwirkung (MIK_K und MIK_D).

2. Weitere Differenzierung nimmt die sogenannte "Raffinerie-Richtlinie Nordrhein-Westfalen", die sich als praxisorientiert erwiesen hat und daher in allen deutschen Bundesländern herangezogen wird, vor. Neben Grenzwerten speziell für Schadstoffkonzentrationen in einzelnen Arbeitsbereichen von Raffinerien enthält sie die in (<u>Bild 5</u>) dargestellte, offiziell anerkannte Methode zur mathematischen Vorherermittlung von Immissionsprognosen unter Einbeziehung von Schadstoffvorbelastungen in der Luft nach einem bestimmten Ausbreitungsmodell. Die Richtlinie unterscheidet Immissionsgrenzkonzentrationen mit belästigender und solche mit gesundheitsgefährdender Wirkung. Die ersteren dürfen an max. 4 % der Jahresbetriebsstunden, die letzteren nur 1 % der Jahresbetriebsstunden überschritten werden, um Schädigungen und Beeinträchtigungen wichtiger Funktionen des menschlichen Organismus nach heutiger Erkenntnis auszuschließen.

Diese Grenzwerte werden bei Rechtsstreitigkeiten über Immissionsfragen vor den Gerichten der Bundesrepublik Deutschland als Beurteilungsgrundlage herangezogen, ihre Überschreitung wir objektiv gesehen mit schädlicher Einwirkung auf die Umwelt gleichgesetzt. Die Gerichte werten die TA-Luft und die VDI-Richtlinien daher als "Antizipiertes Sachverständigengutachten".

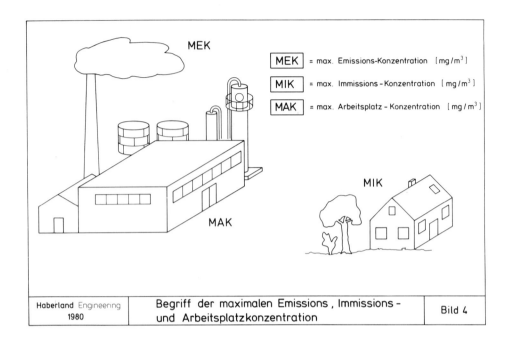

Haberland Engineering 1980 — Begriff der maximalen Emissions-, Immissions- und Arbeitsplatzkonzentration — Bild 4

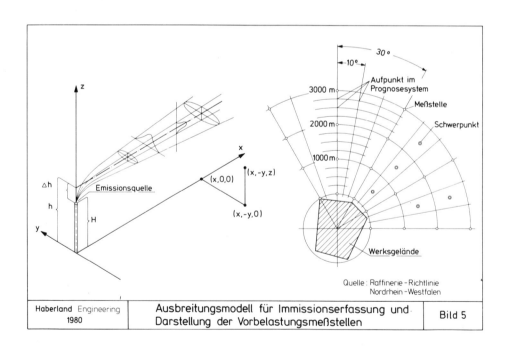

Haberland Engineering 1980 — Ausbreitungsmodell für Immissionserfassung und Darstellung der Vorbelastungsmeßstellen — Bild 5

Zur Überwachung der zulässigen Immissionskonzentrationen in der Luft werden automatisch arbeitende, computerunterstützte Meß- und Auswertungssysteme, vornehmlich in gefährdeten Ballungs- und Siedlungsräumen eingesetzt. Das nebenstehende (Bild 6) zeigt eine typische Meßstation als Glied solcher Systeme.

Als praktisches Beispiel möchte ich Ihnen zum Schluß noch Werte wiedergeben, wie sie in der Umgebung einer großen Zweitraffinerie in der Bundesrepublik Deutschland gefunden wurden. Die Messungen und ihre Auswertung erfolgte durch die zuständige Landesbehörde nach den im Vortrag genannten Regeln.

Die Überprüfung erfolgte durch Immissionsmessungen in genau festgelegten Aufpunkten in konzentrisch geordneten Entfernungen von 250 - 500 m vom Mittelpunkt des Raffineriegeländes. Auch im Gelände selbst waren Meßpunkte eingerichtet. Berücksichtigt wurden außerdem Schadstoffvorbelastung aus fremden Emissionsquellen, Windrichtungen sowie Schleichleckagen aus Apparaten, Rohrleitungen und sonstigen Quellen zur rechnerischen Gegenkontrolle der zu messenden Werte anhand einer Immissionsprognose nach dem vorhin gezeigten Ausbreitungsmodell. Rechnerisch wurden nur Kohlenwasserstoffe und Schwefeldioxyd betrachtet. Erfaßt wurden Schwefeldioxyd, Stäube, Stickoxyde, Summe der Kohlenwasserstoffe, Merkaptane sowie Benzinfraktionen mit modernsten Meß- und Analysenmethoden, u.a. Photometrie, Olfaktometrie usw.

Wie die in (Tabelle VIII) dargestellten Werte zeigen, unterschreiten sämtliche Meßwerte die zulässigen Immissionskonzentrationswerte der TA-Luft bzw. der Raffinerie-Richtlinie NRW erheblich. Dies gilt auch für Geruchsgrenzwerte, wie bei Merkaptanen.

Die Ergebnisse zeigen, daß von Zweitraffinerien keinerlei Gefährdung der Umwelt ausgeht, wenn wichtige Kriterien eingehalten und beachtet werden.

Bild 6

Schadstoffbezeichnung	Meßergebnisse [mg/m³]		zulässige Grenzwerte [mg/m³] TA-Luft		Zahl der $\frac{1}{2}$-h Einzelmessungen
	J 1	J 2	JW1	JW2	
Schwefeldioxid	0,03	0,1	0,14	0,4	1691
Stickoxide	0,019	0,15	0,10	0,3	nicht bekannt
Staub	0,057	0,153	—	—	1970
	96%-Wert	99%-Wert	zulässige Grenzwerte Raffinerierichtlinie 96%-Wert	99%-Wert	
Kohlenwasserstoffe (Summe)	0,43	0,65	2,0	5,0	1931
	94%-Wert	99%-Wert			
Benzin	<0,2	≪0,3	5,0	0,3 (Benzol)	nicht bekannt
Merkaptan	<0,02 >0,005		0,005	0,06	↓
Gesamtorganische Gase (berechnet als Kohlenstoff)	≪2,0	≪5,0	2,0	5,0	

Quelle: Aus Angaben der Landesbehörde

Haberland Engineering 1980	Ergebnisse von Immissionsmessungen im Bereich einer Zweit- raffinerie mit Beurteilung nach Richtlinien des BImSchG	Tab VIII

Allerdings muß deutlich gesagt werden, daß Umweltschutz Geld kostet. Jede nationale Institution muß daher neben dem Erlaß von Vorschriften und Gesetzen den Unternehmen, die sich mit der Aufarbeitung von Altöl befassen, auch die Möglichkeit belassen, das hierfür nötige Kapital zu verdienen. Schließlich sind diese Unternehmen wichtige Treuhänder im Interesse eines einheitlichen Zieles, welches heißt: Erhaltung des natürlichen ökologischen Gleichgewichtes.

Literaturnachweis:

[1] Autorengemeinschaft BP Aktiengesellschaft "Das Buch vom Erdöl", Hamburg (1967).

[2] Umweltbundesamt Berlin, Bericht der Arbeitsgruppe "Mineralölhaltige Rückstände", November 1979.

[3] Düwel, L., Zündorf, Otto-Josef, "Emissionen luftfremder Stoffe aus Industriebetrieben", Schriftreihe Umweltschutz, Band 4, (1974).

UMWELTSCHUTZ BEI DER RAFFINATION VON GEBRAUCHTÖLEN.
VERFAHREN BEI DER WIEDERVERWENDUNG VON SÄURETEEREN
PROTECTION OF THE ENVIRONMENT DURING THE PROCESS OF RE-REFINING USED
OILS.
THE TREATMENT OF ACID SLUDGE
LA PROTECTION DE L'ENVIRONNEMENT DANS LA REGENERATION DES HUILES USEES.
PROCEDE DE RECUPERATION DES BOUES HUILEUSES
LA PROTEZIONE DELL'AMBIENTE NELLA RIRAFFINAZIONE DEGLI OLI USATI.
PROCESSO DI RICUPERO DELLE MELME OLEOSE

C. VERSINO
Istituto di Chimica-Fisica, Università di Torino
C. MOLINO
O.M.A., Rivalta di Torino

RIASSUNTO

Nella riraffinazione all'acido degli oli lubrificanti usati, si ottengono, oltre alla base idrocarburica riraffinata, sottoprodotti quali melme oleose acide e le terre decoloranti esauste, il cui smaltimento crea seri problemi anche di tipo ecologico. Illustrata la validità chimica dell'olio da riraffinazione ed impostati i bilanci di massa, energia ed energia equivalente, si descrive in questa relazione una proposta di processo per lo smaltimento, con recupero, delle melme oleose.
Perfezionando la ricerca, iniziata alcuni anni or sono, si è giunti ad una tecnica di trattamento delle melme, mediante emulsioni oleose esauste, tecnica che raggiunge il duplice ri

sultato di allontanare l'acido solforico della melma, e recuperare nel contempo la parte idrocarburica dell'emulsione esausta. La melma è dapprima sottoposta ad un processo meccanico combinato di compressione e stiramento, denominato tribocompressione, che già allontana il 25% dell'acido presente. Successivamente si tratta il residuo con emulsioni esauste sottoponendo poi la fase idrocarburica alla tribocompressione che ne riduce drasticamente il contenuto in acqua. La composizione media finale è del 94% in composti idrocarburici e derivati, 5% in acqua ed 1% in zolfo.
Si descrive infine l'impianto proposto e lo schema chimico-fisico-biologico di trattamento delle acque di risulta.

ZUSAMMENFASSUNG

Bei der Zweitraffination von Gebrauchtöl nach dem Säureverfahren fallen Nebenprodukte wie Säureteere und verbrauchte Bleicherden an, deren Ableitung ernste Probleme, auch ökologischer Art, aufwirft. Nach Veranschaulichung der chemischen Brauchbarkeit des Zweitraffinats und nach Ansatz der Massen-, Energie- und Energieäquivalent-Bilanzen wird in dem Bericht ein Vorschlag für einen Prozess zur Ableitung - und Wiederverwendung - der Säureteere beschrieben.
Bei der Verbesserung der vor einigen Jahren aufgenommenen Forschung wurde ein Verfahren zur Behandlung der Teere mittels erschöpfter Ölemulsionen gefunden, das zu dem doppelten Ergebnis führt, die Schwefelsäure aus dem Teer zu entfernen und gleichzeitig den Kohlenwasserstoffanteil der erschöpften Emulsion rückzugewinnen. Der Teer wird zunächst einem als "Tribokompression" bezeichneten mechanischen Prozess der kombinierten Kompression und Dekompression unterzogen, bei dem bereits 25 % der vorhandenen Säure entfernt werden.
Der Rückstand wird mit erschöpften Emulsionen behandelt und die Kohlenwasserstoffphase wird dann der Tribokompression unterzogen, wodurch der Wassergehalt drastisch verringert wird. Am Ende beträgt die mittlere Zusammensetzung 94 % Wasserstoffverbindungen und Derivate, 5% Wasser und 1% Schwefel.
Schliesslich werden die vorgeschlagene Anlage und das chemisch-physikalisch-biologische Schema der Abwasserbehandlung beschrieben.

SUMMARY

When re-refining spent lubricating oils by means of acids, apart from the re-refined hydrocarbon base, one gets such by products as acid sludges and spent bleaching earths the elimination whereof is a serious source of trouble with particular regard to environment.
This report proposes a particular process of sludge recuperation after having pointed out the chemical validity of re-refining oil and set forth any data covering mass, energy and equivalent energy.

A further improvement of this research, which was started a few years ago, has brought to such a technique of sludge treatment by means of exhausted emulsions to attain the double goal of eliminating sulphuric acid from sludge, and of recuperating the hydrocarbon part of exhausted emulsion. Sludge is first submitted to a combined mechanical process of compression and stretching which is called tribo-compression and eliminates 25% amount of acid. Later on, it is treated with exhausted emulsion while the hydrocarbon phase is submitted to tribo-compression thus reducing its water content in a drastic way. Its final average composition reads 94% hydrocarbon compounds and derivatives, 5% water and 1% sulphur.

There is, moreover, a description of the proposed equipment and the chemical-physical-biological scheme for treatment of water.

RESUME

La régénération à l'acide des lubrifiants usés fournit, outre la base d'hydrocarbures régénérée, des sous-produits tels que les boues huileuses acides et les terres décolorantes usagées, dont l'élimination pose de sérieux problèmes, notamment du point de vue écologique. Après avoir décrit la valeur chimique de l'huile de régénération et établi les bilans de masse, d'énergie et d'énergie équivalente, l'auteur expose dans le présent rapport une proposition de procédé d'élimination, avec récupération, des boues huileuses.

La poursuite des recherches entreprises il y a quelques années a permis de mettre au point une technique de traitement des boues, à l'aide d'émulsions huileuses usagées, qui permet à la fois d'éliminer l'acide sulfurique des boues et de récupérer les hydrocarbures de l'émulsion usagée. Les boues sont d'abord soumises à un processus mécanique combiné de compression et d'étirement, dénommé "tribocompression", qui élimine déjà 25% de l'acide présent.

Le résidu est ensuite traité avec des émulsions usagées, en soumettant ensuite la phase d'hydrocarbures à la tribocompression, qui réduit dans des proportions considérables la teneur en eau. La composition moyenne finale est de 94% d'hydrocarbures et dérivés, de 5% d'eau et de 1% de soufre.

L'auteur décrit enfin l'installation proposée et la méthode chimico-physico-biologique de traitement des eaux résiduelles.

Introduzione

La moderna tecnologia, basata sull'impiego di macchine sempre più complesse e veloci, richiede una loro efficace lubrificazione, al fine di prolungarne nel tempo le elevate prestazioni richieste.

Veicoli e macchine utensili sono accomunati in questa esigenza ed è necessario un periodico ricambio del lubrificante, olio o emulsione oleosa che sia, per mantenere elevati livelli di prestazione a cui si contrappone il deterioramento del fluido di lubrificazione.

Per molti anni l'abbondanza ed il basso costo dei prodotti petroliferi rendeva non economico il loro recupero. D'altra parte, solo recentemente la legislazione dei vari Stati ha imposto limiti severi allo scarico dei prodotti di scarto. Di qui la necessità di utilizzare quanto prima era indiscriminatamente sversato, ricercando adatte tecnologie di recupero onde evitare, nel contempo, di violare la legge.

In questa situazione si è meglio organizzata la raccolta dei lubrificanti usati, oli ed emulsioni, sebbene a tutt'oggi il livello raggiunto non possa certo definirsi soddisfacente.

Il materiale raccolto deve trovare una sua collocazione in uno smaltimento che non offenda l'ambiente e comporti possibilmente un recupero di materia e/o di energia.

Gli oli usati possono, con evidenti vantaggi di vario tipo, essere avviati alla riraffinazione e quindi ritornare in ciclo. La riraffinazione, però, come un qualunque altro processo industriale, secondo le tecnologie disponibili, genera sottoprodotti che creano a loro volta un nuovo problema: il processo all'acido produce, tra l'altro, una melma oleosa, lo sludge acido (SA), costituito per il 52% da idrocarburi e derivati, per il 40% da acido solforico, per l'8% da acqua.

La parte idrocarburica, riferita all'olio usato, costituisce il 20,7%, mentre circa l'1% è conglobato nelle terre decoloranti esauste (TDE).

Validità chimica degli oli riraffinati (R)

Oltre a vantaggi economici ed energetici, la riraffinazione degli oli usati (U), sia con il classico processo all'acido che con processi di estrazione ed idrogenazione, presenta un risvolto significativamente positivo nella qualità del prodotto riraffinato ottenuto (R). Nella fig. 1 si riportano i cromatogrammi di un olio nuovo (N) e del suo riraffinato (R), dopo esercizio su motore di 4.000 km.

La cromatografia su colonna di SiO_2 condotta secondo la metodica già descritta [1] [2] [3], separa il campione in frazioni caratterizzate quali-quantitativamente per spettro-fotometria IR e per pesata [4] [5] [6]. Nella fig. 1 compaiono; in ascissa il numero progressivo delle frazioni ed in ordinata il loro peso in grammi. Le linee tratteggiate indicano il campo di eluizione dei vari solventi (n-eptano, cloroformio, acetone, metanolo). Si può osservare che i due cromatogrammi sono pressochè sovrapponibili, e d'altra parte gli spettri I.R. degli eluiti con n-eptano (fig. 2) indicano ancora una analoga composizione chimica: si tratta in ogni caso prevalentemente di idrocarburi paraffinico-naftenici con peso molecolare medio attorno a 480.

Nella fig. 3 si riportano i risultati dell'analisi strutturale grafico-statistica, nota come metodo n-d-m [7], condotta sulle frazioni cromatografiche dei due campioni. In ascissa appare la sommatoria della percentuale delle frazioni, ed in ordinata la percentuale di carbonio in struttura paraffinica, naftenica ed aromatica talchè risulti:

$$C_P\% + C_N\% + C_A\% = 100$$

Anche questa tecnica conferma l'equivalenza chimica dell'olio R rispetto all'olio N.

Bilancio di massa

Utilizzando i dati medi di origine industriale a disposizione, si può ritenere che, immessa nel propulsore una quantità unitaria di olio lubrificante (N), solo il 60% sarà recuperabile come olio usato (U), mentre il restante 40% è da ritenersi irrecuperabile (C), poichè consumato durante l'esercizio, o perso nelle fasi di svuotamento coppa, raccolta, etc. [8].

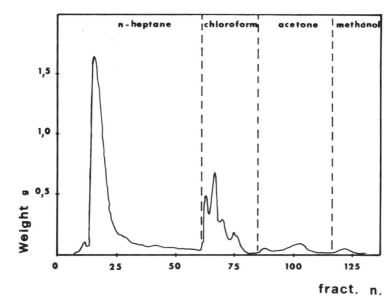

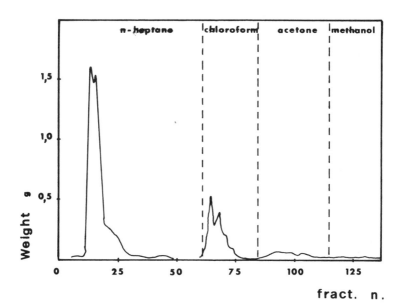

Figure 1

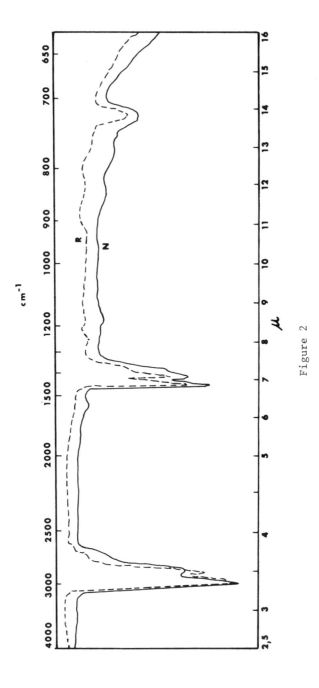

Figure 2

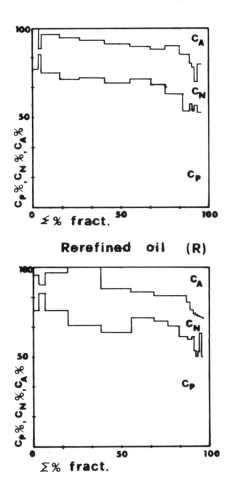

Figure 3

Il bilancio di massa è quindi:
$$L_N = L_C + L_U = (0,40 + 0,60) \, L_N \qquad (1)$$
essendo L_N L_C ed L_U rispettivamente le quantità di olio nuovo (N) immesso nel propulsore, dell'olio consumato o perso (C) e dell'olio usato recuperato (U).
Dalle analisi condotte su un campione medio industriale di U si è dedotta la seguente composizione (Tab. 1).
Si deduce quindi l'eq. di bilancio (2):
$$L_U = (L_R + L_F + L_P + L_c + L_i + L_a) \, 0,6 \cdot L_N \qquad (2)$$
Poichè attualmente le tecniche prevalenti di smaltimento degli U sono la combustione diretta e la riraffinazione, si hanno i seguenti bilanci di massa:
- alla combustione:
$$L_R + L_F + L_P + L_c = 0,565 \cdot L_N \qquad (3)$$
- alla riraffinazione:
$$L_R + L_F + L_P = 0,560 \cdot L_N \qquad (4)$$
I coefficienti 0,565 e 0,560 rappresentano i fattori di recuperabilità di massa degli U, riferiti ad N, nella combustione e nella riraffinazione.

Bilancio di energia ed energia equivalente

Il PCS medio degli U è di 9.800 Kcal/kg e rappresenta il contenuto energetico intrinseco quale ottenibile nella combustione completa degli U.
Analogamente ai fattori di recuperabilità di massa, anche i contenuti energetici intrinseci degli U, alla combustione ed alla riraffinazione, devono essere pressochè coincidenti [9]. Esiste tuttavia una netta differenziazione economica tra i risultati nei due processi: la combustione è distruttiva e restituisce solo il contenuto energetico intrinseco di U; la riraffinazione non solo non intacca questo contenuto ma, attraverso la selezione in gruppi di composti chimici, ne innalza il valore economico. La trattazione completa, dal punto di vista teorico, è già stata ampiamente illustrata [9]; ci limitiamo, in questa sede, ad illustrarne i concetti essenziali.

TABELLA I

Composizione media di U.

Idr. paraffinico-naftenici (R)	66%	L_R =	0,66
Fraz. idroc. leggere (F)	5,6%	L_F =	0,056
Fraz. idroc. pesanti (P)	21,7%	L_P =	0,217
Polvere di carbone (c)	0,8	L_c =	0,008
Composti inorganici (i)	1,2	L_i =	0,012
Acqua (a)	4,7	L_a =	0,047
	100,0		1,000

TABELLA II

Parametri economici, economico-energetici ed equivalenti.

	Costo L/kg	PCS Kcal/kg	f_{ec}	f_{en}	f	f_{eq} (Kcal)$_e$/kg
Olio comb. denso	150	10600	1	1	1	10600
U (dalla comb.)	120	9800	0,8	0,92	0,74	8480
R	600	10600	4,0	1,0	4,0	42400
F dalla riraff.	250	11000	1,67	1,04	1,74	17702
P	150	9700	1,0	0,92	0,92	10600

Si assume, come standard di riferimento economico-energetico, l'olio combustibile denso (OCD) (18-20°E a 50°C) a L. 150/kg e con un PCS di 10.500 Kcal/kg.
Il rapporto dei due valori fornisce il costo della Kcal standard $(Kcal)_S$:

$$1 \ (Kcal)_S = 0{,}014 \ L. \qquad (5)$$

Effettuando i rapporti tra costo unitario di un prodotto suscettibile di combustione e quello dello standard, si ottiene il fattore di equivalenza economica f_{ec}:

$$\frac{\text{costo in L/kg del prodotto}}{\text{costo in L/kg dello standard}} = F_{ec} \qquad (6)$$

L'analogo rapporto fra i PCS unitari fornisce il fattore di equivalenza energetica:

$$\frac{\text{PCS Kcal/kg del prodotto}}{\text{PCS Kcal/kg dello standard}} = f_{en} \qquad (7)$$

Il prodotto dei due fattori dà come risultato il fattore di equivalenza economico-energetico di quel prodotto rispetto allo standard:

$$f_{ec} \times f_{en} = f \qquad (8)$$

Nella Tab. II sono riportati i valori relativi ai vari prodotti già elencati.
Poichè R, F, P sono il risultato della riraffinazione, tenendo conto della (4) e dei dati di Tab. II, si ottiene:
$(4 \times 0{,}66)+(1{,}74 \times 0{,}056)+(0{,}92 \times 0{,}217)= 2{,}937 \qquad (9)$
Il rapporto con il corrispondente valore alla combustione dà:

$$\frac{2{,}937}{0{,}74} = 3{,}97 \qquad (10)$$

ed indica che la riraffinazione produce un vantaggio economico-energetico lordo di circa quattro volte superiore a quello ottenibile dalla combustione diretta degli U.
Poichè questa trattazione è applicabile a materiali vari, a costi amministrativi e fiscali, a costi gestionali etc, di cui non esiste il PCS, con alcune modificazioni conduce al concetto Kcal equivalente $(Kcal)_e$.
Infatti, f_{ec} indica il valore economico del composto o del costo o del beneficio, rispetto allo standard. Il prodotto

di f_{ec} con il PCS dello standard dà un numero di Kcalorie equivalenti $(Kcal)_e$ che esprimono ancora un significato econo mico-energetico:

$$f_{ec} \times 10600 \; Kcal/kg = f_{eq} \; (Kcal)_e/kg \qquad (11)$$

Alcuni valori sono riportati nell'ultima colonna di Tab. II. Utilizzando questi valori, e quelli di Tab. I, si ottiene:

$$(42400 \times 0,66)+(17702 \times 0,056)+(10600 \times 0,217) \; (Kcal)_e/kg =$$
$$= 31275 \; (Kcal_e)/kg \qquad (12)$$

Il rapporto col corrispondente valore alla combustione dà:

$$\frac{31275 \; (Kcal)_e/kg}{8480 \; (Kcal)_e/kg} = 3,69 \qquad (13)$$

e dimostra ancora la convenienza del trattamento di riraffinazione degli U rispetto alla loro semplice combustione.

La trattazione è basata su prezzi che sono assai variabili e fornisce indicazioni lorde. Sicuramente il processo di riraffinazione è più costoso della combustione, ma oltre ad un innegabile beneficio economico, comporta i vantaggi a cui già si è accennato.

Trattamenti di SA e delle terre decoloranti esauste

La frazione idrocarburica pesante P presente negli U la ritroviamo, a valle della riraffinazione, in SA (per il 20,7%) e nelle TDE (per l'1% circa).

Per queste ultime non è apparso conveniente il trattamento di estrazione e rigenerazione poichè trovano attualmente un più razionale smaltimento in cementificio /10/.

SA ha la seguente composizione media:

H_2SO_4	40%		x_o	= 0,40
P	52%	z_o	y_o	= 0,52
H_2O	8%		w_o	= 0,08

Le EE esaminate hanno la seguente composizione media:

H_2SO_4	0,00%		0,000	= x_o'
P	9,40%	z_o'	0,094	= y_o'
H_2O	90,60%		0,906	= w_o'

Il contenuto in zolfo, dello 0,15% analiticamente rilevato, non è stato preso in considerazione.

P rappresenta la parte da recuperare, poichè combustibile.

Le ricerche condotte hanno dimostrato che il trattamento di idrolisi di SA, abbinata alla tribocompressione, conduce ad un prodotto combustibile con contenuto in zolfo inferiore al 1'1% ed un potere calorifico superiore all'intorno di 9.700 Kcal/kg.

Il trattamento d'idrolisi di SA è stato dapprima effettuato con acqua in diversi rapporti di SA/acqua, nelle condizioni rivelatesi ottimali: T = 90 C; tempo di contatto in agitazione: 30 min. [11].

Il risultato di questa ricerca indicava poi ancora come ottimale il rapporto $SA/H_2O = 1/3$, in tre trattamenti successivi. Un ulteriore avanzamento degli studi suggeriva di sottoporre la fase P', in uscita dall'idrolisi, ad un trattamento meccanico composto di tribocompressione in cui si attuavano contemporaneamente uno stiramento ed una compressione di P'. Si realizzava quindi una maggior espulsione della fase acquosa S dalla massa di P'.

In successive esperienze si sostituivano all'acqua, come agente idrolizzante, le emulsioni esauste (EE) raggiungendo un duplice risultato pienamente soddisfacente: ottenere la fase idrocarburica P" di SA e nel contempo recuperare e conglobare in P" la fase idrocarburica presente in EE [12].

Più recentemente si è pensato di sottoporre SA ad un trattamento preliminare di tribocompressione (DTC) al fine di esplorare la possibilità di estrarre una parte dell'acido senza intervento di fase acquosa: è risultato un SA' della seguente composizione:

H_2SO_4	34,5%		$0,345 = x'$
P	58,4%	z'	$0,584 = y'$
H_2O	7,1%		$0,071 = w'$

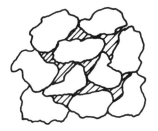

Figure 4

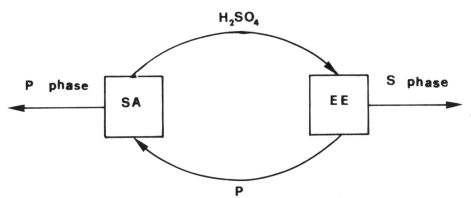

Figure 5

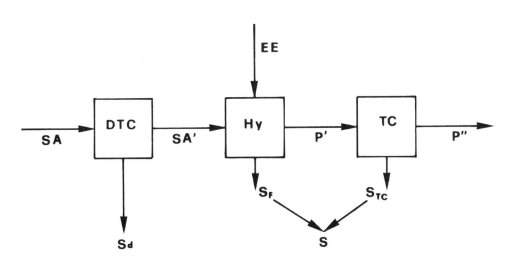

Figure 6

Sottoponendo poi SA' alla sequenza solita di idrolisi-tribocompressione, si ottenevano risultati veramente eccellenti. Così, già per un rapporto SA/EE = 1/2, in due soli trattamenti successivi, si otteneva una fase P" della seguente composizione:

H_2SO_4	5,3%		0,053	= x"
P"	87,3%	z"	0,873	= y"
H_2O	7,4%		0,074	= w"

e col rapporto 1/3 dopo tre trattamenti successivi:

H_2SO_4	1,4%		0,014	= x"
P	93,4%	z"	0,934	= y"
H_2O	5,2		0,052	= w"

Esprimendo la % di H_2SO_4 come zolfo si ottiene, nei due casi:

$$S = 1,7\% \qquad\qquad S = 0,46\%$$

Discussione

Il comportamento di SA alla tribocompressione diretta (DTC) fa pensare che la sua "struttura" possa essere quella rappresentata in fig. 4, in cui le zone tratteggiate rappresentano la fase $H_2SO_4 \cdot H_2O$ libera ed estraibile per tribocompressione diretta. La restante parte di $H_2SO_4 \cdot H_2O$ è in qualche modo legata alla parte idrocarburica e può essere estratta con trattamento a caldo con una fase acquosa.

Il meccanismo d'idrolisi, schematizzato in fig. 5, si attua attraverso le seguenti tappe:

1) fluidificazione di SA per effetto termico
2) trasferimento di H_2SO_4 dalla fase P alla fase S
3) destabilizzazione di EE da parte di H_2SO_4
4) separazione della fase idrocarburica P di EE
5) trasferimento della fase idrocarburica P di EE alla fase idrocarburica P di SA.

Nella fig. 6 si schematizza per blocchi l'intero processo.
Noi riteniamo che lo stadio d'idrolisi (Hy) sia uno stadio di equilibrio e che quindi la fase acquosa che si separa spontaneamente, S_F, abbia la stessa composizione della fase acquosa S_{TC} che verrà espulsa da P' per tribocompressione.
I bilanci di massa e le reazioni di trasferimento per le varie tappe dell'intero processo possono essere così espresse:

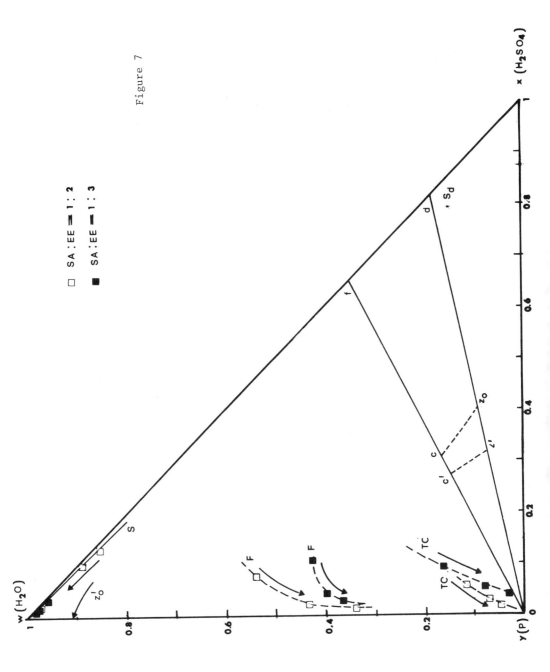

Figure 7

-tribocompressione diretta (DTC):
$$SA = SA' + S_d \qquad (14)$$
$$(a\ H_2SO_4 \cdot b\ P \cdot c\ H_2O)_{SA} = (a'\ H_2SO_4 \cdot b'\ P \cdot c'\ H_2O)_{SA'} +$$
$$+ (a - a')\ H_2SO_4 + (c - c')\ H_2O_{Sd} \qquad (15)$$

-idrolisi (Hy):
$$SA' + EE = P + SF \qquad (16)$$
$$(a'\ H_2SO_4 \cdot b'P \cdot c'\ H_2O)_{SA'} + (b''\ P \cdot c''\ H_2O)_{EE} = (m\ H_2SO_4 \cdot q\ P \cdot$$
$$n\ H_2O)_P, + \bigl[(a' - m)\ H_2SO_4 \cdot (b' + b'' - q)$$
$$P \cdot (c' + c'' - n)H_2O\bigr]_{SF} \qquad (17)$$

-tribocompressione, dopo idrolisi (TC):
$$P' = P'' + S_{TC} \qquad (18)$$
$$(m\ H_2SO_4 \cdot qP \cdot n\ H_2O)_{P'} = (m'\ H_2SO_4 \cdot q'\ P \cdot n'\ H_2O)_{P''} +$$
$$+ \bigl[(m - m')H_2SO_4 \cdot (q - q')P \cdot (n-n')\ H_2O\bigr]_{S_{TC}} \qquad (19)$$

Per le condizioni di equilibrio:
$$(a' - m):(b' + b'' - q):(c' + c'' - n)=(m - m'):(q - q'):$$
$$:(n - n') \qquad (20)$$

La (20) esprime l'eguale composizione di S_F ed S_{TC}.
L'intero processo sopra descritto può essere convenientemente descritto con un diagramma di stato triangolare quale quello riportato in fig. 7.
Ai vertici si indicano con x, y, w rispettivamente H_2SO_4, P, H_2O. Il punto z_o dà la composizione ternaria di SA, il punto z' la composizione ternaria di SA' ed il punto z_o' la composizione binaria di EE.
La zona x-y-d è il luogo dei punti di composizione di sistemi ternari quali gli SA usuali. Il punto z_o giace infatti su y-d.
Questi sistemi possono espellere, per tribocompressione diretta, $H_2SO_4 \cdot H_2O$ nello stesso rapporto con cui sono presenti in SA originale: il punto z' giace infatti ancora sulla retta y-d.
La zona d-y-f è il luogo dei punti di sistemi ternari che acquisita la fase acquosa, non la ricedono spontaneamente.
I punti c e c' sono tipici di SA ed SA' e giacciono sulla y-f.

La zona f-y-w è la zona di lavoro e in essa si collocano i punti dei sistemi ternari S, in equilibrio con le fasi idrocarburiche P' e P".

La composizione di S_d è data dal punto S_d vicino al lato x-w. Questa composizione equivale all'incirca ad un rapporto molecolare $H_2SO_4/H_2O = I/I$ dell'acido solforico monoidrato.

I punti sperimentali dei tre trattamenti con i rapporti SA'/EE pari a I/2 ed I/3 sono raggruppati su tre curve:

- curva S: luogo dei punti di composizione ternaria delle fasi acquose
- curve F: luogo dei punti di composizione ternaria delle fasi P' da idrolisi
- curve TC: luogo dei punti di composizione ternaria delle fasi P" da tribocompressione.

Come si vede le fasi S sono addossate al lato x-w, e verso il vertice w, mentre le fasi P" tendono al vertice y ad indicare l'elevato contenuto in fase idrocarburica ed il ridotto tenore in H_2O ed H_2SO_4.

Sotto il profilo economico-energetico è rilevante che il trattamento di SA con EE conduce allo smaltimento dei due sottoprodotti ed al recupero, oltrechè della fase idrocarburica di SA, anche di quella contenuta in EE.

Con riferimento al 3 trattamento col rapporto SA/EE = I/3, si ha un incremento della fase idrocarburica originale del I63%, come è facilmente verificabile col calcolo.

Il potere calorifico della P" finale è risultato di 9700 Kcal/kg. Trattando allora 1 t di SA con 9 t di EE si ottengono circa 1,37 t di fase P" e quindi:

$$9700 \text{ Kcal/kg} \times 10^3 \times 1,37 \text{ t} = 13289000 \text{ Kcal}$$

che, valutate al costo dell'olio combustibile denso, equivalgono circa a:

188000 L. di utile lordo

Sommando poi quanto si percepisce per la raccolta delle EE, cifra davvero non trascurabile, il beneficio lordo sopra indicato va moltiplicato almeno per un fattore 2.

Ovviamente dovranno detrarsi le spese d'impianto, d'ammortamento, gestione e così via, ed in particolare le spese per il trattamento delle acque di risulta.

Schema d'impianto

Nella fig. 8 è schematizzato l'impianto per il trattamento di idrolisi e tribocompressione di SA con EE.

I serbatoi 1 e 2, in cui sono stoccati SA ed EE, sono riscaldati con serpentina di vapore. Mediante la pompa P_1, SA passa alla tribocompressione diretta in DTC.

La fase fluida S_d è estratta continuamente e riciclata al trattamento di U, mentre SA' è convogliato, attraverso la lama raschiante L, al reattore d'idrolisi e tribocompressione Hy-TC in cui converge EE, dosata dalla pompa P_2. E' previsto un riscaldamento a vapore.

Dopo un tempo di contatto di 30 min., estratta la fase acquosa S_F, si procede alla tribocompressione: eliminata la fase acquosa S_{TC}, la fase P", mediante la lama raschiante L', è avviata allo stoccaggio nel serbatoio 4, previa eventuale aggiunta di flussante dal serbatoio 3, tramite pompa P_3.

La viscosità di P" è infatti di circa 80°E a 50°C.

Trattamento delle acque ($S_F + S_{TC}$)

Le acque provenienti da idrolisi e tribocompressione sono caratterizzate da basso valore di pH ed elevato valore di COD.
Nella Tab. III si riportano alcuni valori dei parametri più importanti.

Nella 3^a colonna si riportano i valori per l'acqua equalizzata di risulta dal processo (rapporto I/3, tre trattamenti successivi); nella 4ª colonna i valori dopo trattamento di coagulazione-flocculazione con calce e polielettroliti a pH=8,5 [13]; nella 5ª colonna dopo trattamento biologico su impianto a dischi porosi rotanti e nella 6ª colonna dopo trattamento a pH = 11. Come si può osservare già il trattamento a pH 8,5 porta i valori di tutti i parametri a livello accettabile ad eccezione di COD e PO_4^{---}.

Per abbattere il valore del COD si è utilizzato un dispositivo a dischi porosi rotanti [14] che ha fornito risultati soddisfacenti. Il pH viene corretto al valore 7 dopo il trattamento di coagulazione-flocculazione. L'acqua è immessa nel reattore biologico con tempi di permanenza dell'ordine di 24 ore: partendo da valori di COD di 3.400 mg/l si ha in uscita un valore di 380 mg/l, già accettabile per la tabella C della Legge Italiana n.319.

TABELLA III

		$S_F + S_{TC}$	dopo trat. a pH =8,5	dopo trat. biologico	dopo trat. a pH =11
pH	mg/l	2,3	8,5	7	11
COD	"	5200	3400	380	345
Estraib.	"	53	8	6	2
SS	"	0,5	0,0	0,0	0,0
Al	"	68	0,8	0,8	0,8
Ba	"	2,6	0,4	0,4	0,3
Cr	"	2,8	0,0	0,0	0,0
Cu	"	0,6	0,0	0,0	0,0
Fe	"	350	3	1,8	0,2
Ni	"	8,2	2,3	2,3	1,4
Mn	"	1,6	0,5	0,5	0,4
Pb	"	3,5	0,3	0,0	0,0
Zn	"	220	1,2	0,6	0,6
N tot.	"	32	24	15	8
PO_4^{---}	"	76	28	18	3

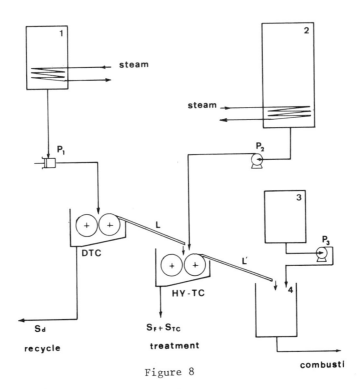

Figure 8

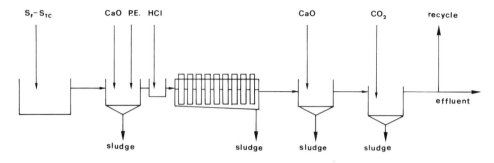

Figure 9

Il processo biologico ha abbattuto solo in parte la concentrazione dei fosfati, ed il risultato accettabile è ottenuto portando il valore del pH ad 11 mediante calce. Questo ultimo trattamento porta a norma i fosfati e migliora ulteriormente il valore di COD. Nella fig. 9 è schematizzato nell'insieme l'impianto di trattamento.

Conclusioni

Nel corso della relazione si è evidenziato il vantaggio economico-energetico della riraffinazione degli oli usati rispetto alla loro combustione diretta. Il processo permette di ottenere un riraffinato R di notevole pregio, e frazioni idrocarburiche leggere facilmente combustibili.

Come sottoprodotti si ottiene lo sludge acido che contiene, per il 52%, composti idrocarburici e derivati suscettibili di combustione.

Mediante tribocompressione diretta di SA si ottiene un sistema SA' che sottoposto a Hy e TC fornisce un combustibile a basso tenore in S ed in H_2O con un PCS di 9700 Kcal/kg.

Il processo raggiunge quindi un doppio obiettivo: recupero della parte idrocarburica di SA e di EE, con evidenti vantaggi di natura economica ed ecologica.

L'aspetto più pesante è il trattamento delle acque di risulta, che, se anche ha trovato una sua soluzione tecnica in laboratorio, comporta evidentemente problemi di costo. Inoltre si dovrà provvedere allo smaltimento dei fanghi che però, essendo costituiti prevalentemente da solfato di calcio, non dovrebbero presentare grosse difficoltà.

Vale la pena di ricordare, in chiusura, che nella valutazione globale del bilancio costi-benefici, è indispensabile tenere in debito conto tutti gli aspetti e tutte le voci che concorrono a formare il costo ed a definire il beneficio.

BIBLIOGRAPHY

[1] C.Versino "Frazionamento di oli min. lub." Annuali di Chimica 53, 136, (1963.

[2] C.Versino, F. Giaretti, L. Fogliano "Cromat. ed eluizione lineare di idrocarburi su SiO2". Nota I. La Riv. dei Combust. XX, 86-90, (1966).

[3] Idem Nota II. La Riv. dei Combust. XXI,389,395,(1967).

[4] C.Versino, C. Del Sole, A Bachiorrini. "Spettrofotometria IR di un olio min. lubrif." La Riv. dei Comb. XXVI, 9,321 (1972).

[5] Idem "Spettrofotometria IR di un olio min. lubrif. esausto". La Riv. dei Combust. XXVI, 12,421,(1972)

[6] Idem "Spettri IR ed UV di oli min. lubrif. trattati al l'acido". La Riv. dei Combust. XXVII, 6, 309 (1973).

[7] C.Versino "Fraz. cromatog. di oli minerali lub." Atti Acc. Scienze di Torino 97,129-1970 (1963).

[8] C.Versino, B Ruggeri, C. Molino, E. Volpi "La riraffinazione degli oli usati" nota I.ICP, Ind. Chim. petr. VII,5,107-II (1979).

[9] Idem Nota II. ICP. Ind. Chim. Petr. VII,7 (1979).

[10] C.Versino, C. Molino. "Una moderna industria a salvaguardia dell'ambiente: la riraffinazione degli oli minerali lubrificanti usati" Environmental 77. 24-29 aprile - Torino (1977).

[11] C.Versino, B Ruggeri, C. Molino, E. Volpi. "La riraffinazione degli oli usati". Nota III. ICP. Ind. Chimica Petr. VIII, 3,97-104 (1980).

[12] Idem Nota IV In stampa su ICP. Ind. Chim. Petr.

[13] C.Versino, C. Folonari, L. Fogliano, "Treatment of water from used metalworking fluids". III Research Institute, Chicago, USAn June 27-29 (1979).

[14] C.Versino."Trattamento delle acque di scarico nella riraffinazione degli oli usati". Conv. IRSA 10-11 maggio 1978 - Roma.

WARUM WIEDERAUFARBEITUNG VON GEBRAUCHTÖLEN ?
ENERGETISCHE UND ÖKOLOGISCHE ASPEKTE DES PROBLEMS
WHY REGENERATE USED OILS ?
ENERGY AND ECOLOGICAL ASPECTS OF THE PROBLEM
POURQUOI REGENERER LES HUILES USAGEES ?
ASPECTS ENERGETIQUES ET ECOLOGIQUES DU PROBLEME
PERCHE RIGENERARE GLI OLI USATI ?
ASPETTI ENERGETICI ED ECOLOGICI DEL PROBLEMA

C. RICHARD
Compagnie Française de Raffinage, Gonfreville-L'Orcher, France

RESUME

La réglementation française a récemment défini les conditions de la collecte des huiles usagées, mais laisse le choix pour l'élimination de ces huiles entre le brûlage dans des installations agréées ou la régénération.

L'auteur montre que la régénération est la meilleure solution, à la fois par la réduction des polluants et par l'économie de combustible. Ces affirmations sont basées sur l'expérience industrielle acquise par les Etablissements MATTHYS dans leur usine de Lillebonne et par la Compagnie Française de Raffinage dans sa raffinerie de Gonfreville-l'Orcher.

SUMMARY

Regulations were recently issued in France establishing the conditions for collecting used oils. However, they leave open the choice for disposal of these oils between burning in suitable installations and regeneration.
The author shows that regeneration is the better solution, both because of the reduction in pollution and the saving of fuel. These claims are based on industrial experience gained by the Matthys company in its factory at Lillebonne and the Compagnie Française de Raffinage at its refinery in Gonfreville-l'Orcher.

ZUSAMMENFASSUNG

In Frankreich wurden vor kurzem Rechtsvorschriften für die Bedingungen der Sammlung von Gebrauchtöl festgelegt; was die Beseitigung dieses Öls anlangt, so kann jedoch frei zwischen dem Verbrennen in zugelassenen Anlagen oder der Zweitraffination gewählt werden.
Der Verfasser legt dar, dass die Zweitraffination wegen der Schadstoffverringerung und der Brennstoffeinsparung die bessere Lösung ist. Die Argumente basieren auf der von MATTHYS in ihrer Anlage in Lillebonne und von der Compagnie Française de Raffinage in ihrer Raffinerie von Gonfreville-l'Orcher gewonnenen Erfahrung.

RIASSUNTO

La regolamentazione francese ha recentemente definito le condizioni di raccolta degli oli usati ma lascia la scelta per quanto riguarda l'eliminazione di tali oli fra la combustione all'interno di impianti autorizzati e la rigenerazione.
L'autore fa presente che la rigenerazione è la soluzione migliore sia per la riduzione dell'inquinamento sia per il risparmio di combustibile. Tali affermazioni sono basate sull'esperienza industriale acquisita dagli "Etablissements Matthys" nel loro impianto di Lillebonne e dalla Compagnia francese di raffinazione nella raffineria di Gonfreville-L'Orcher.

En me confiant la dernière conférence de ces deux journées d'études, Monsieur le Président BRASSART m'a aimablement tendu un piège. En effet, en écoutant les orateurs qui m'ont précédé, je me rappelais une réflexion d'un penseur français du XVIIème siècle, qui écrivait :

" Nous arrivons trop tard. Tout a déjà été dit. "

Excusez-moi donc, Monsieur le Président, si mon exposé reprend des idées qui ont déjà été exprimées au cours de ce Congrès. Dans la mesure où toutes ces communications ont été rédigées séparément, ces répétitions auront au moins l'avantage de prouver la communauté des pensées et la convergence des raisonnements dans l'ensemble des pays ici représentés.

Le Gouvernement Français, suivant la recommandation de la Directive Européenne du 16 Juin 1975, a défini par le décret 79-981 du 21 Novembre 1979, la réglementation de la récupération des huiles usagées.

Ce texte précise que "les huiles usagées collectées sont préférentiellement destinées à être éliminées par régénération ou recyclage".

Tout en n'interdisant pas le brûlage dans des installations agréées, nous voudrions expliquer les raisons qui ont incité le Gouvernement à indiquer sa préférence. Dans ce but, nous comparerons les conséquences pour la collectivité publique, d'une part de la régénération, d'autre part du brûlage d'une même quantité d'huiles usagées.

Nous appuierons notre démonstration sur des rendements mesurés :

- à l'usine de Lillebonne de la Société MATTHYS, en ce qui concerne la régénération des huiles usagées, pour l'unique raison que cette usine traite à elle seule plus de 50 % des tonnages français et peut donc être considérée comme représentative du marché;

- à la Raffinerie de Gonfreville l'Orcher, de la COMPAGNIE FRANCAISE DE RAFFINAGE, en ce qui concerne la fabrication d'huiles de base, cette Raffinerie elle aussi étant, de son côté, la plus importante de France.

Avant d'aborder le point de vue énergétique de la comparaison, il est utile de considérer d'abord la composition des huiles usagées arrivant aux usines de régénération.

I - Composition des huiles usagées.

Il n'existe pas une seule sorte d'huile usagée, mais de nombreux types dépendant des origines de la collecte. Nous ne citerons ici que trois types qui constituent l'essentiel des tonnages recueillis par les récupérateurs :

- les huiles moteurs représentant environ 50 % des tonnages.

- les huiles industrielles représentant environ 25 %.

- les boues huileuses représentant environ 25 %; mais après séparation de la phase aqueuse elles ne fournissent que le quart de leur poids en huiles régénérables.

Les huiles moteurs représentant donc un peu plus de 60 % des hydrocarbures récupérables à la régénération, nous baserons sur une huile moteur l'exemple de notre démonstration.

Une analyse typique d'huile moteur usagée (après collecte) est la suivante (voir Tableau I) :

- Hydrocarbures
  - légers   1,0 %
  - gazole   2,8 %
  - constituants des huiles de base   79,3 %

- Additifs (essentiellement des composés organo-métalliques)   7,0 %

- Solvants chlorés   0,9 %

- Sédiments (métaux d'usure, imbrûlés, poussières)   1,0 %

- Solution aqueuse   8,0 %

Les additifs sont de compositions très diverses; de plus, les huiles moteurs contiennent des quantités notables de particules d'oxydes de plomb, provenant de la combustion des alkyles de plomb des carburants.

Tableau I

Analyse-type d'une huile moteur usée.

(en poids)

- Hydrocarbures
  - Hydrocarbures légers — 1,0 %
  - Gazole — 2,8 %
  - Constituants des huiles de base — 79,3 %
- Additifs (eesentiellement sels de métaux lourds) — 7,0 %
- Solvants chlorés — 0,9 %
- Sédiments (métaux d'usure, imbrûlés, poussières) — 1,0 %
- Solution aqueuse — 8,0 %

Teneur en métaux (parties par million, en poids)

Fer :	135
Cuivre :	28
Plomb :	2.050
Zinc :	635
Calcium :	913
Magnésium :	88
Aluminium :	24
Sodium :	75
Potassium :	25
Chrome :	5

Par exemple, on peut donner la décomposition suivante des teneurs en métaux :

- Fer   :   135 ppm   (en poids)
- Cuivre   :   28 ppm
- Plomb   :   2.050 ppm
- Zinc   :   635 ppm
- Calcium   :   913 ppm
- Magnésium :   88 ppm
- Aluminium :   24 ppm
- Sodium   :   75 ppm
- Potassium :   25 ppm
- Chrome   :   5 ppm

II - Régénération des huiles par les procédés MATTHYS.

Le schéma de procédé présenté à la Figure 1 est une simplification des opérations effectuées à l'usine de Lillebonne des Etablissements MATTHYS pour régénérer les huiles. Sur ce schéma, nous avons indiqué le rendement moyen obtenu à partir de 100 tonnes d'huiles usagées, tel qu'il résulte de la production de l'année 1979.

On voit que le procédé commence par un tri à la réception des camions. Un échantillon de chaque camion est analysé par absorption d'un rayonnement infra-rouge mettant en évidence la constitution des principales molécules. Ce tri permet d'identifier, par lecture du spectre, les principales impuretés contenues dans l'échantillon et de classer le camion correspondant dans l'une des quatre catégories :

- huiles moteurs.

- huiles industrielles.

- émulsions qui devront subir un pré-traitement séparant l'eau (les huiles extraites rejoignent les huiles industrielles en vue de leur régénération).

- huiles usagées dont la teneur en huiles de base, après régénération, ne justifie pas le traitement. (En pratique, cette teneur doit être de l'ordre de 50 % si l'on veut éviter des consommations excessives de réactifs chimiques).

Après ce tri, les huiles de chacune des **quatre** catégories sont traitées séparément. L'expérience a montré, en effet, que le mélange d'huiles moteurs avec les huiles industrielles provoque des réactions chimiques entre additifs nuisant à la régénération.

Figure 1   Société MATTHYS: Schéma de Procédé et Bilan Pondéral

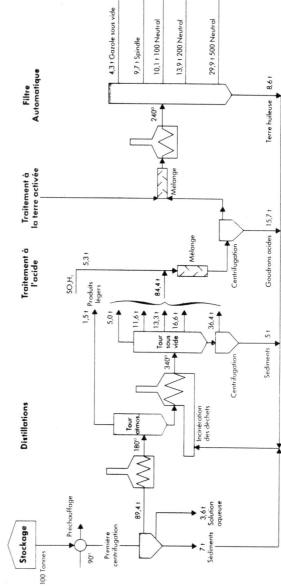

On voit, d'après ce schéma, que ces procédés font très largement appel à la centrifugation à toutes les étapes du traitement. Les centrifugeuses utilisées ont été mises au point en vue de cette utilisation particulièrement sévère, due surtout à la présence de sédiments abrasifs dans les huiles usagées.

Ces centrifugeuses, construites par la GARAP, permettent d'obtenir une accélération de 7.500 g, supportent des températures de 250° C et séparent les mélanges qu'on y charge en trois phases :

- une phase hydrocarbure
- une phase aqueuse
- un produit pâteux contenant les sédiments solides avec une très faible teneur en huile.

Selon la composition de la charge, de telles centrifugeuses peuvent traiter entre 1.500 et 3.000 litres/heure de charge.

Ces procédés présentent plusieurs autres particularités :

1°) ils ne nécessitent l'apport d'aucun combustible extérieur, les calories étant entièrement fournies par l'incinération des sédiments, des goudrons acides et des terres huileuses produites dans le traitement des huiles.

2°) les incinérateurs, fours rotatifs horizontaux, sont conçus de manière à agglomérer les sels métalliques contenus dans les sédiments sous forme de billes de sels inertes non volatiles qui sont recueillis sur la sole du four. Le taux de récupération de ces métaux dépasse 90 %.

Enfin, la lecture de la figure 1 montre un rendement pondéral en hydrocarbures de 67,9 %, dont 63,6 % sont des bases de constitution d'huiles ayant des caractéristiques de :

- couleur
- stabilité thermique
- viscosité
- indice de viscosité

comparables à celles des huiles de base produites directement à partir des pétroles bruts par distillation, extraction et déparaffinage.

Ainsi qu'on peut le voir, le procédé actuel utilisé par MATTHYS est un dérivé du procédé à l'acide.

Même si la réduction des quantités d'acide consommées dépasse 50 % par rapport au procédé à l'acide classique (compte tenu de la nature des huiles usagées traitées) il reste que la

Figure 2   C.F.R. Unités d'huiles: Schéma Simplifié et Bilan Pondéral

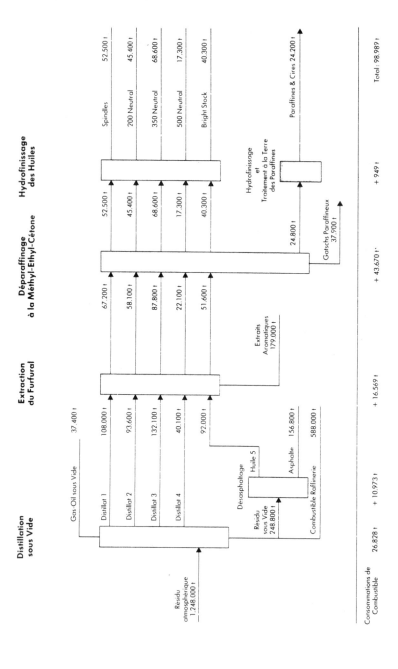

Société MATTHYS s'est fixée comme objectif de supprimer complètement l'utilisation d'acide.

A cette fin, elle a actuellement deux variantes, toutes deux basées sur l'hydrogénation, l'une des deux ayant recours au procédé d'ultrafiltration.

Ce dernier procédé est actuellement étudié en liaison avec la COMPAGNIE FRANCAISE DE RAFFINAGE et le C.E.A. qui fournit les barrières.

Ce procédé en est au stade du pilote et les résultats sont excellents.

Le deuxième, qui est 100 % MATTHYS, en est au stade des essais industriels.

Pour autant qu'on puisse en juger, les prix de revient des deux procédés seraient équivalents et un peu plus favorables que le procédé actuel MATTHYS, qui est déjà parmi les plus performants.

On notera que l'intérêt de ces différents procédés - qu'il s'agisse du procédé actuel ou des deux procédés à l'étude - réside dans le fait qu'ils se distinguent par la possibilité qu'ils offrent de traiter, non seulement les huiles moteurs mais, sans aucun problème, la totalité des huiles industrielles usagées, sous la seule réserve que les huiles moteurs et les huiles industrielles aient été préalablement ségréguées (jusqu'ici le mélange d'huiles industrielles entre elles n'a posé aucun problème).

III-Qu'advient-il en cas de combustion des huiles usagées ?

Pour pouvoir être brûlée comme combustible, l'huile usagée doit subir une filtration afin que ses sédiments ne viennent, ni se décanter dans les cuves de stockage, ni obstruer les brûleurs.

La combustion d'une huile usagée telle que celle dont nous avons examiné la composition au paragraphe I, fournit une quantité de chaleur de l'ordre de 8.000 Kcal/kg, en raison de la vaporisation inévitable de ses fractions **aqueuses**.

Enfin, dans cette combustion, les métaux contenus dans les additifs sont dispersés dans les fumées sous forme d'oxydes ou de sels métalliques. Le constituant le plus nocif est le plomb.

Pour une huile moteur ayant la composition citée, il faudra la diluer dans 40 fois son volume d'un combustible exempt de plomb, pour que la combustion soit conforme à l'arrêté du 7 Juin 1980, limitant à 5 milligrammes par thermie la teneur en plomb dans les fumées.

Il est peu probable que la plupart des utilisateurs disposent des équipements nécessaires pour effectuer un tel mélange.

IV - Fabrication d'huiles de base.

Nous allons maintenant déterminer la quantité de fond de colonne atmosphérique nécessaire pour la fabrication d'huiles de base par les procédés classiques comportant :

- distillation sous vide

- désasphaltage du résidu sous vide

- extraction au furfurol des fractions à bas indice de viscosité

- déparaffinage à la méthyl-éthyl-acétone

- raffinage à l'hydrogène.

Répondre à cette question est relativement compliqué. En effet, les rendements et la consommation de chaque unité dépendent d'abord de la qualité de la charge, donc des pétroles bruts dont sont issus les résidus atmosphériques.

Rendements et consommations varient également selon la qualité de l'huile finie, en viscosité, indice de viscosité et point d'écoulement.

Enfin, la fabrication de ces huiles est physiquement inséparable de la fabrication des cires et paraffines qui en sont extraites et, seules, des situations particulières permettent d'attribuer à ces cires et paraffines une quote-part du combustible consommé.

Pour résoudre cette triple difficulté, nous avons choisi de prendre des chiffres moyens en examinant la production de toutes les huiles de base, cires et paraffines produites à Gonfreville l'Orcher pendant un semestre.

Comme nous l'avons fait pour présenter la régénération des huiles, nous avons schématisé sur la figure 2 l'ensemble des unités de fabrication des huiles de base. Nous avons également porté sur cette figure les tonnages chargés par chaque unité et les consommations de combustible équivalent correspondant à chaque unité.

Enfin, nous admettrons arbitrairement que la fabrication d'une tonne de cire ou paraffine requiert la même quantité d'énergie que la fabrication d'une tonne d'huile de base.

Cette convention arbitraire ne saurait d'ailleurs introduire une grosse erreur puisque le tonnage de cires et paraffines représente un peu moins de 10 % du total des fabrications.

Nous insistons sur le fait que ce schéma est extrêmement simplifié, puisque la COMPAGNIE FRANCAISE DE RAFFINAGE dispose à Gonfreville l'Orcher de trois chaînes d'huiles parallèles et que les résidus atmosphériques comptabilisés dans la figure 2 proviennent de trois pétroles bruts différents. Les chiffres de la figure 2 ne représentent que le cumul d'une série d'opérations ayant chacune son rendement propre et produisant des huiles d'une qualité précise.

Vous voyez que, partant d'une charge de 1.248.000 tonnes de résidus atmosphériques, on a obtenu :

24.200 tonnes de cires et paraffines

et 224.100 tonnes d'huiles de base

en consommant 99.000 tonnes de combustible.

En résumé, la fabrication d'une tonne d'huile de base nécessite la consommation, par combustion, de 0,4 tonne de fuel.

V - Bilan énergétique.

Revenons aux 100 tonnes d'huiles usagées de la figure 1.

Si nous les brûlons, nous avons vu qu'elles équivalent à 80 tonnes de combustible.

Si nous les régénérons, les 67,9 tonnes obtenues représentent :

- Gazole .... 4,30 T.
- Fuel oil équivalent nécessaire pour la production des huiles de base correspondantes ...... 63,6 x 1,4 = 89,04 T.

Soit .... 93,34 T.

Le gain énergétique est donc de l'ordre de 13 T. pour chaque régénération de 100 T. d'huiles usagées (puisque la combustion de ces 100 T. n'équivaut qu'à 80 T. de combustible).

Conclusions.

A la fois sur le plan de la protection de la qualité de l'air et sur le plan énergétique, la régénération des huiles usagées, avec les techniques existantes, est préférable au brûlage.

Il est donc souhaitable que les prix de reprise des huiles usagées en vue de leur brûlage ne détournent pas une part considérable des huiles collectées de ce qui devrait être leur destination normale.

SCHLUSSWORT
CONCLUSIONS
CONCLUSIONS
CONCLUSIONI

P. BRASSART

Präsident des Kongresses
President of Congress
Président du Congrès
Presidente del Congresso

Minister,

May I first present to you our Association, the European Union of Independent Lubricant Producers.

It represents 7 countries in all (tomorrow there will be 8) and 450 companies which employ some 35,000 people across the Community.

Within this association, the European Regeneration Commission plays a very important role.

It has had the honour - and also the heavy responsibility - of organizing this Congress.

The Regeneration Industry, which was born before the war, experienced enormous difficulties after it in all the European countries, due in particular to the fact that in view of the very low price of crude oil no-one thought much about saving raw materials, or indeed about protecting the environment.

It finally needed a directive from the European Communities' Commission to remind member states that they should concern themselves with such things.

In the course of the last decade, I have attended a number of congresses, both in Europe and in the United States, on the subject of the elimination of used oils.

Under the pressure of events, I have been aware during the recent past of certain upheavals which an attentive observer could interpret as predicting a revival for this industry.

You will understand how delighted I am, Minister, to see, after more than twenty years of constant struggle to defend an Industry then in poor shape - firstly under the wing of my father, then alone, and finally with the help of a growing number of friends - that during this Congress it has suddenly emerged into the limelight, and has become powerful and sophisticated where once it was only artisanal.

The exhibition that your colleagues have visited, and the technical discussions to which they have listened, are proof of this ; but the advantages to be derived from this form of recycling compared with any others have been especially revealed by the particularly brilliant exposés we have heard from Mr. Thomson, of the Bartlesville Energy Technology Center, and Mr. Richard, Past Chairman of Concawe.
You will understand, Minister, how proud I am as a Frenchman to see that your Government, by including new regulations in the law on energy saving, is moving in the direction that we felt necessary.

I know how strongly you feel about environmental protection.

I had the honour of attending the press conference you gave in the context of the campaign launched by ANRED, which declared with emphasis that used oils must be collected for recycling, and I have also had occasion to note to what powerful effect Mr. Giraud, our Minister for Industry, has intervened in this matter ; and I would like to ask his representative, who is present here tonight, to be good enough to convey our gratitude to him.

However, Minister, allow me to conclude by saying to you that regulations can only be, and only are, good to the extent that those whose responsibility it is to apply them take into account the fact that the best results are always obtained when a maximum of freedom is left to the people in the field.

In this context, I would like to draw your attention to the fact that since used oil is a waste product, it cannot by definition have fixed and unchanging characteristics.

This means that to obtain optimum efficiency in the regeneration operations there must be careful segregation at the time of collection, in order that products whose mixture in a common storage system would lead to secondary reactions, which would impede subsequent treatment, should be kept apart.

It is a remarkable thing that this need has been emphasized by all the speakers who have touched on this subject.

The consequence is that only the rerefiners can lay down, at one and the same time :
- the criteria for this segregation,
- the measures to be taken to obtain it,
- the checks that are necessary to ensure that they are applied.

If this necessity is not borne in mind, we shall not be bearing in mind the conditions for success.

Minister, you have before you representatives of 35 countries, - a gathering that extends widely beyond the context of Europe.

They have attended all the lectures with remarkable assiduity, despite the marvellous weather we are having in Paris, a city that many of them had never visited before.

In spite of this temptation, there have been very few abstentees among the 400 people who have registered ; this demonstrates both the high level of the exposés we have listened to and also the faith that they have in the future of our Industry.

ALLOCUTION DE MONSIEUR F. DELMAS

Secrétaire d'Etat à l'Environnement

Monsieur le Président,
Monsieur le Député,
Messieurs les Directeurs,
Mesdames, Messieurs,

C'est avec plaisir que j'ai répondu il y a quelques semaines favorablement à la demande du Président Brassart de venir clôturer, au nom de Monsieur Michel d'Ornano, Ministre de l'Environnement et du Cadre de Vie, ce deuxième Congrès Européen sur le Recyclage des Huiles Usagées - C'est en effet un domaine qui nous tient particulièrement à coeur, et qui se caractérise depuis un an par des initiatives françaises décisives.

Comme vous l'avez rappelé il y a un instant, Monsieur le Président, le Gouvernement Français, l'un des premiers parmi les pays européens, a mis en place, en novembre 1979, un dispositif réglementaire complet visant à une collecte exhaustive des huiles usées et à une valorisation de ces produits.

J'estime donc particulièrement opportune la tenue de ce congrès à quelques semaines de l'entrée en vigueur de ce dispositif et me réjouis d'autant plus de son patronage par la Commission des Communautés Européennes, qui a joué à cet égard un rôle incitateur que je me plais à souligner, en proposant dès 1975 la directive européenne sur l'utilisation des huiles usagées, directive qui se trouve désormais traduite dans le droit français.

Quel était l'enjeu ?

Il s'agit en premier lieu, pour moi, représentant le Ministre chargé de l'Environnement, d'éliminer des pollutions réellement dangereuses pour l'homme et son environnement. Je pense notamment aux cours d'eau et aux nappes souterraines ; mais je songe également à la suppression de pollutions atmosphériques inacceptables liées au brûlage d'huiles usagées dans les conditions trop souvent insuffisantes au regard de certaines émissions, de métaux lourds par exemple.

Il s'agit en second lieu, - et je me retourne à ce stade vers Monsieur Jean-Pierre Capron, Directeur des Hydrocarbures représentant le Ministre de l'Industrie, - de valoriser convenablement une ressource nationale non négligeable au plan énergétique.

Le parlement français a voté il y a près de trois mois, dans le cadre de la loi relative aux économies d'énergie et à la récupération de la chaleur, des dispositions indiquant très clairement que valorisation signifiait régénération de préférence à incinération, même si cette dernière est parfaitement contrôlée.

On peut ainsi espérer que le dispositif français, fondé d'une part sur un agrément exclusif de collecte au niveau de chaque département, d'autre part sur une meilleure organisation de valorisation s'appuyant sur des installations performantes en rendement et en prévention des pollutions, permettra d'augmenter de plus de 50% la récupération actuelle des huiles usagées, en portant à 15% environ la part des huiles commercialisées en provenance de circuits de régénération.

Cet objectif est réaliste : je suis en effet persuadé que ce système correspond à une réalité économique dans un contexte de renchérissement croissant des matières premières ; je suis confirmé dans cette idée par les résultats exprimés à l'occasion de ce Congrès, notamment par l'ancien Président du groupe professionnel des pétroliers européens pour la protection de l'atmosphère et de l'eau (concawe), - dont chacun connaît le sérieux des analyses - et par le Directeur du "Bartles energy' technology center" aux Etats-Unis, résultats qui expriment finalement un bilan global positif en faveur de la régénération.

Cet objectif peut donc être tenu au plan économique ; il doit l'être aussi au plan technique. La valorisation des huiles usées relève en effet d'opérations parfois complexes et technologiquement sophistiquées, très sensibles à la qualité des approvisionnements. Il importe donc que tout au long de la chaîne de production de collecte et de traitement, la ségrégation des produits soit assurée de la manière la plus efficace possible. Ceci sera obtenu, j'en suis sûr, par une étroite collaboration entre ramasseurs et éliminateurs.

Je voudrais, pour conclure, insister sur l'originalité d'un dispositif qui, malgré les difficultés que nous ne sous-estimons pas, doit pouvoir être étendu à l'ensemble des pays de la Communauté dans le cadre tracé par la Directive Européenne de 1975.

Le progrès est en effet considérable ;
- au plan énergétique, c'est évident,
- au plan de la protection de l'Environnement,
il faut bien comprendre que le système mis en place permettra de soustraire des milieux naturels précisément les rejets sauvages qui leur faisaient le plus de mal.

Ce gain viendra donc conformer les efforts développés par la France pour redonner à tous les cours d'eau une qualité satisfaisante, efforts qui ont déjà conduit à une régression de la pollution de 5% par an.

Enfin mon propos serait incomplet si je ne soulignais pas le rôle important et décisif qu'ont joué dans la mise en place du dispositif français, l'Agence Nationale pour l'Elimination et la Récupération des Déchets (ANRED), établissement public mis en place en 1977 par mon département avec l'aide du Ministre de l'Industrie et qui constitue le plaque tournante de ce dispositif, et aussi les représentants des professionnels qui ont été associés depuis le début à sa mise en place et sans lesquels nous n'aurions pu élaborer un système fondé sur la capacité de l'industrie française de la collecte et de la régénération des huiles usagées de mener, au plan technique, cette mission nationale.

TEILNEHMERLISTE
LIST OF PARTICIPANTS
LISTE DES PARTICIPANTS
LISTA DEI PARTECIPANTI

BELGIQUE

BERG N. G.
CASTROL N. V.
HELMSTRAAT 107
2200 BORGERHOUT

BRAT H.
FINA S. A.
VAN SERVERLAAN 28
1970 WEZEMBEER OPPEM

DEQUENNE J.
BELGIAN SHELL
CANTERSTEEN 47
1000 BRUXELLES

DONCKERWOLCKE M.
I. H. M. B.
49 SQUARE MARIE LOUISE
1040 BRUXELLES

FABRY R.
MINISTERE DES AFFAIRES
ECONOMIQUES
ADMINISTRATION DE
L'ENERGIE
30 RUE DE MOT
1040 BRUXELLES

GRARD P.
S. A. COTRIOL
CHAUSSEE DE BRUXELLES 2a
7458 MASIERES LEZ MONS

HATRY P.
FEDERATION PETROLIERE
BELGE
RUE DE LA SCIENCE 4
1040 BRUXELLES

JACQUEMIN
FACULTES UNIVERSITAIRES
DE NAMUR
3 REMPART DE LA VIERGE
5000 NAMUR

KEMPENEERS R.
S. A. FABRIQUE NATIONALE
HERSTAL
RUE VOIE DE LIEGE 33
4400 HERSTAL

LOWEY BALL A. H.
PHILLIPS PETROLEUM CHE-
MICALS
STEENWEG OP BRUSSEL 355
1900 OVERIJSE

MATTHISSEN R.
I. C. D. I.
335 AVENUE E. MASCAUX
6001 MARCINELLE

MESSENS F.
CHEVRON OIL BELGIUM
166 AVENUE MARIE LOUISE
1050 BRUXELLES

MULKENS J.
FEDERATION PETROLIERE
BELGE
RUE DE LA SCIENCE 4
1040 BRUXELLES

NOEL A.
S. D. R. W.
5 RUE GRAFE
5000 NAMUR

ORBAN A.
ESSOCHEM EUROPE INC
NIEUWE NIJVERHEIDSLAAN 2
1950 MACHELEN

PLONSKER L.
ETHYL S. A.
AVENUE LOUISE 523
BOITE 19
1050 BRUXELLES

RENGUET J.
S. A. MOTTAY PISART
STEENKAI 42
1800 VILVOORDE

SORGO P. M.
PHILLIPS PETROLEUM
CHEMICALS
355 STEENWEG OP BRUSSEL
1900 OVERIJSE

VAN HECK F.
S. A. COPPEE RUST N. V.
TAVERNIERKAAI 2
BOITE 3
2000 ANVERS

VERTONGEN M.
AMOCO MARKETING BELGIUM
LIPPENSLAAN 90
BOITE 1
8300 KNOKKE

VOLRAL
I. C. D. I.
335 AVENUE E. MASCAUX
6001 MARCINELLE

BUNDESREPUBLIK DEUTSCHLAND

BRANDT G.
LUBRIZOL GMBH HAMBURG
C/O LUBRIZON FRANCE
CEDEX 7
92080 PARIS LA DEFENSE

COENEN H.
F. KRUPP GMBH
KRUPP FORSCHUNGSINSTITUT
MÜNCHENER STRASSE 100
4300 ESSEN 1

DECKER M.
NORDKLIMA

DEGEN H.
FUCHS MINERALLWERKE GMBH
FRIESENHEIMER STRASSE 15
6800 MANNHEIM

DINSE O.
PINTSCH ÖL GMBH
LIEBIGSTRASSE 41
2000 HAMBURG 74

EBERT O.
LUDWIG HERZOG GMBH & CO KG
HAIDGRAGEN 9
8012 OTTOBRUNN

EDER R.
AMMRA
GURLITTSTRASSE 31
2000 HAMBURG 1

EGGE W.
ESSO AG
KAPSTADTRING
2 HAMBURG 60

DR EICKE H.
SÜD CHEMIE AG
PASTFACH 22 22 40
8000 MÜNCHEN 2

ERDWEG K.
LEYBOLD HERAEUS GMBH
WILHEM RHON STRASSE
0645 HANAU

DR FUCHS M.
RUDOLF FUCHS GMBH & C°
FRIESENHEIMER STRASSE 15
6800 MANNHEIM

FUHSE H.
HORST FUHSE MINERALÖLRAFF.
HALSKESTRASSE 40
2000 HAMBURG 40

GIERE F.
HABERLAND & C°
BAHNHOFSTRASSE 82
3161 DOLLBERGEN

GROH F.
UNITI
BUCHSTRASSE 10
2000 HAMBURG 76

HAAKE R.
BERND MEINKEN
ZUM STEVERTAL 8
4358 HALTERN

MME HABERLAND M.
HABERLAND & C°
BAHNHOFSTRASSE 82
3161 DOLLBERGEN

HANGSTEIN H. C.
NORDDEUTSCHE LANDESBANK
GEORGSPLATZ 1
3000 HANNOVER 1

HAVEMANN R.
HABERLAND & C°
BAHNHOFSTRASSE 82
3161 DOLLBERGEN

HEIDLER F.
HARPENER AG
SILBERSTRASSE 22
4600 DORTMUND

HIRSCHELMANN G.
RHEINISCHE MOTORÖL
KRABBENKAMP 11
4100 DUISBURG 12

HOLLMANN G.
EDELHOFF
POSTFACH 500
5860 ISERLOHN

IGEL K.
SUDÖL MINERALÖL RAFF.
EISLINGEN

IVERSEN B.
ÖLWERKE J. SCHINDLER GMBH
NEUHOFER BRÜCKENSTRASSE 127
2102 HAMBURG

KALK T.
RHEINISCHE MOTORÖL
KRABBENKAMP 11
4100 DUISBURG 12

KÄSEBERG J.
WENZEL &WEIDMANN GMBH
JÜLISCHER STRASSE
5180 ESCHWEILER

DR KNOBLOCH T. P.
UMWELTBUNDESAMT
BISMARCKPLATZ 1
1000 BERLIN 31

DE KOEHN H.
MINERALÖL RAFF. WEISS&C°
SCHMIDTS BREITE 3
2102 HAMBURG 93

KOPNER D.
KARL POHLMANN GK
AM KNIEP 2
3540 KORBACH

KUHN P.
SÜDÖL MINERALÖL RAFF.
GMBH
SCHLOSSSTRASSE 19
7332 EISLINGEN

DR LAFRENZ CH.
HABERLAND & C°
BAHNHOFSTRASSE 82
3161 DOLLBERGEN

LIERSCH K. M.
ABWASSERTECHNISCHE
VEREINIGUNG
LOHFELD WEG 23
3000 HANNOVER

MEINKEN B.
ING. BUREAU MEINKEN
P. O. 216
HALTERN

MELLO H.
ESSO CHEMIE GMBH
BAHNHOFSTRASSE 1-9
5000 KÖLN 1

NAGEL W.
DEUTSCHE TOTAL/GARAP
KIRCHFELDSTRASSE 61
4000 DÜSSELDORF

DR NEULING
PAUL NEULIN KG
BERLIN

NIEMAX J.
DEUTSCHE PENTOSIN
WERKE GBMH
INDUSTRIESTRASSE 39-43
2000 WEDEL

PAULER
RUDOLF FUCHS GMBH & C°
FRIESENHEIMERSTRASSE 15
6800 MANNHEIM

SACHSE H.
ALBERT SCHÖNING INH. H.
SACHSE
POSTFACH 7002
5810 WITTEN 7

SCHULZE-SCHWINKING
WESTFALIA SEPARATOR AG
4740 OELDE A
WESFALEN

STUDDERS H. W.
U. E. I. L.
POSTACH 1709
2160 STADE

GRAF VON STAUFFENBERG G.
PINTSCH ÖL GMBH
SAARSTRASSE 5
6450 HANAU

WEDEPOHL E.
INSTITUT FÜR ERDÖLFORS-
CHUNG
AM KLEINEM FELDE
HANNOVER

POHLMANN K. FR.
KARL POHLMANN KG
AM KNIEP 2
3540 KORBACH

SAMBALE D.
KLÖCKNER INDUSTRIE
ANLAGEN GMBH
NEUDORFERSTRASSE 3-5
4100 DUISBURG

SCHÜTT W.
OROGIL FRANKFURT/MAIN
AM SCHOLLENGARTEN 1
6231 SCHWALBACH/TS

PROF. THOME-KOZMIENSKY
UNIVERSITE TECHNIQUE DE
BERLIN

VON THADEN G.
MINARALÖLRAFF. WEISS & C°
SCHMIDTSBREITE 3
2101 HAMBURG 93

DR WILDERSOHN M.
WENZEL & WEIDMANN GMBH
JÜLISCHER STRASSE
5180 ESCHWEILER

DR RICHTER F.
ÖLWERKE SCHINDLER GMBH
NEUHÖFER BRÜCKENSTRASSE
127-152
2102 HAMBURG 93

SENK J.
ÖLWERKE SCHINDLER GMBH
NEUHÖFER BRÜCKENSTRASSE
127-152
2102 HAMBURG 93

STEINMETZGER G.
HABERLAND & C°
BAHNHOFSTRASSE 82
3161 DOLLBERGEN

TRUM R.
VERBAND PRIVATER STÄDTE-
REINIGUNGSBETRIEBE
POSTFACH 900845
5000 KÖLN 90

WEBER D.
FIRMA DIETER WEBER
IM DUGENDORF
7335 SALACH

(CEE) - COMMISSION DES COMMUNAUTES EUROPEENNES

CARPENTIER
COMMISSION DES COMMUNAUTES
EUROPEENNES
200 RUE DE LA LOI
1049 BRUXELLES

KRAEMER M.
SERVICE ENVIRONNEMENT
COMMISSION DES COMMUNAUTES
EUROPEENNES
200 RUE DE LA LOI
1049 BRUXELLES

HANSEN T. M. SC.
TECHNOLOGICAL INSTITUTE
2630 TASTRUP

ALEXALINE J.
SRRHU
159 QUAI AULAGNIER
92600 ASNIERES

MME STALPAERT J.
COMMISSION DES COMMUNAUTES
EUROPEENNES
200 RUE DE LA LOI
1049 BRUXELLES

RISCH B. W. K.
COMMISSION DES COMMUNAUTES
EUROPEENNES
200 RUE DE LA LOI
1049 BRUXELLES

DANMARK

MULLER K.
HANDELSHOJSKOLEN I KØBENHAVN
SANKT KNUDS VEJ 42
1903 KØBENHAVN

FRANCE

ANDRE P.
HUILES MINERALES PRECY
45220 CHATEAU RENARD

ANTOINE
MATTHYS LUBRIFIANTS
84 RUE DE VILLIERS
92300 LEVALLOIS PERRET

BARTAGNON CH.
SOCIETE FRANCAISE DES
PETROLES BP
10 QUAI PAUL DOUMER
92410 COURBEVOIE CEDEX

BAUDASSE M.
DAFFOS & BAUDASSE
61 RUE DECOMBEROUSSE
69100 VILLEURBANNE

BELLOIN J. C.
WESTFALIA SEPARATOR FRANCE
B. P. 11
02400 CHATEAU THIERRY

BERNARD R.
LUWA
14 AVENUE DE LA PLAGE
94340 JOINVILLE

BONIS-CHARANCLE F.
ESSO SAF
6 AVENUE A. PROTHIN
92400 COURBEVOIE

BOUGRAT A
S. F. S.
3 AVENUE DE LA REPUBLIQUE
60000 BEAUVAIS

BOURIOT C.
MINISTERE DE L'ENVIRONNEMENT
14 BOULEVARD DE GAL LECLERC
92521 NEUILLY SUR SEINE

BOUSSUGE
SOCIETE COOPERATIVE D'INVES-
TISSEMENTS
2 RUE GOUNOD
75017 PARIS

BRASSART P.
PRESIDENT SNFRHG
44 RUE LA BOETIE
75008 PARIS

BRAULT J. P.
SUPERFINEST
Z. I. 36100 ISSOUDUN

BRET CH.
ELF FRANCE
137 RUE DE L'UNIVERSITE
75007 PARIS

BRUN A.
ESSO S. A. F.
COURBEVOIE

BRUN A.
OROGIL
47 RUE DE VILLIERS
92527 NEUILLY

MME CAILLAUD J.
RAFF. D'OLERAT
B. P. 25
LA ROCHEFOUCAULD

CASAMAYOU J. M.
PROCON FRANCE
TOUR FIAT CEDEX 16
92084 PARIS LA DEFENSE

CHAMPAGNE D.
SODETEG
9 AVENUE REAUMUR
92350 PLESSIS ROBINSON

CHAUVEAU F.
ESSO S. A. F.
6 AVENUE A. PROTHIN
92400 COURBEVOIE

COUDREAU A.
CSNCRA
6 RUE LEONARD DE VINCI
75116 PARIS

COURTOIS J.
SOCIETE FRANCAISE DES
PETROLES BP
10 QUAI PAUL DOUMER
COURBEVOIE

CROSA
UNION DES CHAMBRES SYNDI-
CALES DU PETROLE
16 AVENUE KLEBER
75116 PARIS

CROSLEBAILLY H.
SOPALUNA
119 BOULEVARD FELIX FAURE
93305 AUBERVILLIERS CEDEX

CROZIER A.
HEURTEY INDUSTRIES
30 RUE GUERSANT
75017 PARIS

DATCHARRY D.
A. N. R. E. D.
2 SQUARE LA FAYETTE
B. P. 406
49004 ANGERS CEDEX

DAUNY J.-CH.
AMOCO CHEMICALS FRANCE
22 RUE DE MARNES
92380 GARCHES

DELABRECHE B.
THEMEROIL
71240 VARENNES LE GRAND

DELBENDE J.
ETS GEERAERT & MATTHYS
48-58 RUE DES FORTS
59210 COUDEKERQUE BRANCHE

DELLOYE E.
SOCIETE DES HUILES
LEMAHIEU
26 RUE GAY LUSSAC
Z. I. 59147 GONDECOURT

DENISAN J.
SOPALUNA
119 BOULEVARD FELIX
FAURE
93305 AUBERVILLIERS

DORDE
RODOR
VILLENEUVE ST GEORGES

MME DORDE
RODOR
VILLENEUVE ST GEORGES

DUCATILLON J.
RAFF. IMPERATOR
59780 BAISIEUX

DUCATILLON Y.
SOLUNOR
59780 BAISIEUX

DUDAY CH.
A. N. R. E. D.
2 SQUARE LA FAYETTE
B. P. 406
49004 ANGERS

DUMORTIER J.
MATTHYS LUBRIFIANTS SA
84 RUE DE VILLIERS
92300 LEVALLOIS PERRET

DU PASSAGE H.
OROGIL
47 RUE DE VILLIERS
92527 NEUILLY

DUBREUIL
SOCIETE COOPERATIVE
D'INVESTISSEMENTS
2 RUE GOUNOD
75017 PARIS

EMPIS G.
CENTRE PROFESSIONNEL
DES LUBRIFIANTS
4 AVENUE HOCHE
75008 PARIS

FAUVET A.
S. R. R. H. U.
159 QUAI AULAGNIER
92600 ASNIERES

FISNOT S.
UNION FRANCAISE DES
PETROLES
2 AVENUE DU GAL DE GAULLE
54380 DIEULOUARD

FORTIN J.
SOLUNOR
RUE DE BREUZE
59780 BAISIEUX

GALANT G.
SOCIETE DES HUILES YACCO
42 AVENUE DE LA GRANDE
ARMEE
PARIS

GALLET M.
SOFRALUB
17 SENTE ST DENIS
95000 CERGY

GANIER M.
CETIM
10 RUE BARROUIN
SAINT ETIENNE

GENDREAU F.
DEUTZ FRANCE SA
185 AVENUE DE NEUILLY
92200 NEUILLY SUR SEINE

GENTIL P.
SITREM
64-66 RUE DE PARIS
93130 NOISY LE SEC

GRIGNON DUMOULIN A.
LITWIN SA
10 RUE JEAN JAURES
92407 PUTEAUX

DR GUINAT E.
SOPHOS
7 RUE AMPERE
92801 PUTEAUX

HAMMEL M.
SONOLUB
18 RUE DE LA MARNE
76410 ST AUBIN LES ELBEUF

HARANP
INSTITUT FRANCAIS DU PETROLE
CENTRE DE DOCUMENTATION
1 ET 4 AVENUE DU BOIS PREAU
92506 RUEIL MALMAISON

HOURNAC R.
TECHNIP
CEDEX 23
92090 PARIS LA DEFENSE

HYRON J.
SUPERFINEST
Z. I. 36100 ISSOUDUN

HAMMEL A.
SONOLUB
18 RUE DE LA MARNE
ST AUBIN LES ELBEUF

IMBEAUX E.
COMPAGNIE FRANCAISE
DE RAFFINAGE
84 RUE DE VILLIERS
92300 LEVALLOIS PERRET

JUREDIEU
IGOL FRANCE
130 BOULEVARD PEREIRE
75017 PARIS

JUSTE A.
BURMAH FRANCE
66 ROUTE DE SARTROUVILLE
78230 LE PECQ

KACHLER R.
S. N. F. R. H. G.
44 RUE LA BOETIE
75008 PARIS

KAESER G.
LUBRIZOL FRANCE
CEDEX 7
92080 PARIS LA DEFENSE

KARLEN R.
LUBRIZOL FRANCE
CEDEX 7
92080 PARIS LA DEFENSE

KATZ J.
SOPALUNA
119 BOULEVARD FELIX FAURE
93305 AUBERVILLIERS CEDEX

KERN G.
SOCIETE FRANCAISE DES
PETROLES B. P.
10 QUAI PAUL DOUMER
92412 COURBEVOIE CEDEX

KROLL W.
KROLL SARL
3 RUE DU MOULIN BATEAU
94380 BONNEUIL SUR MARNE

LANGENFELD F. H.
NYCO S. A.
51 RUE DE PONTHIEU
66 CHAMPS ELYSEES
75008 PARIS

LAURET J.
UNION EUROPEENNE DES INDE-
PENDANTS EN LUBRIFIANTS
33 RUE MARBEUF
75008 PARIS

LECLERC J.
PROCON FRANCE
TOUR FIAT CEDEX 16
92084 PARIS LA DEFENSE

LEMEE M.
INSTITUT FRANCAIS DE
PETROLE
1 ET 4 AVENUE DU BOIS
PREAU
92506 RUEIL MALMAISON

LEPERCQ B.
GERLAND
13 RUE DE LA MONTJOIE
93 LA PLAINE ST DENIS

LOPES J.
SOPALUNA
119 BOULEVARD F. FAURE
93305 AUBERVILLIERS CEDEX

MARCHAND
MATTHYS LUBRIFIANTS
84 RUE DE VILLIERS
92300 LEVALLOIS PERRET

MASCARELL M.
SHELL CHIMIE
27 RUE DE BERRI
75397 PARIS CEDEX 08

MATTHYS P.
MATTHYS S. A.
84 RUE DE VILLIERS
92300 LEVALLOIS PERRET

MAURAN A.
HUILERIE GENERALE DU
MIDI
ODARS
31450 MONTGISCARD

MAURIN J.
COMPAGNIE FRANCAISE DE
RAFFINAGE
CENTRE DE RECHERCHES
TOTAL
B. P. 27
76700 HARFLEUR

MAITRE MAURO
PARIS

MAYRAS P.
SHELL FRANCAISE
29 RUE DE BERRI
75008 PARIS

MERLIN
KROLL SARL
3 RUE DU MOULIN BATEAU
94380 BONNEUIL SUR MARNE

MERLIN U.
KROLL SARL
3 RUE DU MOULIN BATEAU
94380 BONNEUIL SUR MARNE

MICHEL P.
SNPECE
37-39 AVENUE LEDRU ROLLIN
75012 PARIS

MILHENCH S.
BURMAH CASTROL EUROPE
66 ROUTE DE SARTROUVILLE
B. P. 9
78730 LE PECQ

MOAL O.
S. I. C. A. "FLEURS"
KERISNEL
29250 ST POL DE LEON

MORVAN A.
S. I. C. A. "FLEURS"
KERISNEL
29250 ST POL DE LEON

DR OFFTERDINGER
KROLL SARL
3 RUE DU MOULIN BATEAU
94380 BONNEUIL SUR MARNE

OHANA
PROCON FRANCE
2 RUE DESCAMPS
PARIS

PAIX M.
C. S. N. I. L.
33 RUE MARBEUF
75008 PARIS

PANCHOUT A.
COMPAGNIE FRANCAISE DE
RAFFINAGE
84 RUE DE VILLIERS
92300 LEVALLOIS PERRET

PICOT J.
COMPAGNIE FRANCAISE DE
RAFFINAGE
22 RUE BOILEAU
75781 PARIS CEDEX 16

PICOT J.
CONDAT S. A.
16 RUE FREDERIC MISTRAL
38670 CHASSE SUR RHONE

MINISCLOUX J.
SOLUNOR
RUE DE BREUZE
59780 BAISIEUX

PORRINI M.
CITEPA
28 RUE DE LA SOURCE
75016 PARIS

PULVIN P.
MOBIL OIL FRANCAISE

PUTOUD A.
ELF FRANCE
PARIS

RASZKIEWICZ
ELF FRANCE
PARIS

RATYNSKI W.
SOCIETE DES HUILES
RENAULT-ELF FRANCE
55-65 RUE C. DESMOULINS
92130 ISSY LES MOULINEAUX

RAULINE A.
ESSO S. A. F.
CENTRE DE RECHERCHES
B. P. 6
76130 MONT ST AIGNAN

RENARD J. J.
MOREM
RUE PASTEUR PROLONGEE
94400 VITRY SUR SEINE

RICHARD C.
PAST PRESIDENT CONCAWE
PARIS

RICHARD G.
CETIM
B. P. 67
60304 SENLIS

MME RICHE R.
C. F. D. E.
11BIS RUE L. JOUHAUX
75010 PARIS

ROBERT J. P.
CHRYSO S. A.
ROUTE D'ORLEANS
B. P. 1
91380 CHILLY MAZARIN

SCHVARTZ J.
DEPUTE DE LA MOSELLE

ROLLAND R.
FINA FRANCE
19 RUE DU GAL FOY
75008 PARIS

ROSSARIE J.
COMPAGNIE FRANCAISE DE
RAFFINAGE
B. P. 27
76700 HARFLEUR

SAINTVILLE F.
CENTRE PROFESSIONNEL DES
LUBRIFIANTS
4 AVENUE HOCHE
75008 PARIS

SIGNORET - CIE FRANCAISE
DES NAPHTES
124 BOULEVARD DE PLOMBIERES
13337 MARSEILLE CEDEX 3

STORCK E.
PETROCARBOL ASSAINISSEMENT
2 AVENUE DU GAL DE GAULLE
54380 DIEULOUARD

TROESCH A.
SOPALUNA
119 BOULEVARD F. FAURE
93305 AUBERVILLIERS CEDEX

VALIQUETTE G.

WEINER S.
SOCIETE DE FABRICATION
D'ELEMENTS CATALYTIQUES
B. P. 33
84500 BOLLENE

ZAUGG E.
MOTUL S. A.
119 BOULEVARD F. FAURE
93305 AUBERVILLIERS

ITALIE

d'AGOSTINA E.
AGIP PETROLI SPA
449 VIA LAURENTINA
ROMA

ASSOM G.
FIAT AUTO SPA
C. SO G.AGNELLI 200
TORINO

BORZA M.
ASSORENI
VIA R. FABIANI 1 S. DONATO
MILANESE

BOTTARO P.
EUROFILTRI
VIA DEI SANSONE 4
16128 GENOVA

CACCIOLA P.
IP. INDUSTRIA ITALIANA
PETROLI
PIAZZA DELLA VITTORIA 1
GENOVA

CAMPI R.
INDUSTRIA ITALIANA PETROLI
PIAZZA DELLA VITTORIA 1
GENOVA

CARRIERO P. A.
SIRIUS SPA
20070 PIEVE FISSIGARA
(MILANO)

CAUDA F.
FIAT LUBRIFICANTI SPA
VIA SANTENA 1
10026 VILLASTELLONE

CHIOZZI P.
AGIP SPA
PLACE E. MATTEI 1
00144 ROMA

CONDELLO S.
ITALSIDER SPA
CENTRO SIDERURGICO DI
TARANTO

CONTARDI C.
REGIONE PIEMONTE

CONTINI G.
RONDINE AZ PETROCHIMICA SPA
STRADA STATALE SEMPIONE 17
PERO (MILANO)

CORSI L.
CONTRO SPETIMENTALE METAL-
LURGICO
VIA DEL CASTEL ROMANO
00129 ROMA

DR DE VITA F.
FRATELLI DE VITA SPA
VIA DEL PONTASSO 5
16015 CASELLA S. (GE)

DI LORETO V.
VERGA ENGINEERING
PIAZZA DUCA D'AOSTE 14
20124 MILANO

DI SERIO A.
AGIP PETROLI
VIA LAURENTINA 449
00142 ROMA

DR DONNABELLA P.
RA. M. OIL SPA
VIA FILICHITO N. 16/A
80013 CASALNUOVO (NA)

DR FERRANTE A.
RIVOL SPA
VIA SOSTEGNO A
27010 SPESSA PO (PAVIA)

FRANCHI G.
R. O. L. RAFFINERIA OLII
LUBRIFICANTI
VIA DE NOTARIS 50
MILANO

GIORGI G.
ASSOC. NAZ. INDUSTRIA
CHIMICA
VIA T. VALFRE 12
00165 ROMA

GUIDICI A.
REGIONE LOMBARDI-ASS.
ECOLOGICA
VIA F. FILZI 20
MILANO

PR LEOCI B.
RIVOL SPA
VIA SOSTEGNO 1
27010 SPESSA PO

LOMBARDI G.
TECHNIPETROL
ROMA

MAJOLO F.
ENTE NAZIONALE IDOR-
CARBURI
PIAZZA E. MATTEI 1
00144 ROMA

MALTONI R.
AGIP SPA
PIAZZALE E. MATTEI 1
00144 ROMA

MARALDI O.
AGIP PETROLI
ROMA

MARIOTTI C.
ICEP

MARMORALE G.
AGIP PETROLI SPA
ROMA

MENEGAZZO G.
TECHNIPETROL
ROMA

MERELLO G.
VISCOLUBE ITALIANA SPA
20070 PIEVE FISSIRAGA
(MI)

MINERVINI T.
AGIP PETROLI SPA
ROMA

MODENISI A.
SNAMPROGETTI
CIALE DE GASPERI
20100 MILANO

DR MOLINA C. O
O. M. A. SNC
VIA PAPINI N° 53/55
TORINO

NINI
COMMUNE DI MILANO
MILANO

DR ORLANDI C.
IP. INDUSTRIA ITALIANA
PETROLI
PIAZZA VITTORIA 1
GENOVA

PAPALE I.
AGIP PETROLI SPA
VIA LAURENTINA 449
00 142 ROMA

RICCO'A.
CAFFARO SPA
VIA PRIVATA VASTO 1
20121 MILANO

DR ROBOTTI G.
SIRO - SOC. IND. RAFF.
OLI
VIA MONTE ROSA
13 MILANO

RUOTA F.
COMMUNE DI MILANO
MILANO

SANTELLI G.
AGIP PETROLI SPA
VIA LAURENTINA 449
ROMA

DR SCHIEPPATI R.
PRESIDENT GRUPPE AZIENDE
INDIPENDENTI LUBRIFICANTI
MILANO

SECHI P.
ASCHIMICI G. A. I. L.
VIA FATEBENEFRATELLI 10
20121 MILANO

SECLI A.
AGIP PETROLI SPA
ROMA

TACCHINO E.
MIN. INDUSTRIA DGFE
VIA MOLISE 2
ROMA

TREMOLADA
I. C. E. P.

TRILLO
CLIPPER OIL ITALIANA

VERSINO C.
UNIVERSITA DI TORINO
CORSE M. D'AZEGLIO 48
TORINO

DR ZAVATTI M.
FINA ITALIANA SPA
VIA ROSSINI 6
20122 MILANO

NEDERLAND

DE IONGH H.
KON. NEDLLYOD GROEP NV
HOUTLAAN 21
3016 ROTTERDAM DA

DE BLIECK J. L.
KINETICS TECHNOLOGY INTERNAT.
B. V.
26 BREDEWATER
2715 CA ZOETERMEER

DAVID J.
STICHTING CONCAWE
VAN HOGENHOUCKLAAN 60
2596 TE THE HAGUE

CROENENDAAL
VERDO OIL RECYCLING
BERGEN OP ZOOM

BUYS T. A.
MIN. OF HEALTH AND ENVI-
RONMENTAL PRETECTION
POSTBUS 439
2260 AK LEIDSCHENDAM

WESTENBRINK J.
KINETICS TECHNOLOGY INTER-
NATIONAL
B. B.
26 BREDEWATER
2715 CA ZOETERMEER

VERHOEVEN
VERDO OIL RECYCLING
BERGEN OP ZOOM

VAN DER TORN W.
NEDLLOYD GROEP
HOUTLAAN 21
3016 DA ROTTERDAM

TERLOUW G.
CHEVRON CENTRAL LABORA-
TORIES
P. O. BOX 7970
3000 HZ ROTTERDAM

ZWIJNDRECHT W.
VERDO OIL RECYCLING
BERGEN OP ZOOM

UNITED KINGDOM

BARCLAY D. P. A.
BURMAH CASTROL IND LTD
BURMAH HOUSE PIPERS WAY
SWINDON

BEESLEY W. E.
HENRY BEESLEY & SONS LTD
EXCHANGE WORKS KELVIN WAY
WEST BRONWICH
WEST MIDLANDS B707JN

BILSLAND A.
WASTE LUBRICATING OILS LTD
CHEMICAL LANE TUNSTALL
STOKE ON TRENT

BRETT A. R.
BRETT'S OILS LTD
PIPEWELLGATE GATESHEAD
TYNE AND WEAR NE8 2BN

BUCHANAN A.
J. O. BUCHANAN & C° LTD
THE WHARF
RENFREW PA4 8SN

BURDEN R. A.
BRETT'S OILS LTD
PIPEWELLGATE GATESHEAD
TYNE AND WEAR NE8 2BN

BUTTERFIELD J.
ROBERT OWEN HOLDINGS LTD
DORSET

CROCKER F.
WASTE OILS (SOUTHERN)
211 CLIFFE ROAD
STROOD ROCHESTER KENT

DAVITT R. M.
CURRAN OILS LTD
HURMAN STREET
CARDIFF

DICK A.
FOSECO MINSEP LTD
LONG ACRE NECHELLES
BIRMINGHAM B7 5JR

DOMINEY J. A.
BP OIL LIMITED
BP HOUSE VICTORIA STREET
LONDON SW1

DROUIN J. M.
ESSO CHEMICAL LTD
P. O. BOX 1
ABINGDON OXON
21600 ABINGDON

EVISON
SHELL INT. PETRO. LTD
LONDON

FROST S.
BURMAH CASTROL CY
SWINDON

HATELEY E. D.
EDGAR VAUGHAN & C° LTD
LEGGE STREET
BIRMINGHAM B4 7EU

HUMPHRYS
SHELL LTD PETR. LTD
LONDON

HUSSEY PH.
HALES INDUSTRIAL
SERVICES LTD
ARMOURY ROAD
SMALLHEATH
BIRMINGHAM

HUNTER T. C.
HENRY BALFOUR & C° LTD
LEVEN FIFE KY8 4RW
SCOTLAND

JONES S. F.
CENTURY OILS GROUP LTD
P. O. BOX 2 CENTURY WORKS
HANLEY STOKE ON TRENT

JONES
SOUTHERN ECONOMY FUELS LTD
WARE HERTS

LITTLEWOOD P.
BUSCH U. K.
2 BENSHAM LANE
CROYDON SURREY CRO 2RQ

MATTHEWS T. H.
CENTURY OILS GROUP LTD
P. O. BOX 2
CENTURY WORKS
HANLEY STOKE ON TRENT

McCARTHY J. M.
BURMAH CASTROL C°
SWINDON

McKELVIE
TOTAL OIL GB LTD
33 CAVENDISH SQUARE
LONDON MIM OJE

MITCHELL CH. H.
CENTURY OILS GROUP LTD
P. O. BOX 2
CENTURY WORKS
HANLEY STOKE ON TRENT

NASH G. H. S.
SOUTHERN WASTE OILS
86 COBBAM ROAD
FERNDONN
IND. EST. WIMBORNE DORS.

NICHOLAS G. H.
SHELL INT. PETR. LTD
LONDON

OWEN R. G. F.
ROBERT OWEN (HOLDINGS) LTD.
18 MARKET STREET
POOLE DORSET BH15 INF

PARKER D. J.
BP C° LTD
BRITANNIC HOUSE
MOOR LANE
LONDON EC2Y 9BU

RATCLIFFE C. J.
ANGLO PENSYLVANIAN OIL
C° LTD
C/O CENTURY OILS GROUP
LTD P. O. BOX 2
CENTURY WORKS
HANLEY STOKE ON TRENT

ROBERTSON J. A.
SHELL U. K. OIL
SHELL MEX HOUSE
STRAND
LONDON WC2R ODX

ROBSON R.
ESSO CHEMICALS
PARAMINS DIVISION
P. O. BOX 1
ABINGDON OXFORDSHIRE
OX13 6BB

ROSE G. S.
DALTON AND C° LTD
SILKOLENE OIL REFINERY
BELPER DERBY

BIGGIN R.
LUBRIZON INT. LAB.
DERBY

SCOOT CAMPBELL G.
GULF OIL (GB) LTD
THE QUADRANGLE IMPERIAL
SQUARE
CHELTENGAM GLO GLSO ITF

SHAMA Y.
PROCON GB
GREATER LONDON HOUSE
HAMPSTEAD ROAD
LONDON NWI 7SB

THORNEYCROFT B.
ROBERT OWEN HOLDINGS LTD
18 MARKET STREET
POOLE DORSET BH15 INF

VAN VESSEM J. L.
CALTEX U. K. LTD
30 OLD BURLINGTONSTREET
LONDON WIX 2AR

VICKERS J. F.
BENJ. R. VICKERS & SONS LTD
5 GROSVENOR ROAD
LEEDS LS6 2EA

WALKER W. B.
BRAYBROOKE CHEMICAL SERVICES
LTD
P. O. 2 CENTURY WORKS
HANLEY STOKE ON TRENT

WARNER C. H. R.
FILTRATE LTD
P. O. BOX 67
LEEDS LSI ILS

WELLS M. N.
LONDON AND COASTAL OIL
WHARVES
CANVEY ISLAND
ESSEX SS8 ONR

AUTRES PAYS
===========

AUSTRALIE

SALUSINSZKY A. L.
A. L. SALUSINSZKY & ASS.
PRY.LTD
27 BERTRAM STREET
BURWOOD
VICTORIA 3125

DR BRAITHWAITE R. L.
SOUTH AUSTRALIAN INSTITUTE OF TECHNOLOGY

AUSTRIA

BUCHINGER G.
GESELLSCHAFT FÜR UMWELTSCHUTZ
SALESIANERSTRASSE 4
1030 WIEN

DOPPLER F.
DOPPLER MINERALÖL GMBH
VOGELWEIDER STRASSE 8
4600 WELLS

HIRSCH M.
ÖVG-ÖLVERWELTUNGSGES GMBH
WIESINGERSTRASSE 3/20
1010 WIEN

HÜBL F.
BADEN/WIEN

WALTHER P.
ÖMW AKTIENGESELLSCHAFT
OTTO WAGNER PLATZ 5
1091 WIEN

BRAZIL

GERAISSATE A.
LUBRINASA LUBRIFICANTES NACIONAIS SA
AVENIDA SENADOR QUEIROZ 279
9° ANDAR

CANADA

ABRACEN S.
LES HUILES NORCO LTEE
230 RUE NORMAN
QUEBEC

DON LOISELLE
LOISELLE PETROLEUM LTD
2124 RUE CHATEAUGUAY
MONTREAL QUEBEC H3K IL4

STRIGNER P.
NATIONAL RESEARCH
COUNSIL OF CANADA

BERNARD L. H.
SHELL CANADA LTD
505 UNIVERSITY AVENUE
TORONTO

FISHER D. A.
SHELL CANADA LIMITED
505 UNIVERSITY AVENUE
TORONTO ONTARIO

WITTENBERG S.
IMPERIAL OIL LTD
BOX 3004
SARNIA ONTARIO N7T 7M5

BROUWER B. H.
IMPERIAL OIL LIMITED
BOX 3004
SARNIA ONTARIO N7T 7M5

STREET R. J.
MOHAWK LUBRICANTS LTD
130 FORESTER STREET
NORTH VANCOUVER B. C.

CHINE

DEWEI ZHENG
CHINA NATIONAL PETROLEUM
CORPORATION

PAN LIMIN
CHINA NATIONAL PETROLEUM
CORPORATION

ALGERIE

DJATOU G.
SONATRACH
ARZEN

SARKA
SONATRACH

FAID M. K.
SONATRACH
ALGER

TAXEKA
SONATRACH

FINLAND

PIIRILA E.
TECHNICAL RESEARCH CENTRE OF FINLAND
BIOLOGINKUJA 3-5
SF 02150 ESPOO 15

GABON

WIELESYNSKI
GOUVERNEMENT GABONAIS
B. P. 884
LIBREVILLE

GREECE

ANTONOPOULOS N.
NAFTA S. A.
10 RUE VISSARIONOS
ATHENES 135

LIPOVATZ N.
ORYLIP LTD
17 RUE HAIDARIOU
LE PIREE

LERIOS N.
MOTOR OIL HELLAS

HUNGARY

SASS L.
INSTITUT HONGROIS DES HYDROCARBURES
2443 SZAZHALOMBATTA
PF 32

INDIA

BHARGAVA M. K.
CHEMOLEUMS PRIVATE LTD
THIRUNEERMALAI ROAD
CHROMEPET
MADRAS 6000444

BHATNAGAR S. R.
CASTROL LTD INDIA
C/O CASTROL LTD
91 WALKESHWAR ROAD
BOMBAY 400006

GOEL P. K.
INDIAN OIL CORPORATED LTD
SECTOR 13
FARIDABAD 121002

MUTHIAH M. CT.
CHEMOLEUMS PRIVATE LTD
THIRUNEERMALAI ROAD
CHROMEPET
MADRAS 600044

IRELAND

DUFFY F. M. J.
ATLAS OIL RECYCLERS LTD
32 NASSAU STREET
DUBLIN 2

ISRAEL

KATZ J.
PRESSURE LUBRICANTS (ISRAEL) LTD
P. O. BOX LOD 31

HEIFETZ Y.
"DELEK" THE ISRAEL FUEL CORP. LTD
"BEIT GIBOR"
6 PROF. KAUFMAN STREET
POB 1831 TEL AVIV

MOROCCO

HADDI M.
SADOIL
42 BOULEVARD DE LA RESISTANCE
CASABLANCA

LAHLOU S.
SADOIL
42 BOULEVARD DE LA RESISTANCE
CASABLANCA

NEW ZEALAND

HILLAM D. B.
DOMINION OIL REFINING C° LTD
P. O. BOX 13044 ONEHUNGA
AUCKLAND

PHARE J. D.
DOMINION OIL REFINING C° LTD
P. O. BOX 13044 ONEGUNGA
AUCKLAND

NORWAY

KISE T.
REGINOL A/S NORSK OLJERAFFINERI
P. O. BOX 35 LEIRDAL
OSLO 10

MYRVOLD H.
REGINOL S/A NORSK OLJERAFFINERI
P. O. BOX 35 LEIRDAL
OSLO 10

PEDERSEN T.
NORVEGIAN PETROLEUM INSTITUTE

SIMONSEN B. S.
REGINAL S/A NORSK OLJERAFFINERI
P. O. BOX 35 LEIRDAL
OSLO 10

JAPAN

OHBA T.
TIWDA
11-23 NAGATA CHO 1 CHOME CHIODAKU
TOKYO

YOSHIDA A.
KAKOKI ENVIRONMENT SERVICE LTD
1-11 SHIM BASHI 6
CHOME MINATOKU
TOKYO

SPAIN

CASTRO DEL RIO D. J.
ULIBARRI S. A.
PASEO DE LA CASTELLANA 174
MADRID 16

DAMETO COTONER F.
DOR MINERALOIL REFINERIE ESPAÑOLA SA
PADRE NADAL 4

GOMEZ MIÑANA J. A.
ULIBARRI S. A.
PASEO DE LA CASTELLANA 174
MADRID 16

HURTADO F.
LUDESA
CALABRIA 175   ENTLO 1°
BARCELONA

MARTINEZ SANTOS D. V.
ULIBARRI S. A.
PASEO DE LA CASTELLANA 174
MADRID 16

SWITZERLAND

DIEM A.
ASSOCIATION SUISSE DES IMPORTATEURS
D'HUILES DE GRAISSAGE VSS
LÖWENSTRASSE 1
8001 ZÜRICH

HAURI F. W.
LUWA AG
ANEMONENSTRASSE 40
8057 ZÜRICH

LÄMMLE L.
GARANTOL AG
BLÄSIMÜHLE
8321 MADETSWIL

RITZ W. K.
ASEOL AG
EFFINGERSTRASSE 17
3001 BERN

RUEGG D.
ASSOCIATION SUISSE DES IMPORTATEURS
D'HUILES DE GRAISSAGE
LÖWENSTRASSE 1
8001 ZÜRICH

TUNISIA

FTOUH A.
SOCIETE TUNISIENNE DE LUBRIFIANTS
9 RUE NOUVELLE DEHLI
TUNIS

PORTUGAL

PEREIRA J. M. R. S.
PETROGAL PETROLEOS DE PORTUGAL EP
PATIO DO PIMENTA 25
1200 LISBOA

REPUBLIC OF SOUTH AFRICA

COWAN W. F. H.
DUROL OIL CO (PTY) LTD
P. O. BOX 371
0200 ROSSLYN
PRETORIA

MEYER J. H.
SOUTH AFRICAN BUREAU OF STANDARDS
PRIVATE BAG X 191
PREROTIA 0001

ROUMANIE

NICA V.
CENTRAL MANAGEMENT FOR DELIVERY OF
PETROLEUM PRODUCTS "PECO"
GRAL BUDISTEANU 11 BIS
BUCHAREST SR ROMANIA

SENEGAL

MBACKE A.
TOTAL SENEGAL
15 AVENUE DE LA REPUBLIQUE
DAKAR

SWEDEN

KJNGSTRÖM H.
AB NYNÄS PETROLEUM
BOX 5842
102-48 STOCKOLM

UNITED STATES OF AMERICA

ANDRETICH G.
ASSOCIATION OF PETROLEUM RE-REFINERS

BAUER J. F.
RESOURCE TECHNOLOGY INC
809 SOUTH MN ST.
KANSAS CITY KANSAS

BECKER D. A.
NATIONAL BUREAU OF STANDAND
RECYCLED OIL PROGRAM
WASHINGTON DC 20234

BOTH JR
BOTH OIL
ASSOCIATION OF PETROLEUM RE-REFINERS

BOTH III
BOTH OIL
ASSOCIATION OF PETROLEUM RE-REFINERS

BRONSON D.
LAKEWOOD OIL SCE
ASSOCIATION OF PETROLEUM RE-REFINERS

DAVIS G.
DAVIS REFINING CORP.
ASSOCIATION OF PETROLEUM RE-REFINERS

DEHAAN J.
TOTAL PETROLEUM INC
EAST SUPERIOR STREET
ALMA MICHIGAN

DUPUIS
ASSOCIATION OF PETROLEUM RE-REFINERS

FEAGAN J.
RESOURCE TECHNOLOGY
ASSOCIATION OF PETROLEUM RE-REFINERS

GETTINGER G.
MIDWEST OIL REFINING
ASSOCIATION OF PETROLEUM RE-REFINERS

HAMBLIN R. J.
PROCON INTERNATIONAL
50 VOP PLAZA
DES PLAINES ILLINOIS 61616

HUNTER J.
RESOURCE TECHNOLOGY
ASSOCIATION OF PETROLEUM RE-REFINERS

KALMACOFF J.
HUB OIL
ASSOCIATION OF PETROLEUM RE-REFINERS

KAUFMANN H. B.

KERRAN C.
DOUBLE EAGLE LUB.
ASSOCIATION OF PETROLEUM RE-REFINERS

KIPATSA D.
RESOURCE TECHNOLOGY INC
P. O. BOX 5187
KANSAS CITY

McBAIN J.
EXECUTIVE DIRECTOR OF ASSOCIATION OF
PETROLEUM RE-REFINERS

MEMBER OF CONGRESS
WASHINGTON

MORRIS K.
CAM OR INC.
ASSOCIATION OF PETROLEUM RE-REFINERS

ODENWALD R.
ASSOCIATION OF PETROLEUM RE-REFINERS

ODENWALD S.
ASSOCIATION OF PETROLEUM RE-REFINERS

REINTJES J.
ASSOCIATION OF PETROLEUM RE-REFINERS

SAVORY R.
EXXON RESEARCH & ENGINEERING
P. O. BOX 101
FLORHAM PARK
NEW JERSEY 07932

SAWVELL LR
S AND R OIL
ASSOCIATION OF PETROLEUM RE-REFINERS

SPARKS T.
RESOURCE TECHNOLOGY
ASSOCIATION OF PETROLEUM RE-REFINERS

SWASEY R.
S AND R OIL
ASSOCIATION OR PETROLEUM RE-REFINERS

THOMSON
BARTLESVILLE ENERGY TECHNOLOGY CENTER

WALKER S. R.
SUN INTERNATIONAL INC
200 W LANCASTER AVENUE
WAYNE PENNSYLVANIA

WARDEN A.
WARDEN OIL
ASSOCIATION OF PETROLEUM RE-REFINERS

MRS DANA WILSON
ASSOCIATION OF PETROLEUM RE-REFINERS

VENEZUELA

AZNAR R.
CORPOVEN S. A.
EDIF. PETROLEOS DE VENEZUELA
TORRE OESTE APTDO. 61373

MARSICOBETRE D.
CORPOVEN S. A.
EDIF. PETROLEOS DE VENEZUELA
TORRE OESTE APTDO. 61373

YUGOSLAVIA

ILIC
NAPHTAGAS
BELGRADE

JOUANOVIC
NAPHTAGAS
BELGRADE

JURKOVIC
PETROLINVEST
RUE TVORNICKA
B. P. 343
71000 SARAJEVO

PINDOVIC
NAPHTAGAS
BELGRADE

TERZIMEHIC
PETROLINVEST
RUE TVORNICKA
B. P. 343
71000 SARAJEVO

ZAGRADNIK J.
PETROL RAFFINERIE MARIBOR
LINHARTOVA 17/A
62000 MARIBOR

ZIMAKISEVIC
RAFFINERIE DE MODRICA

ZEDIC
RAFFINERIE DE MODRICA